Microextraction Techniques in Analytical Toxicology

Microextraction Techniques in Analytical Toxicology

Edited by
Rajeev Jain and Ritu Singh

CRC Press is an imprint of the
Taylor & Francis Group, an **informa** business

First edition published 2022
by CRC Press
6000 Broken Sound Parkway NW, Suite 300, Boca Raton, FL 33487-2742

and by CRC Press
2 Park Square, Milton Park, Abingdon, Oxon, OX14 4RN

© 2022 selection and editorial matter, Rajeev Jain and Ritu Singh; individual chapters, the contributors

CRC Press is an imprint of Taylor & Francis Group, LLC

Reasonable efforts have been made to publish reliable data and information, but the author and publisher cannot assume responsibility for the validity of all materials or the consequences of their use. The authors and publishers have attempted to trace the copyright holders of all material reproduced in this publication and apologize to copyright holders if permission to publish in this form has not been obtained. If any copyright material has not been acknowledged please write and let us know so we may rectify in any future reprint.

Except as permitted under U.S. Copyright Law, no part of this book may be reprinted, reproduced, transmitted, or utilized in any form by any electronic, mechanical, or other means, now known or hereafter invented, including photocopying, microfilming, and recording, or in any information storage or retrieval system, without written permission from the publishers.

For permission to photocopy or use material electronically from this work, access www.copyright.com or contact the Copyright Clearance Center, Inc. (CCC), 222 Rosewood Drive, Danvers, MA 01923, 978-750-8400. For works that are not available on CCC please contact mpkbookspermissions@tandf.co.uk

Trademark notice: Product or corporate names may be trademarks or registered trademarks and are used only for identification and explanation without intent to infringe.

Library of Congress Cataloguing-in-Publication Data
Names: Jain, Rajeev, 1986- editor. | Singh, Ritu, 1986- editor.
Title: Microextraction techniques in analytical toxicology /
edited by Rajeev Jain and Ritu Singh.
Description: First edition. | Boca Raton : CRC Press, 2022. |
Includes bibliographical references and index.
Identifiers: LCCN 2021027444 (print) | LCCN 2021027445 (ebook) |
ISBN 9780367651947 (hardback) | ISBN 9780367651954 (paperback) |
ISBN 9781003128298 (ebook)
Subjects: LCSH: Analytical toxicology--Technique. | Extraction (Chemistry)
Classification: LCC RA1223.E98 M53 2022 (print) |
LCC RA1223.E98 (ebook) | DDC 615.9/07--dc23
LC record available at https://lccn.loc.gov/2021027444
LC ebook record available at https://lccn.loc.gov/2021027445

ISBN: 978-0-367-65194-7 (hbk)
ISBN: 978-0-367-65195-4 (pbk)
ISBN: 978-1-003-12829-8 (ebk)

DOI: 10.1201/9781003128298

Typeset in Times
by MPS Limited, Dehradun

Dedicated with love to: my parents, wife Richa & son Animesh

(Rajeev Jain)

Contents

Preface ix
Editors' Biographies xi
List of Contributors xiii

1   Microextraction Techniques in Analytical Toxicology: An Overview 1
    Rajeev Jain, Ritu Singh, and Abuzar Kabir

2   Application of Solid-Phase Microextraction in Analytical Toxicology 11
    Rakesh Roshan Jha and Rajeev Jain

3   Applications of Micro Solid-Phase Extraction in Analytical Toxicology 23
    Shahram Seidi and Maryam Rezazadeh

4   Stir Bar Sorptive Extraction in Analytical Toxicology Studies 61
    Natasa P. Kalogiouri, Elisave-Ioanna Diamantopoulou, and Victoria F. Samanidou

5   Microextraction by Packed Sorbent 71
    Tiago Rosado, Eugenia Gallardo, Duarte Nuno Vieira, and Mario Barroso

6   Thin-Film Solid-Phase Microextraction: Applications in Analytical Toxicology 117
    Lucas Morés, Josias Merib, and E. Carasek

7   Application of Single Drop Microextraction in Analytical Toxicology 131
    Archana Jain, Manju Gupta, and Krishna K. Verma

8   Applications of Liquid-Phase Microextraction in Analytical Toxicology 147
    María Ramos-Payán, Samira Dowlatshah, and Mohammad Saraji

9   Dispersive Liquid-Liquid Microextraction and Its Variants 159
    Rakesh Roshan Jha, Rabindra Singh Thakur, and Rajeev Jain

10   Electromembrane Extraction in Analytical Toxicology 171
    Marothu Vamsi Krishna, Kantamaneni Padmalatha, and Gorrepati Madhavi

11   Fabric Phase Sorptive Extraction in Analytical Toxicology 183
    Natalia Manousi, Abuzar Kabir, and George A. Zachariadis

12   Sorbent-Based Microextraction Using Molecularly Imprinted Polymers 193
    Cecilia Ortega-Zamora, Gabriel Jiménez-Skrzypek, Javier González-Sálamo, and Javier Hernández-Borges

13   Applications of Ionic Liquids in Microextraction 205
    Devendra Kumar Patel, Neha Gupta, Sandeep Kumar, and Juhi Verma

14	**Deep Eutectic Solvent-Based Microextraction** 221
	Amir M. Ramezani, Yadollah Yamini, and Raheleh Ahmad
15	**Hyphenation of Derivatization with Microextraction Techniques in Analytical Toxicology** 239
	Abhishek Chauhan and Utkarsh Shukla

Index 249

Preface

Analytical toxicologists encounter difficulties with the continuous addition of new drugs, pesticides, and other substances, which present novel challenges in the analysis and interpretation of results. The accurate measurement of xenobiotics in complex biological matrices greatly depends upon the sample treatment and extraction techniques used prior to instrumental analysis. Therefore, sample preparation is the most critical and challenging step in any analytical methodology as it determines the quality of the results obtained.

In the past few years, significant advances have been made in the development and application of microextraction techniques and are becoming alternatives to classical methods such as solid-phase extraction. Some of the basic driving factors for the development of microextraction techniques are elimination of sample treatment steps, minimizing sample amount, and a significant reduction in consumption of hazardous reagents and solvents. Microextraction techniques came into existence with the introduction of solid-phase microextraction (SPME) in the 1990s and gained much attention from the scientific community leading toward its commercialization. In later years, other microextraction techniques based on solid and liquid formats have emerged. These microextraction techniques are characterized by better extraction efficiencies, higher enrichment factors, and reduced consumption of toxic organic solvents in comparison to classical extraction methods.

Despite the complexity of biological matrices encountered in analytical toxicology laboratories, microextraction techniques have been applied for various analytes, ranging from gaseous poisons, volatile organic chemicals, drugs of abuse and metabolites, therapeutic drugs, pesticides, alkaloids, and endogenous compounds. No doubt, these microextraction techniques will be routinely used in analytical laboratories in the future and will replace traditional extraction techniques; however, it will take time.

This book aims to provide principles and practical information about technical know-how and implementation of microextraction techniques in laboratories – for the analysis of drugs, poisons, and other relevant analytes in biological specimens, especially pertaining to analytical toxicology. The book itself is structured around the robust anatomy of the subject. Following a basic introduction (Chapter 1), which includes a brief theory and overview of microextraction techniques from the perspective of analytical toxicology, chapters onward (Chapters 2–6) are dedicated to applications of sorbent-based microextraction techniques in analytical toxicology. This part includes solid-phase microextraction, micro solid-phase extraction, stir bar sorptive extraction, microextraction by packed sorbents, and thin-film solid-phase microextraction. Liquid-phase microextraction techniques are compiled in Chapters 7–10, where single drop microextraction (SDME), liquid-phase microextraction, dispersive liquid-liquid microextraction (DLLME), and electromembrane extraction are presented. SDME is the first liquid-based microextraction technique that has reduced the volume of a solvent to a single drop of microliter level, whereas DLLME is the most popular, easy, and cost-effective microextraction technique in this format. Chapter 11 covers a relatively new microextraction technique, introduced in 2014, named fabric phase sorptive extraction (FPSE). Here, extraction of target analytes takes place on a FPSE membrane coated with sol-gel derived sorbent. The porous surface of fabric in the FPSE membrane offers a high primary contact surface area between analyte and sorbent material. Further, Chapters 12–15 compile some special topics such as molecularly imprinted polymer-based microextraction, green solvent-based microextraction (ionic liquids and deep eutectic solvents), and microextraction techniques coupled with a derivatization approach.

The primary readership is expected to be forensic and clinical toxicologists, researchers, and academicians. The secondary readership is anyone curious about analytical toxicology, including undergraduates and professionals in other fields. It is designed to equip the reader with the ability to coherently appraise the merits or otherwise of any of the analytical techniques.

The authors of the chapters are pioneers in their field. Their expertise, wisdom, knowledge, and active collaboration have made this book possible. As editors, we feel pleased and honored to work with such distinguished scientists and academicians.

We would like to express our sincere thanks to CRC Press for their trust and support during the entire preparation process.

Dr. Rajeev Jain
Central Forensic Science Laboratory
Chandigarh, India

Dr. Ritu Singh
Central University of Rajasthan
Rajasthan, India

Editors' Biographies

Dr. Rajeev Jain is Senior Scientist of Forensic Toxicology at Central Forensic Science Laboratory, Ministry of Home Affairs, Govt. of India. He has more than ten years of research experience in the field of analytical toxicology. During his tenure as Forensic Toxicologist, he has examined post-mortem samples of more than 500 medico-legal cases pertaining to drug overdose and poisoning. He obtained his Ph.D. degree in analytical toxicology from CSIR-Indian Institute of Toxicology Research (India). His research work is focused on the development of simple, rapid, eco-friendly, cost-effective, sensitive, and selective analytical methods based on microextraction techniques (e.g., SPME, DLLME) for the determination of chemical analytes for forensic toxicological, clinical, and environmental importance. He has published many research papers, review papers, and book chapters in various refereed journals on this subject. He was one of the first scientists to explore the possibility of coupling dispersive liquid-liquid microextraction with injector port silylation for rapid analysis of polar toxicants. He is also acting as invited reviewer in various refereed journals and is on the editorial board of various open access journals of toxicology and chromatography.

Dr. Ritu Singh is an Assistant Professor in the Department of Environmental Science, School of Earth Sciences, Central University of Rajasthan, Rajasthan, India. Her research interests include analytical method development, microextraction techniques, nano-remediation, environmental monitoring, and assessment. She received her doctorate degree in environmental science from Babasaheb Bhimrao Ambedkar University, Lucknow. She worked as research fellow in CSIR-Indian Institute of Toxicology Research for five years and made a significant contribution to the field of nano-remediation. She has published her work in several international journals of repute and has contributed several chapters to international and national books. She also has one international book to her credit. She is the recipient of the 'DEF-Young Scientist Award' in the field of environmental science. She is an editorial board member of the *Journal of Nanoscience, Nanoengineering and Applications*, the *Journal of Water Pollution & Purification Research*, and the *Eurasian Journal of Soil Science,* and she is an invited reviewer of several refereed journals. She is also serving as a life member of various academic bodies.

List of Contributors

Abuzar Kabir
Department of Chemistry and Biochemistry, Florida International University
Florida, USA

Rajeev Jain
Central Forensic Science Laboratory, Directorate of Forensic Science Services, Ministry of Home Affairs, Govt. of India
Chandigarh, India

Ritu Singh
Department of Environmental Science, School of Earth Sciences, Central University of Rajasthan
Ajmer, RJ, India

Rakesh Roshan Jha
Centre of Analytical Bioscience, School of Pharmacy, University of Nottingham
Nottingham, UK

Shahram Seidi
Department of Analytical Chemistry, Faculty of Chemistry, K. N. Toosi University of Technology
Tehran, Iran

Maryan Rezazadeh
Chemical Analysis Laboratory, Geological Survey of Iran
Tehran, Iran

Natasa P. Kalogiouri
Laboratory of Analytical Chemistry, Department of Chemistry, Aristotle University of Thessaloniki
Thessaloniki, Greece

Elisavet-Ioanna Diamantopoulou
Laboratory of Analytical Chemistry, Department of Chemistry, Aristotle University of Thessaloniki
Thessaloniki, Greece

Victoria F. Samanidou
Laboratory of Analytical Chemistry, Department of Chemistry, Aristotle University of Thessaloniki
Thessaloniki, Greece

Tiago Rosado
Centro de Investigação em Ciências da Saúde, Faculdade de Ciências da Saúde da Universidade da Beira Interior (CICS-UBI), Covilhã, Portugal, Laboratório de Fármaco-Toxicologia-UBI Medical, Universidade da Beira Interior, Covilhã, Portugal, C4 - Cloud Computing Competence Centre, Universidade da Beira Interior
Covilhã, Portugal

Eugenia Gallardo
Centro de Investigação em Ciências da Saúde, Faculdade de Ciências da Saúde da Universidade da Beira Interior (CICS-UBI), Covilhã, Portugal Laboratório de Fármaco-Toxicologia-UBI Medical, Universidade da Beira Interior
Covilhã, Portugal

Duarte Nuno Vieira
Faculdade de Medicina, Universidade de Coimbra
Coimbra, Portugal

Mario Barroso
Serviço de Química e Toxicologia Forenses, Instituto de Medicina Legal e Ciências Forenses - Delegação do Sul
Lisboa, Portugal

Lucas Mores
Universidade Federal de Santa Catarina
Brazil

Josias Merib
Universidade Federal de Ciências da Saúde de Porto Alegre
Brazil

Eduardo Carasek
Universidade Federal de Santa Catarina
Brazil

Archana Jain
Department of Chemistry, Rani Durgavati University
Jabalpur, MP, India

Krishna K. Verma
Department of Chemistry, Rani Durgavati University
Jabalpur, MP, India

María Ramos Payán
Department of Analytical Chemistry, Faculty of Chemistry, University of Seville
Seville, Spain

Samira Dowlatshah
Department of Chemistry, Isfahan University of Technology
Isfahan, Iran

Mohammad Saraji
Department of Chemistry, Isfahan University of Technology
Isfahan, Iran

Rabindra Singh Thakur
Analytical Chemistry Laboratory, Regulatory Toxicology Group, CSIR-Indian Institute of Toxicology Research (CSIR-IITR), Lucknow, UP, India, Academy Council of Scientific and Innovative Research (AcSIR), CSIR-IITR Campus
Lucknow, UP, India

Marothu Vamsi Krishna
Vijaya Institute of Pharmaceutical Sciences for Women
Vijayawada, AP, India

Kantamaneni Padmalatha
Vijaya Institute of Pharmaceutical Sciences for Women
Vijayawada, AP, India

Gorrepati Madhavi
Vijaya Institute of Pharmaceutical Sciences for Women
Vijayawada, AP, India

N. Manousi
Laboratory of Analytical Chemistry, Department of Chemistry, Aristotle University of Thessaloniki
Thessaloniki, Greece

G. Zachariadis
Laboratory of Analytical Chemistry, Department of Chemistry, Aristotle University of Thessaloniki
Thessaloniki, Greece

Cecilia Ortega-Zamora
Departamento de Química, Unidad Departamental de Química Analítica, Facultad de Ciencias, Universidad de La Laguna (ULL). Avda. Astrofísico Fco. Sánchez, San Cristóbal de La Laguna
España

Gabriel Jiménez-Skrzypek
Departamento de Química, Unidad Departamental de Química Analítica, Facultad de Ciencias, Universidad de La Laguna (ULL). Avda. Astrofísico Fco. Sánchez, San Cristóbal de La Laguna
España

Javier González-Sálamo
Departamento de Química, Unidad Departamental de Química Analítica, Facultad de Ciencias, Universidad de La Laguna (ULL). Avda. Astrofísico Fco. Sánchez, San Cristóbal de La Laguna, España, Instituto Universitario de Enfermedades Tropicales y Salud Pública de Canarias, Universidad de La Laguna (ULL). Avda. Astrofísico Fco. Sánchez, San Cristóbal de La Laguna
España

Javier Hernández-Borges
Departamento de Química, Unidad Departamental de Química Analítica, Facultad de Ciencias, Universidad de La Laguna (ULL). Avda. Astrofísico Fco. Sánchez, San Cristóbal de La Laguna, España, Instituto Universitario de Enfermedades Tropicales y Salud Pública de Canarias, Universidad de La Laguna (ULL). Avda. Astrofísico Fco. Sánchez, San Cristóbal de La Laguna
España

Devendra Kumar Patel
Analytical Chemistry Laboratory, CSIR-Indian Institute of Toxicology Research Lucknow, UP, India, Academy of Scientific & Innovative Research (AcSIR)
Ghaziabad, UP, India

Neha Gupta
Analytical Chemistry Laboratory, CSIR-Indian Institute of Toxicology Research
Lucknow, UP, India
Academy of Scientific & Innovative Research (AcSIR)
Ghaziabad, UP, India

List of Contributors

Sandeep Kumar
Analytical Chemistry Laboratory, CSIR-Indian Institute of Toxicology Research
Lucknow, UP, India
Academy of Scientific & Innovative Research (AcSIR)
Ghaziabad, UP, India

Juhi Verma
Analytical Chemistry Laboratory, CSIR-Indian Institute of Toxicology Research
Lucknow, UP, India

Amir M. Ramezani
Healthy Ageing Research Centre, Neyshabur University of Medical Sciences
Neyshabur, Iran

Yadollah Yamini
Department of Chemistry, Faculty of Sciences, Tarbiat Modares University
Tehran, Iran

Raheleh Ahmadi
Healthy Ageing Research Centre, Neyshabur University of Medical Sciences
Neyshabur, Iran

Abhishek Chauhan
Polymer R&D Division, North Site, Atul Ltd.
Valsad (Gujarat), India

Utkarsh Shukla
Polymer R&D Division, North Site, Atul Ltd.
Valsad (Gujarat), India

1
Microextraction Techniques in Analytical Toxicology: An Overview

Rajeev Jain[1], Ritu Singh[2], and Abuzar Kabir[3]
[1]Central Forensic Science Laboratory, Directorate of Forensic Science Services, Ministry of Home Affairs, Govt. of India, India
[2]Department of Environmental Science, School of Earth Sciences, Central University of Rajasthan, NH8, Bandarsindri, Kishangarh, India
[3]Department of Chemistry and Biochemistry, Florida International University, Miami, USA

CONTENTS

1.1 Introduction to Analytical Toxicology 2
1.2 Nature of Specimens in Analytical Toxicology 2
    1.2.1 Blood 2
    1.2.2 Urine 2
    1.2.3 Saliva / Oral Fluid 3
    1.2.4 Hair and Nails 3
    1.2.5 Vitreous Humour (VH) 3
    1.2.6 Liver 3
    1.2.7 Stomach Contents 3
    1.2.8 Other Tissues 3
1.3 Microextraction Techniques in Analytical Toxicology: Classification, Theory, and Practical Applications 4
    1.3.1 Classification of Microextraction Techniques Used in Analytical Toxicology 4
    1.3.2 Theoretical Considerations 4
        1.3.2.1 Solvent-Based Microextraction 4
        1.3.2.2 Sorbent-Based Microextraction 6
    1.3.3 Sorbent-Based Microextraction Techniques 6
        1.3.3.1 Solid-Phase Microextraction (SPME) 6
        1.3.3.2 Micro Solid-Phase Microextraction (µSPE) 6
        1.3.3.3 Stir Bar Sorptive Extraction (SBSE) 7
        1.3.3.4 Microextraction by Packed Sorbent (MEPS) 7
        1.3.3.5 Electromembrane Extraction (EME) 7
        1.3.3.6 Fabric Phase Sorptive Extraction (FPSE) 8
        1.3.3.7 Molecularly Imprinted Polymer-Based Microextraction 8
    1.3.4 Solvent-Based Microextraction Techniques 8
        1.3.4.1 Single Drop Microextraction (SDME) 8
        1.3.4.2 Liquid-Phase Microextraction (LPME) 8
        1.3.4.3 Ionic Liquid-Based Microextraction 9
        1.3.4.4 Deep Eutectic Solvent-Based Microextraction 9
1.4 Conclusion and Future Trends 9
References 9

1.1 Introduction to Analytical Toxicology

Analytical toxicology involves detection, identification, and quantification of xenobiotic compounds (exogenous compounds), such as drugs, pesticides, poisons, pollutants, and their metabolites in various complex sample matrices, such as ante- and postmortem blood, urine, tissue, or vitreous humour (VH), or alternative samples, such as hair, nail, meconium, sweat, oral fluid, etc. Analytical toxicologists play an important role in diagnosis, management, and prevention of poisoning by detecting, identifying, and measuring the unknown drug or poison in the biological specimens (Maurer 2007; Maurer 2010). In most cases of analytical toxicology and doping control, the nature of the target analyte is usually unknown prior to analysis. Additionally, the presence of endogenous biomolecules and other xenobiotic compounds makes the matrix more complex, which raises the need for highly selective and sensitive analytical methods to determine unknown toxicants. Moreover, drugs and their metabolites are generally present at very trace levels in biological fluids, which further makes the whole analysis a daunting task. Since the analysis is usually untargeted and sample availability is also limited, sample preparation methodologies that require the least amount of sample and are capable of removing insoluble residues and interfering compounds are of the utmost importance in analytical toxicology (Flanagan 2007; Jain and Singh 2016).

1.2 Nature of Specimens in Analytical Toxicology

Various disciplines, such as clinical toxicology, forensic toxicology, therapeutic drug monitoring (TDM), screening of drugs of abuse, as well as occupational and environmental toxicology are covered under the aegis of analytical toxicology. However, there is considerable overlap between all the disciplines. Therefore, the specimens commonly encountered in analytical toxicology are basically of biological origin obtained under different conditions, which may range from liquid (e.g., pure solutions of a drug, blood, urine, cerebrospinal fluid, oral fluid) to semi-solid and solid material (e.g., tissue and pharmaceutical tablets). Analysis of liquid samples is generally easier in comparison to solid samples, which generally require homogenization, digestion, and protein precipitation.

1.2.1 Blood

Blood is the sample of choice in living humans as analyte concentrations in blood are closely related with their dose and biological effect. Beside blood, plasma and serum are also used for analysis of drugs. In postmortem toxicology, two blood specimens are collected: one from the heart and another from a peripheral site, e.g., femoral or ileac vein. These specimens may be significantly decomposed or contaminated from chest fluid, pericardial fluid, and gastric contents in the case of traumatic death (Jones 2008; Kerrigan and Levine 2020). Beside quantification, blood samples are also useful for screening of xenobiotics if their concentration is high enough. Postmortem blood has a high degree of haemolysis, and therefore direct analysis of whole blood is preferred.

1.2.2 Urine

Urine is an important specimen for targeted and non-targeted comprehensive screening of drugs and xenobiotic compounds as it represents a major route for their elimination from the body. Additionally, the collection process of urine samples is non-invasive, and the concentration of drugs is relatively high. Analysis of a urine specimen is also relatively simple as it comprises more than 99% water and is devoid of lipids, circulating serum proteins, and large molecular weight compounds due to the glomerular filtration process, which facilitates its analytical investigation by immunoassay, spot-tests, or sample preparation for instrumental analysis (Dinis-Oliveira et al. 2010). However, in forensic postmortem toxicology, urine is available only in 50% of deaths as the bladder usually voids during the dying process (Jones 2008).

1.2.3 Saliva / Oral Fluid

There has been growing interest in using saliva as a diagnostic medium of drug abuse since it can be obtained quickly and non-invasively without privacy violation, unlike urine sample collection. Saliva contains the free form of the drug, and its concentration can be correlated to the free drug concentration in plasma. For many drugs, only free fraction is physiologically active; therefore, saliva can better indicate the state of intoxication (Schramm et al. 1992).

1.2.4 Hair and Nails

Analysis of hair and nails is particularly useful for retrospective information of drug abuse and metal poisoning. The circulating drugs in the blood stream get incorporated into the cells of the hair and nails and get trapped when they are keratinized. The advantage of hair and nail testing is their non-invasive and easy collection, storage at room temperature, and small sample size requirement for analysis. The growth rate of nails is slower, which makes them suitable for retrospective analysis of drug abuse. Various drugs of abuse (e.g., amphetamines, cannabinoids, benzodiazepines, morphine, heroin, cocaine), trace elements (e.g., arsenic), and doping substances (e.g., ephedrine), etc., can be detected in hair and nails and can establish their chronic exposure (Daniel et al. 2004).

1.2.5 Vitreous Humour (VH)

VH is located between the lens and the retina of the eye and fills the eye chamber. VH is basically a salt solution that consists 99% of water and contains very little protein. Hence, any drug and metabolite present in VH can be easily extracted. VH is resistant to putrefactive changes as it resides in an anatomically isolated area; therefore, it has been used widely for estimation of ethanol and other drugs in postmortem forensic toxicology. The main drawback of VH is its small volume, i.e., up to 3–4 mL in each eye.

1.2.6 Liver

The liver is one of the most important and primary solid tissue used in postmortem toxicology for the analysis of drugs and poisons. The liver is the main metabolic organ of the body, where a higher concentration of basic drugs can be found in comparison to other body organs. The collection and sample preparation of the liver is easier; it is available in sufficient quantities for analysts, and unlike blood, it is not affected by postmortem redistribution as the concentration of drugs is relatively stable after death (Jones 2008; Dinis-Oliveira et al. 2010).

1.2.7 Stomach Contents

Stomach or gastric contents are mainly important for qualitative analysis in the case of oral overdose of drugs and poisons, especially when the specimen is obtained soon after the intoxication. The concentration of drug after oral ingestion may be high in the stomach contents; therefore, it is suitable for toxicological screening of xenobiotics. The drugs that are difficult to be detected in blood due to their extensive distribution can be easily detected in their parent form in stomach contents. In some cases where death occurred within a short time after oral ingestion, unabsorbed tablets or capsules may be detected in their intact form (Jones 2008; Dinis-Oliveira et al. 2010).

1.2.8 Other Tissues

When administration of drugs or poisons takes place by inhalation or intravenous routes, as in the case of solvent abuse, their high concentrations may be detected in lung specimens. It also depends on the properties of xenobiotics. Some specific poisons, such as paraquat, are accumulated in lung tissues in high quantities.

The kidney is a useful body organ for identification of drugs and poisons as most of them are passed through the kidney and then excreted in urine. In cases of heavy metal poisoning, kidney specimens are particularly useful as heavy metal accumulation takes place in the kidney. Quantitative analysis of xenobiotics in the kidney is not of much significance and is only important in accessing the overall body burden of a xenobiotic.

The brain is a relatively protected and isolated organ and remains unaffected by postmortem redistribution of drugs, unlike centrally located organs (e.g., liver). Therefore, the brain is a useful specimen in cases of death due to trauma to the chest and abdomen. Additionally, the brain is relatively lesser prone to decomposition, which also makes it a specimen of choice for detection and quantitation of xenobiotics in decomposed bodies (Rohrig and Hicks 2015). Some other organs, such as the spleen, are used as a secondary specimen. The spleen is generally rich in blood and therefore is particularly useful in the analysis of carbon monoxide and cyanide poisoning. Quantitation of xenobiotics in such specimens generally contributes to the assessment of their overall body burden.

1.3 Microextraction Techniques in Analytical Toxicology: Classification, Theory, and Practical Applications

1.3.1 Classification of Microextraction Techniques Used in Analytical Toxicology

Microextraction techniques were developed as a green alternative to classical extraction techniques, such as solid-phase extraction (SPE) and liquid-liquid extraction (LLE). As the name implies, microextraction techniques employ a very small volume of the extraction phase compared to the volume of the sample (Lord and Pawliszyn 2000). Subsequent to the introduction of solid-phase microextraction (SPME) by Professor Janusz Pawliszyn and his group in 1987, a large number of microextraction techniques were introduced during the last three decades. Microextraction techniques can be classified into two major classes based on the nature of the extracting phase: (a) sorbent-based microextraction techniques and (b) solvent-based microextraction techniques (Figure 1.1). Noteworthy members of sorbent-based microextraction techniques include: (1) SPME; (2) micro SPE (µSPE); (3) stir bar sorptive extraction (SBSE); (4) microextraction by packed sorbent (MEPS); (5) thin-film microextraction (TFME); (6) electro membrane extraction (EME); (7) fabric phase sorptive extraction (FPSE); and (8) molecularly imprinted polymer-based microextraction. Major members of the solvent-based microextraction family include: (1) single drop microextraction (SDME); (2) liquid-phase microextraction (LPME); (3) ionic liquid-based microextraction; and (4) deep eutectic solvent-based microextraction. It is noteworthy to mention that, unlike extraction techniques, microextraction techniques are governed by the equilibrium between the donor phase (primarily aqueous sample) and the acceptor phase (the extracting phase). Due to their green nature and miniaturized format, microextraction techniques have found many new applications in the field of analytical toxicology, where the available sample volume is often limited.

1.3.2 Theoretical Considerations

1.3.2.1 Solvent-Based Microextraction

Solvent-based microextraction techniques have numerous variants. One popular technique is a two-phase LPME technique consisting of one donor phase and one acceptor phase. Jeannot and Cantwell proposed a general model for equilibrium and mass transfer in a two-phase system. The rate constant K for the equilibrium is given by the equation:

$$\frac{1}{\beta_{oo}} = \frac{1}{\beta_o} + \frac{K_{ow}}{\beta_w} \quad (1.1)$$

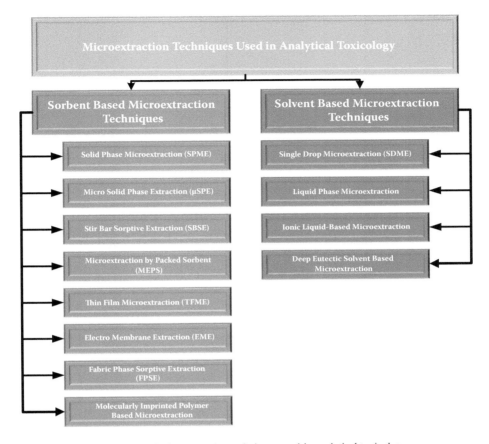

FIGURE 1.1 Classification scheme of microextraction techniques used in analytical toxicology.

$$K = \frac{Ai\beta oo \left[Kow. \left(\frac{Vo}{Vw}\right) + 1 \right]}{Vo} \tag{1.2}$$

where Ai = Interfacial area between the organic and the aqueous phase

β_{oo} = Overall mass transfer coefficient for the organic phase in cm/s

β_o = Mass transfer coefficient of the organic phase in cm/s

β_w = Mass transfer coefficient of the aqueous phase in cm/s

V_o = Volume of the organic phase

V_w = Volume of the aqueous phase

K_{ow} = Distribution ratio between the organic and the aqueous phases

As can be deduced from the equations, the time required for accomplishing the equilibrium would be the minimum if:

1. Ai, β_o, and β_w are maximized;
2. V_w is minimized.

As such, if the mass transfer coefficient values (β values) from water to the organic phase are maximized and water volume is kept at the minimum possible value, the equilibrium will be reached faster. However, the absolute mass of the analyte in the organic phase may remain too low for the analytical instrument to be able to detect. Therefore, the overall sensitivity of the technique must be given appropriate consideration.

1.3.2.2 Sorbent-Based Microextraction

Sorbent-based microextraction techniques, such as SPME, SBSE, and TFME, utilize a thin film of a polymeric-extracting phase to selectively isolate and pre-concentrate the analytes of interest from the sample when it is exposed directly to the sample (direct immersion extraction) or the headspace of the sample resides in a confined container (headspace extraction) for a predetermined time. Although not all the microextraction techniques have been designed to extract the analytes in both the headspace extraction mode and the direct extraction mode, in principle all are capable of performing in both the extraction modes. The mass transfer of analytes from the bulk of the sample to the extracting sorbent of the microextraction device begins immediately when the sorbent is exposed to the sample, either directly (direct immersion extraction) or indirectly (headspace microextraction). Since the mass transfer in microextraction techniques is governed by the distribution equilibrium between the sample matrix and the extracting phase, the mass transfer continues until the equilibrium is reached. As such, under equilibrium extraction conditions, the maximum amount of the extractable analyte is fixed and independent of the extraction time once the extraction equilibrium is reached. Therefore, the maximum sensitivity of a microextraction technique can be achieved only when it is performed under equilibrium conditions.

Extracted analytes from the microextraction devices can be desorbed thermally (thermal desorption) or by exposing them to an organic solvent (solvent desorption) for eluting the analytes from the microextraction device and the subsequent introduction into the chromatographic system.

1.3.3 Sorbent-Based Microextraction Techniques

1.3.3.1 Solid-Phase Microextraction (SPME)

SPME undoubtedly deserves credit for the beginning of a new era of microextraction technologies characterized with miniaturization and solvent-free or solvent-minimized sample preparation. SPME extracts analytes by absorption or adsorption into a polymeric coating immobilized on the surface of a fibre (fibre-SPME) or inside of a fused silica capillary (in-tube SPME). Subsequent to the extraction of analyte either in headspace extraction mode or in direct immersion extraction mode, the analytes are desorbed by exposing the SPME fibre into a gas chromatography (GC) inlet for thermal desorption or into a special interface in high-performance liquid chromatography (HPLC) for solvent-mediated desorption. Due to the lack of chemical bonding between the polymeric coating and the fibre, solvent-mediated desorption is not a common practise. The majority of the applications developed with SPME are based on thermal desorption in the GC inlet. Recent introduction of biocompatible SPME coatings has positioned the technique to be better equipped for toxicological analysis. SPME, due to its numerous advantageous features, enjoys enormous popularity in analytical toxicology (Ulrich 2000; Pragst 2007; Kataoka 2015).

1.3.3.2 Micro Solid-Phase Microextraction (μSPE)

μSPE is a miniaturized format of classical SPE that is based on the use of a cartridge (spin column μSPE, SC-μSPE) or pipette tip (pipette tip μSPE, PT-μSPE) packed with different sorbent materials, including C18, C8, etc. As a new format of conventional SPE, μSPE is an exhaustive or near-exhaustive sample preparation technique. The analyte extraction process in μSPE begins by drawing the sample solution into the tip (PT-μSPE), followed by dispensing back into the sample tube. These two cumulative steps are defined as one aspirating/dispensing cycle. By replicated aspirating/dispensing cycles, the extraction procedure will reach to equilibrium. Finally, the absorbed analytes are eluted using an appropriate solvent (Beckett et al. 2021; Napoletano et al. 2012). μSPE enjoys all the advantageous features of SPE, such as simplicity, rapidity, and the ability to achieve a high enrichment factor. The new format also allows handling a small volume of biological samples, an inevitable criterion in analytical toxicology as the available sample volume is often limited.

1.3.3.3 Stir Bar Sorptive Extraction (SBSE)

SBSE was introduced primarily to address the poor sensitivity of SPME attributed to the minuscule amount of extraction sorbent (~0.5 µL). SBSE utilizes a glass tube to house a cylindrical bar magnet, and the extracting sorbent is immobilized on the outer surface of the glass tube. Due to the high volume of the extracting sorbent, SBSE claims to be 1000× more sensitive to SPME fibre. The extraction is carried out by immersing the stir bar inside the sample in a sample vial on a magnetic stirrer. The stir bar rotates at a given rpm and continues collecting the analytes until the equilibrium of the analyte between the sample and the stir bar is reached. The analytes are desorbed using a thermal desorption unit, and the sample vapour enters into the GC via an interface. Due to the viscous nature of the polymeric sorbent (polydimethylsiloxane and polydimethylsiloxane/ethylene glycol mixed), the sample must be clean and particle free. As such, the application of SBSE in analytical toxicology is limited.

1.3.3.4 Microextraction by Packed Sorbent (MEPS)

MEPS is a miniaturized format of SPE that allows sample volume of as little as 10 µL. MEPS utilizes approximately 1–2 mg of the solid sorbent, such as C2, C8, and C18, packed inside a syringe barrel as a plug or between the barrel and the needle as a cartridge. Due to the integration of solid sorbent inside the syringe, MEPS can be connected online to a GC or liquid chromatography (LC) without any modification. The extraction is carried out via draw-eject cycle of the sample through the sorbent using an autosampler. At the end of the pre-determined draw-eject cycle, the sorbent is washed to remove any unwanted compounds or matrix interferents. Subsequently, the adsorbed analytes are eluted with an organic solvent or the mobile phase (if HPLC is used for the analysis), and the eluant is injected into the chromatographic system. The entire process is automated and therefore compatible with high throughput analytical/bioanalytical laboratories (Abdel-Rehim 2010).

1.3.3.5 Electromembrane Extraction (EME)

EME combines the benefits of electroanalysis and LPME. An EME device consists of a hollow fibre impregnated with a supported liquid membrane (SLM), an acceptor phase placed in the lumen of the hollow fibre, and two platinum electrodes – one placed inside the hollow fibre and the other one placed inside the sample. The sample volume typically ranges between 150 and 500 µL. The volume of the acceptor phase depends on the dimension of the hollow fibre. The pH of the acceptor phase is adjusted to a value so that the analytes remain in a charged state. During analyte extraction in EME, the electrodes are connected to a power supply to create an electric field across the SLM, where SLM functions as a resistor. The applied electric potential across the SLM can be kept between 1 and 300 V. Instead of classical alternate current, direct current obtained from common batteries (9 V) can be used as the power source. When extracting cations, the cathode is placed in the acceptor phase, and the anode is placed inside the sample solution. During extraction of anions, electrodes are positioned in the reverse direction. The applied voltage forces the charged analytes to migrate from the sample solution, through the SLM, toward the electrode placed inside the acceptor phase. The inherent advantages of EME have drawn substantial attention from the toxicologists and clinical chemists. The technique has made electro-assisted extraction of acidic and basic (ionized or ionizable compounds) drugs simple, rapid, and convenient (Jamt et al. 2012; Petersen et al. 2011).

1.3.3.6 Fabric Phase Sorptive Extraction (FPSE)

FPSE represents a major breakthrough invention in separation science since the invention of SPME in 1987. FPSE has successfully combined the extraction principle of two major yet competing sample preparation techniques: SPE, governed by the exhaustive extraction principle, and SPME, governed by the equilibrium extraction principle. The integration of the two extraction principles into a single sample preparation technique in FPSE has positioned itself as an inevitable component of modern analytical and bioanalytical laboratories. FPSE utilizes a porous and permeable fabric as the substrate to host sol-gel derived advanced material systems as the extracting sorbent. Instead of the physical coating process used for immobilizing sorbent polymers in classical microextraction techniques, FPSE utilizes the advantages of a sol-gel synthesis process that chemically binds the polymeric network of the extracting sorbent to the fabric support and consequently provides remarkably high thermal, solvent, and chemical stability to the FPSE device. As such, the FPSE membrane can be exposed to high temperatures or any organic or organo-aqueous solvent for desorbing/eluting the adsorbed analyte without compromising the integrity of the device. Due to the flexibility of the FPSE membrane and its planar geometry, the FPSE device can be inserted directly into the biological sample matrix without requiring any sample pre-treatment, such as protein precipitation, filtration, centrifugation, etc. FPSE has also eliminated the post-extraction steps, such as solvent evaporation and sample reconstitution, from the sample preparation workflow. The simplification sample preparation workflow in FPSE not only saves money, solvent, and labour but also minimizes analyte loss and improves the overall quality of the analytical data.

FPSE can be carried out in immersion extraction mode as in SPME fibre. FPSE membrane can be used as an SPE disk, too. FPSE can also be used as a better alternative to a dried blood spot card. The numerous advantages of FPSE have been exploited in many recent toxicological studies (Kabir et al. 2017; Locatelli et al. 2018, 2019, 2020; Taraboletti et al. 2019; Tartaglia et al. 2020).

1.3.3.7 Molecularly Imprinted Polymer-Based Microextraction

Molecularly imprinted polymers (MIPs) are a new class of compounds that have drawn enormous interest in recent years in many fields, including analytical toxicology. MIPs are synthetic antibodies created using one or multiple template molecules. MIPs recognize the template molecules in terms of their shape, size, and functional composition of the templates. MIPs demonstrate very high selectivity and binding capacity toward the template molecule. In addition, MIPs demonstrate high chemical, mechanical, and thermal stabilities. Because of the inherent selectivity toward the template molecule, MIPs are highly favourable in analytical toxicology, where isolation of the target analyte from a highly complex sample matrix is always challenging (Ansari and Karimi 2017).

1.3.4 Solvent-Based Microextraction Techniques

1.3.4.1 Single Drop Microextraction (SDME)

SDME utilizes a single drop of a water-miscible extracting solvent, where the analytes are partitioned between the single drop of solvent and the aqueous solution based on the analyte's partition coefficient between the two liquid phases. The single drop of the solvent, formed at the tip of a GC or LC syringe, is exposed to the aqueous sample during analyte extraction. At the end of extraction, the solvent drop is retracted back into the syringe needle and is injected into the chromatographic system.

1.3.4.2 Liquid-Phase Microextraction (LPME)

LPME, a miniaturized form of liquid-liquid extraction, generally utilizes sub-microliter volume of the solvent for the extraction process. Since its introduction two decades ago, many new formats have emerged, such as single drop microextraction, hollow-fibre microextraction, electromembrane extraction, and dispersive liquid–liquid microextraction. Regardless of the format, analytes are extracted from the aqueous phase to the organic phase based on the partition coefficient of the analyte between the organic phase and the aqueous phase. At the end of the extraction, the organic phase is separated from water and is injected into the chromatographic system.

1.3.4.3 Ionic Liquid-Based Microextraction

Ionic liquids (ILs), often known as liquid salts, are considered to be a promising alternative to the conventional organic solvents traditionally used in liquid-phase extraction and microextraction due to their low volatility, low flammability, and good thermal and solvent stability. To extract the target analytes from the biological sample, a water-immiscible IL is added to the aqueous sample to form two distinct phases. The cloudy sample solution is often vortexed and centrifuged to separate the IL from the aqueous phase. The IL enriched with the analytes is then injected into the chromatographic system.

1.3.4.4 Deep Eutectic Solvent-Based Microextraction

Deep eutectic solvents (DESs), a subclass of ILs, have drawn enormous interest in LPME due to their facile synthesis procedures, low cost, and biodegradability as well as the inclusion of hydrogen bond acceptor (HBA) and hydrogen bond donor (HBD) in their chemical structure. Typically, in microextraction using DES, a water-miscible DES is added to water to form a homogeneous solution. Subsequently, an emulsifier solvent, such as tetrahydrofuran, is added to convert the homogeneous solution into a turbid solution. An ultrasonic bath is often used to disperse the aggregated DES droplets into the aqueous sample. At the end of extraction, the DES containing the analytes is collected by an external magnet, whereas the aqueous phase is decanted. Finally, the analytes are eluted by an appropriate solvent, and an aliquot of the eluant is injected into the chromatographic system.

1.4 Conclusion and Future Trends

Sample preparation remains the most important step in the overall analytical workflow in the broad field of analytical toxicology. Limited and overwhelmingly complex sample composition; the ever-changing list of the target analytes, including designer drugs; ultra-low concentration of the target analytes; and other factors pose unprecedented challenges to analytical toxicologists. Thanks to the development of new microextraction-based sample preparation technologies in recent years, the analytical workflow has been substantially simplified. Many redundant and error-prone steps have been eliminated, and the use of toxic organic solvents has been minimized/eliminated. The recent improvements in the hardware for gas-phase and liquid-phase separation techniques, mass spectral detection, and powerful computing software for rapid data collection and analysis can be fully exploited only when a sample is prepared properly, truly represents the original sample, and is free of matrix interferents and other components that potentially harm the performance of the analytical instrument. Microextraction techniques have successfully demonstrated their inevitability in the progress of analytical toxicology. It is expected that the development of both the solvent- and sorbent-based microextraction techniques will continue in years to come, with a focus on automation and advanced materials.

REFERENCES

Abdel-Rehim, Mohamed. "Recent Advances in Microextraction by Packed Sorbent for Bioanalysis." *Journal of Chromatography A* 1217 (2010): 2569–80.

Ansari, Saeedeh, and Majid Karimi. "Recent Progress, Challenges and Trends in Trace Determination of Drug Analysis Using Molecularly Imprinted Solid-phase Microextraction Technology." *Talanta* 164 (2017): 612–25.

Beckett, Nicola, Rebecca Tidy, Bianca Douglas, and Colin Priddis. "Detection of Intact Insulin Analogues in Post-mortem Vitreous Humour-Application to Forensic Toxicology Casework." *Drug Testing and Analysis* 13 (2021): 604–13.

Daniel, C. Ralph, Bianca Maria Piraccini, and Antonella Tosti. "The Nail and Hair in Forensic Science." *Journal of the American Academy of Dermatology* 50 (2004): 258–61.

Dinis-Oliveira, Ricardo J., Felix Carvalho, Jose A. Duarte, Fernando Remiao, Antonio Marques, Agostinho Santos, and Teresa Magalhaes. "Collection of Biological Samples in Forensic Toxicology." *Toxicology Mechanisms and Methods* 20 (2010): 363–414.

Flanagan, Robert J. *Fundamentals of Analytical Toxicology* [electronic resource]. England: John Wiley & Sons, 2007.

Jain, Rajeev, and Ritu Singh. "Applications of Dispersive Liquid-liquid Micro-extraction in Forensic Toxicology." *Trac-Trends in Analytical Chemistry* 75 (2016): 227–37.

Jamt, Ragnhild Elén Gjulem, Astrid Gjelstad, Lars Erik Eng Eibak, Elisabeth Leere Oiestad, Asbjørg Solberg Christophersen, Knut Einar Rasmussen, and Stig Pedersen-Bjergaard. "Electromembrane Extraction of Stimulating Drugs from Undiluted Whole Blood." *Journal of Chromatography A* 1232 (2012): 27–36.

Jones, Graham R. "Clarke's Analytical Forensic Toxicology." *Postmortem Toxicology* 2 (2008): 191–217.

Kabir, Abuzar, Rodolfo Mesa, Jessica Jurmain, and Kenneth Furton. "Fabric Phase Sorptive Extraction Explained." *Separations* 4 (2017): 21.

Kataoka, Hiroyuki. "SPME Techniques for Biomedical Analysis." *Bioanalysis* 7 (2015): 2135–44.

Kerrigan, Sarah, and Barry S. Levine. *Principles of Forensic Toxicology*. Cham: Springer Nature, 2020.

Locatelli, Marcello, Kenneth G. Furton, Angela Tartaglia, Elena Sperandio, Halil I. Ulusoy, Abuzar Kabir. "An FPSE-HPLC-PDA Method for Rapid Determination of Solar UV Filters in Human Whole Blood, Plasma and Urine." *Journal of Chromatography B* 1118–1119 (2019): 40–50.

Locatelli, Marcello, Angela Tartaglia, Francesca D'Ambrosio, Piera Ramundo, Halil I. Ulusoy, Kenneth G. Furton, and Abuzar Kabir. "Biofluid Sampler: A New Gateway for Mail-in-Analysis of Whole Blood Samples." *Journal of Chromatography B* 1143 (2020): 122055.

Locatelli, Marcello, Nicola Tinari, Antonino Grassadonia, Angela Tartaglia, Daniela Macerola, Silvia Piccolantonio, Elena Sperandio, Christian D'Ovidio, Simone Carradori, Halil Ibrahim Ulusoy, Kenneth G. Furton, and Abuzar Kabir. "FPSE-HPLC-DAD Method for the Quantification of Anticancer Drugs in Human Whole Blood, Plasma, and Urine." *Journal of Chromatography B* 1095 (2018): 204–13.

Lord, Heather and Janusz Pawliszyn. "Evolution of Solid-Phase Microextraction Technology." *Journal of Chromatography A* 885, no. 1–2 (2000): 153–193.

Maurer, Hans H. "Analytical Toxicology." *Analytical and Bioanalytical Chemistry* 388 (2007): 1311.

Maurer, Hans H. "Analytical Toxicology." In *Molecular, Clinical and Environmental Toxicology*. Vol. 2, *Clinical Toxicology*, edited by Andreas Luch, 317–37. Basel: Birkhauser, 2010.

Napoletano, Sabino, Camilla Montesano, Dario Compagnone, Roberta Curini, Giuseppe D'Ascenzo, Claudia Roccia, and Manuel Sergi. "Determination of Illicit Drugs in Urine and Plasma by Micro-SPE Followed by HPLC-MS/MS." *Chromatographia* 75 (2012): 55–63.

Petersen, Nickolaj Jacob, Knut Einar Rasmussen, Stig Pedersen-Bjergaard, and Astrid Gjelstad. "Electromembrane Extraction from Biological Fluids." *Analytical Sciences* 27 (2011): 965–72.

Pragst, Fritz. "Application of Solid-phase Microextraction in Analytical Toxicology." *Analytical and Bioanalytical Chemistry* 388 (2007): 1393–414.

Rohrig, Timothy P., and Charity A. Hicks. "Brain Tissue: A Viable Postmortem Toxicological Specimen." *Journal of Analytical Toxicology* 39 (2015): 137–9.

Schramm, Willfried, Richard H. Smith, Paul A. Craig, and David A. Kidwell. "Drugs of Abuse in Saliva – A Review." *Journal of Analytical Toxicology* 16 (1992): 1–9.

Taraboletti, Alexandra, Maryam Goudarzi, Abuzar Kabir, Bo-Hyun Moon, Evagelia Laiakis, Jerome Lacombe, Pelagie Ake, Sueoka Shoishiro, David Brenner, Albert Fornace Jr, and Frederic Zenhausern. "Fabric Phase Sorptive Extraction - A Metabolomic Pre-processing Approach for Ionizing Radiation Exposure Assessment." *Journal of Proteome Research* 18 (2019): 3020–31. acs.jproteome.9b00142-undefined.

Tartaglia, Angela, Abuzar Kabir, Francesca D'Ambrosio, Piera Ramundo, Songul Ulusoy, Halil I. Ulusoy, Giuseppe M. Merone, Fabio Savini, Cristian D'Ovidio, Ugo De Grazia, Kenneth G. Furton, and Marcello Locatelli. "Fast Off-Line FPSE-HPLC-PDA Determination of Six NSAIDs in Saliva Samples." *Journal of Chromatography B-Analytical Technologies in the Biomedical and Life Sciences* 1144 (2020): 9.

Ulrich, Sven. "Solid-Phase Microextraction in Biomedical Analysis." *Journal of Chromatography A* 902 (2000): 167–94.

2

Application of Solid-Phase Microextraction in Analytical Toxicology

Rakesh Roshan Jha[1,2] **and Rajeev Jain**[3]
[1]*Centre of Analytical Bioscience, School of Pharmacy, University of Nottingham, Nottingham NG7 2RD, UK*
[2]*Analytical Chemistry Laboratory, Regulatory Toxicology Group, CSIR-Indian Institute of Toxicology Research (CSIR-IITR), Vishvigyan Bhawan, UP, India*
[3]*Central Forensic Science Laboratory, Directorate of Forensic Science Services, Ministry of Home Affairs, Govt. of India, India*

CONTENTS

2.1 Introduction 11
2.2 Applications of SPME in Analytical Toxicology 13
    2.2.1 Analysis of Pesticides 13
    2.2.2 Analysis of Benzodiazepines 14
    2.2.3 Analysis of Amphetamines and Related Substances 14
    2.2.4 Analysis of Cannabinoids 15
    2.2.5 Analysis of Cocaine and Its Metabolites 15
    2.2.6 Analysis of Opium Alkaloids and Opiates 16
    2.2.7 Analysis of Therapeutic Drugs 17
    2.2.8 Analysis of Volatile and Other Toxicants 17
2.3 Conclusion 18
Reference 18

2.1 Introduction

Analytical toxicology is a branch of science that deals with qualitative and quantitative determination xenobiotic and toxic compounds in complex biological matrices with the aim of resolving various research questions. Analytical toxicology plays a vital role in the identification and prevention of poisoning (Flanagan et al., 2007; Wille and Lambert, 2007). The most important part of the analytical method of development is sample preparation, and it probably could be treated as the backbone of the analytical toxicology. Sample preparation is the key to resolve the complexity of the matrices, which is always a big challenge for analytical chemists, especially in the case of analytes that are present in traces (Pragst, 2007). Sample preparation methods are categorized as traditional, or old, extraction techniques and modern, or miniaturized, extraction techniques. Liquid-liquid extraction (LLE), Soxhlet extraction, and solid-phase extraction (SPE) are examples of traditional techniques that are excellent sample preparation methods for isolation and determination of chemical entities from different environmental, biological, and food matrices (Jha et al., 2017, 2018a). These techniques are frequently used in various research laboratories for routine analysis of samples, even today. However, there are certain limitations associated with these methods, such as the requirement of (i) large amount of samples, (ii) large amount of extraction solvents (mostly toxic), (iii) lengthy extraction time, and (iv) multi-step procedures. Additionally, these techniques are neither cost-effective nor environmentally friendly. These limitations associated with traditional sample preparation methods are overcome by miniaturized extraction methods. Liquid-liquid microextraction (LLME), dispersive liquid-liquid microextraction (DLLME), single droplet microextraction (SDME), and solid-phase microextraction (SPME) are some examples of modern sample preparation techniques (Jha et al., 2018b; Kumari et al., 2015). These methods require a very small amount of samples (mL or mg and even in μL and μg), for which microliters of extraction solvent are enough to successfully extract the target analytes from matrices. Further, the microextraction method is quick, cost-effective, and eco-friendly, too.

SPME was first introduced in the 1990s by Arthur and Pawliszyn. SPME features simultaneous extraction and pre-concentration of analytes directly from aqueous and gas samples. This technique is very fast, easy to use, portable, and it has high extraction efficiency for the targeted analytes (Frison et al., 2001). The working principle of SPME is the use of the extracting phase in the assistance of solid support, which is kept in contact with the sample phase for a period until equilibrium exists between the sample phase and the extraction phase. SPME in combination with an analytical instrumentation technique, such as gas chromatography-mass spectrometry (GC-MS), becomes a very influential tool for the investigation of the various chemical entities as the SPME fibre laden with the chemical analytes through the extraction process could be directly inserted into the injector port of the GC-MS (Pragst, 2007; Risticevic et al., 2009). The direct injection of the SPME fibre lowers the time of sample preparation and further reduces the analyte loss during the pre-concentration process, offering superior extraction efficiency of the targeted analytes. The coupling of SPME with GC-MS makes it a solvent-free extraction method as none of the solvent is involved in the extraction and pre-concentration process (Dong et al., 2013). Hence, SPME is the most appropriate technique for the analysis of volatile and semi-volatile compounds using GC-MS.

SPME can be categorized into three different groups based on the mode of action, such as (a) direct immersion SPME (DI-SPME), (b) headspace SPME (HS-SPME), and (c) membrane-protected SPME (MP-SPME). In the DI-SPME mode, the analyte of interest is directly transferred to the coated fibre of SPME after insertion into the sample matrix, and the transfer takes place until equilibrium is achieved between the sample matrix and the extractant phase. The DI-SPME mode is assisted with agitation for smooth transfer of analytes to the extractant phase (fibre coating), and the extent of agitation depends on the nature of the samples. In the case of liquid samples, fast agitation may be required, which could be achieved by the stirrer, and the level of agitation could be controlled with the RPM of the stirrer. On the other hand, for gaseous samples, a gentle agitation is enough for complete and smooth transfer of the analyte of interest from the matrix to the fibre. In HS-SPME, the analyte of interest passes through the air barrier prior to reaching the coating of the fibre (Lord and Pawliszyn, 2000; Abdulra'uf et al., 2012). The fibre coated with the organic polymer is subjected to the HS just above the sample, where volatilized targeted analytes are adsorbed on the fibre, and once extraction is completed, the fibre could be directly injected into the analytical instrument of choice for analysis. SPME in HS mode offers a very short time of analysis with high sensitivity of analytes up to femtograms. Further, in the HS-SPME mode, fibres have long durability as they are prevented from higher molecular mass and several interferences. Additionally, factors affecting the extraction efficiency of the analytes, such as salt addition and pH for the sample matrix, could be adjusted without damaging the fibre in the HS-SPME mode. The analytes having low volatility when present in the dirty sample matrix could be extracted using MP-SPME, where a membrane is used to protect the fibre of the SPME from the matrix impurities (Lord and Pawliszyn, 2000). In MP-SPME, the fibre is not in direct contact with the sample matrix, which increases the life span of the fibre, and also the membrane in most of the cases provides further selectivity of the analytes by allowing analytes of interest to reach the fibre through its pores (Zhang et al., 1996).

The SPME fibre is coated with polymeric material, which is designed according to the chemical compounds to be analyzed. The sample solution, where the SPME fibre is immersed for extraction purposes, is continuously stirred for maximum and uniform adsorption of the analyte on the fibre, completing the extraction procedure until equilibrium is attained (Lord and Pawliszyn, 2000). There are several polymeric-coated silica fibres commercially available today, including polyethylene glycol (PEG), divinylbenzene (DVB), polydimethylsiloxane (PDMS), polyacrylate (PA), and carboxen (CW). These coatings are available in different thicknesses and dimensions, which could be used as per the analysis required (Spietelun et al., 2010). Most of the polycyclic aromatic hydrocarbons (PAHs), pesticides, and aromatic amines could be extracted using PDMS coated fibre while VOC and metals can be extracted either with CW or with PDMS. Hence, users have numbers of SPME fibres available for volatile and semi-volatile compounds, and they can choose the fibre as per their requirement. Further, different approaches have been used in the discovery of a new coating of the fibre for extraction of specific analytes for which molecularly imprinted polymer, sol-gel, ionic imprint, on-fibre derivatization, and immunosorbent techniques have been used (Dietz et al., 2006).

There are certain variable parameters that may affect the extraction efficiency of the targeted analytes using SPME such as pH, ionic strength of the sample solution, speed of agitation, time of extraction to attain equilibrium, temperature, fibre coating material, and fibre dimensions (especially fibre thickness) (Spietelun et al., 2013). All these parameters need to be optimized for a set of experiments in order to achieve the best extraction efficiency of the analyte of interest, which can be achieved either using the one time one variable method or some statistical application, such as design of experiments. By adjusting pH and ionic strength, affinity of the analyte toward the SPME fibre can be enhanced, resulting in high extraction of the targeted analytes. Similarly, speed of agitation promotes transfer of analytes from the sample matrix to the extractant phase. In the case of gas samples, gentle agitation provides the smooth transfer of analytes, whereas in the case of liquid samples, medium to high agitation is required to achieve the best extraction efficiency (Lord and Pawliszyn, 2000). Once equilibrium is attained between the sample phase and the extractant phase, further increases in extraction time may decrease the extraction efficiency. Additionally, a less stable or unstable analyte should be immediately subjected for analysis after equilibrium. The choice of fibre and its dimensions are the most significant aspects while performing SPME, as the fibre is generally specific to a class of compounds.

The present chapter is mainly focused on applications of SPME for the analysis of pesticides, drugs of abuse, cannabinoids, cocaine (COC), amphetamines, volatile organic compounds, and therapeutic drugs in various biological matrices. Particular focus is made on protocol parts of the cited studies in order to help the reader get insights on practical aspects of the SPME technique.

2.2 Applications of SPME in Analytical Toxicology

2.2.1 Analysis of Pesticides

SPME has been extensively applied for analysis of pesticides from various matrices, such as fruits, vegetables, soil, and water (Abdulra'uf et al., 2012). However, applications of SPME for the determination of pesticides from biological samples, such as blood and postmortem (PM) tissues, are limited (Pragst, 2007). SPME is suitable for extraction of pesticides from biological samples due to their relatively hydrophobic nature (Pragst, 2007). HS-SPME was used for monitoring 18 organochlorine (OC) pesticides from human serum samples of 1,904 adults. About 1 mL of serum sample was diluted with high-performance liquid chromatography (HPLC) grade water followed by the addition of 0.1 g NaCl (for salting out effect) and 0.02 g of K_2CO_3 (for pH adjustment to 11). SPME fibre with 85 μm of PA coating was exposed to the HS of the sample for 50 min at 90 °C under constant stirring of 500 rpm. This was followed by desorption of analytes into a heated GC-MS injection port at 280 °C for 2 min. Hexachlorobenzene was the most frequently detected OC pesticide in all the tested samples (Kim et al., 2013).

An HS-SPME method in combination with GC-MS-MS was reported for trace-level determination of 11 organophosphorous (OP) and OC pesticides in the whole human blood, such as lindane, hexachlorobenzene (HCB), chlorpyrifos, endosulfan, etc. The procedure was simple and consisted of dilution of the blood sample with ultrapure water (1:1 v/v). The sample was then pre-heated at 90 °C for 30 min, and then HS-SPME was performed using PA fibre for 30 min. This was followed by drying the fibre and desorption of analytes into the GC injection port at 240 °C for 4 min. Limit of detection (LOD) was found to be in the range of 0.02–3 ng mL^{-1} (Hernandez et al., 2002).

Tsoukali et al. developed and validated an analytical method based on HS-SPME followed by GC with nitrogen phosphorous detection (NPD) for determination of methyl parathion (MP) in PM samples, such as whole blood, liver, and kidney. Prior to SPME, tissue samples were homogenized, and 300 μL of homogenate was used for extraction. The sample was pre-incubated for 15 min, followed by HS-SPME with 85 μm PA fibre for 20 min. A small amount of NaCl was added in order to increase the ionic strength of the sample. The method successfully detected MP in the PM blood sample of a 21-year-old women who had committed suicide by injecting MP intravenously. The concentration of MP in the blood sample was found to be 24 μg mL^{-1} (Tsoukali et al., 2004).

2.2.2 Analysis of Benzodiazepines

Benzodiazepines are one of the most frequently prescribed tranquilizers, sedatives, and hypnotic drugs and are commonly encountered in clinical and forensic cases. Five common benzodiazepines, namely oxazepam, diazepam, nordiazepam, flunitrazepam, and alprazolam, were analyzed by SPME-GC-MS in human urine and plasma samples. Before performing DI-SPME, octanol was immobilized on a PA fibre for improved enrichment of benzodiazepines. Extraction was performed in a direct immersion mode for 15 min at room temperature under slightly acidic conditions. SPME parameters were optimized using a design of experiment strategy. The method was found to be sensitive with LODs in the range of 0.01–0.45 μg mol-1 and 0.01–0.48 μg mol^{-1}, respectively, in urine and plasma samples (Reubsaet et al., 1998). Another method was described for determining midazolam in human plasma by SPME-GC-MS. The author first deproteinized plasma samples and performed SPME with 85 μm PA fibre at 50 °C for 10 min. LOD for midazolam was found to be 1 ng mL^{-1} (Frison et al., 2001).

A biocompatible SPME fibre was designed by coating alkyldiol-silica (ADS) on a stainless-steel wire. An epoxy binding agent for immobilization of ADS on stainless steel wire was used. This specially fabricated fibre was able to fractionate the protein and analyte component from the biological sample; hence, no blood protein precipitation was required, resulting in minimized sample preparation time. The ADS-SPME fibre was directly immersed into the blood sample for extraction of diazepam and its major metabolites N-desmethyldiazepam, oxazepam, and temazepam. After extraction, the fibre was rinsed with water and interfaced with LC-MS for desorption and separation of extracted analytes (Walles et al., 2004).

2.2.3 Analysis of Amphetamines and Related Substances

Amphetamines are powerful central nervous system stimulants and the second most commonly used illicit drug worldwide. Some popular examples are amphetamine (AMP), methamphetamine (MA), 3,4-methylenedioxyamphetamine (MDA), 3,4-methylenedioxymethamphetamine (MDMA, or ecstasy), and 3,4-methylenedioxyethamphetamine (MDEA) (Jain and Singh, 2016a). SPME has been applied for determining amphetamines and other drugs of abuse from oral fluid samples. For this purpose, DI-SPME has been applied for extraction of AMP, MA, and MDMA, which has shown greater sensitivity in comparison to HS-SPME. Extraction was performed at room temperature under constant stirring of the sample using PDMS fibre (Fucci et al., 2003). In another similar application, AMP and MA were derivatized using butyl chloroformate directly in oral fluid samples. These derivatized analytes were then extracted using PDMS fibre by directly immersing the fibre into the sample (Yonamine et al., 2003). Both analytical methods have shown similar sensitivity for amphetamines in the range of 1–10 ng mL^{-1}.

Analytical methods for determining amphetamines in urine samples are generally based on their derivatization to improve volatility and decrease polarity. Ugland et al. described a method for derivatization of amphetamines and ecstasy directly in urine samples using propylchloroformate reagent, which produced their water stable carbamate derivatives. Under alkaline conditions, these derivatives were extracted using 100 μm PDMS-coated SPME fibre in DI mode for 16 min. LODs of ecstasy, MA, and MDEA were found to be 5 ng mL^{-1}, whereas they were 15 ng mL^{-1} for AMP and MDA (Ugland et al., 1999). Later on, Huang et al. utilized heptafluorobutyric anhydride and heptafluorobutyric chloride as derivatizing reagents for derivatization of AMP and MA. Here, derivatizing reagents were kept in a glass insert that was subsequently kept in a glass vial where SPME fibre was also exposed. The sample was heated at 100 °C for 20 min, and the vapours of analytes were diffused into the glass insert through the holes. In this way, vaporization, adsorption, and absorption could be achieved in a single step. The method has shown superior sensitivity, with detection limits of 0.3 and 1 ng mL^{-1} for MA and AMP, respectively (Huang et al., 2002).

A different approach of derivatization of AMP and MA was described by Okajima et al. They used pentafluorobenzylbromide (PFBBr) as a derivatizing reagent. PFBBr was added directly into blood samples, and the sample was heated at 90 °C for 30 min. This was followed by HS-SPME of derivatives using 100 μm PDMS fibre for another 30 min and subsequent GC-MS analysis. Low detection limits of 0.5 ng g^{-1} could be achieved by this approach (Okajima et al., 2001).

2.2.4 Analysis of Cannabinoids

Analysis of cannabinoids in various complex samples by microextraction techniques has been extensively reviewed by Jain et al. (Jain and Singh, 2016b). As far as SPME is concerned, oral fluid remains one of the preferred matrices of choice for testing cannabinoids due to its easy availability, non-invasive collection, and relatively low protein content. Anzillotti et al. compared the LC-MS-MS and SPME-GC-MS methods for quantitative analysis of Δ^9-tetrahydrocannabinol (THC) in oral fluid samples for assessing driving under the influence of drugs (DUID). The authors analyzed 70 samples by both techniques and concluded that SPME-GC-MS offered superior sensitivity in comparison to the LC-MS-MS method. For THC, the lower LODs were 0.5 and 2 ng mL^{-1} by SPME-GC-MS and LC-MS-MS, respectively. Additionally, along with THC, cannabidiol (CBD) and cannabinol (CBN) were also detected by SPME-GC-MS (Anzillotti et al., 2014). Recently, a pilot study was conducted for analysis of natural and synthetic cannabinoids, such as THC, CBD, CBN (natural) JWH 250, JWH 019, JWH 122, etc., in oral fluid samples. The authors compared the HS and DI mode of SPME and observed that most of cannabinoids could be extracted satisfactorily with the DI mode. The method was found suitable for confirmation of THC at low concentration levels in oral fluid samples as the LOD offered was 1 ng mL^{-1} against the cut-off limit of 2 ng mL^{-1} (Anzillotti et al., 2019).

A simple and rapid analytical method was reported by Emidio et al. for monitoring cannabinoids (THC, CBD, and CBN) in human hair samples. Initially, hair samples were decontaminated, followed by alkaline digestion with NaOH at 90 °C. Cannabinoids were extracted by HS-SPME using PDMS fibre for 40 min at 90 °C. The method did not require derivatization of cannabinoids for GC-MS analysis. Use of ion trap tandem mass spectrometry offered a very low quantitation limit for THC (0.062 ng mg^{-1}), which was below the cut-off value set by the Society of Hair Testing (Emídio et al., 2010). The protocol is shown in Figure 2.1.

2.2.5 Analysis of Cocaine and Its Metabolites

COC, its major metabolite benzylecgonine (BE), and cocaethylene (CE) (which is a transesterification product of COC formed when it is consumed with ethanol) have been analyzed in biological samples such as urine, hair, and plasma by DI-SPME in combination with GC-MS. Analytes

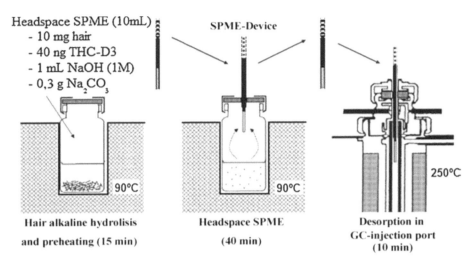

FIGURE 2.1 Procedure for HS-SPME of cannabinoids from hair samples (reproduced with permission from Emídio et al., 2010).

were extracted with PDMS fibre under constant stirring of an alkaline sample solution for 20–25 min. In the case of the hair sample, prior decontamination with dichloromethane followed by long digestion for 18 hours at 50 °C was required in order to liberate the drugs from the matrix. Plasma samples were also subjected to deproteinization with acetonitrile, and supernatant was used for extraction of drugs. These methods were sensitive enough to detect COC, BE, and CE in the range of 5–19 ng mL^{-1} and 0.1–0.5 ng mg^{-1} (de Toledo et al., 2003; Yonamine and Saviano, 2006; Álvarez et al., 2007).

2.2.6 Analysis of Opium Alkaloids and Opiates

A highly sensitive analytical method based on electrically accelerated hollow fibre SPME (EA-HF-SPME) coupled with HPLC was reported by Rihai-Zanjani et al. Ethylenediamine was coated on carbon nanotubes (CNTs). These functionalized CNTs were coated on porous propylene hollow fibre. Adsorption and desorption of morphine from urine samples were aided by a specially designed electric device that could produce an electric voltage in the range of 0–30 V. Adsorbed analytes were desorbed in a washing solution of HPLC by reversing the electric voltage. The method was able to detect morphine in urine samples up to a very low concentration of 0.15 ng mL^{-1}, which is significantly lower than previously reported methods (Riahi-Zanjani et al., 2018). Morphine, codeine, and 6-monoacetylmorphine were extracted using automated HS-SPME from hair samples. Prior to SPME, hair samples were digested by adding methanol and incubated for 18 hours at 50 °C. The methanolic extract was evaporated, and silylation of analytes was performed using bis(trimethylsilyl)trifluoroacetamide containing 1% trimethylchlorosilane (BSTFA + TMCS). The derivatives were then extracted using HS-SPME with PDMS fibre at 125 °C for 25 min. The method was proved to be sensitive and offered LODs in the range of 0.002–0.005 ng mg^{-1} (Moller et al., 2010).

DI-SPME methods have been reported for determining methadone and its metabolite 2-ethylene-1,5-dimethyl-3,3-diphenylpyrrolidine (EDDP) from various biofluids, such as urine, oral fluid, and plasma samples. All methods comprised DI-SPME of these opiate drugs for 30 min at alkaline pH. LODs were achieved in the range of 0.04–6 ng mL^{-1} for methadone (Myung et al., 1999; Bermejo et al., 2000; dos Santos Lucas et al., 2000).

Tramadol and fentanyl were analyzed by HS-SPME in plasma samples in two different applications. PDMS/divinylbenzene (PDMS/DVB, 65 μm) fibre was used for HS-SPME of tramadol at 100 °C for 30 min and analyzed by GC-MS. Tramadol could be detected up to a concentration level of 0.2 ng mL^{-1}

(Sha et al., 2005). Homemade sol-gel based PEG- and Ucon-coated SPME fibres were compared with commercial PDMS fibres for extraction of fentanyl from plasma samples under alkaline conditions. Although PEG- and Ucon-coated fibres exhibited better extraction efficiency for fentanyl, their stability was insufficient due to the presence of etheric functional group, which is susceptible to acidic and alkaline conditions and got exhausted after 20 extractions. The LOD for fentanyl was found to be 0.03 ng mL^{-1} (Bagheri et al., 2007).

2.2.7 Analysis of Therapeutic Drugs

SPME has been widely applied for the analysis of various therapeutic drugs in biological samples. SPME parameters, such as choice of fibre coatings, extraction mode (HS or DI), extraction time, extraction temperature, pH, and ionic strength, have to be carefully optimized for better extraction efficiencies of therapeutic drugs. Some biological samples, such as blood and plasma, require a deproteinization step prior to SPME. For instance, blood samples were deproteinized with perchloric acid prior to DI-SPME for analysis of barbiturates and phenothiazines (Iwai et al., 2004; Kumazawa et al., 2000). However, for local anaesthetic drugs (lidocaine, mepivacaine, prilocaine, etc.), instead of deproteinization, blood samples were directly heated at a high temperature of 120 °C for HS-SPME from blood samples (Watanabe et al., 1998).

Adjustment of pH is also a crucial factor for better extraction of drugs from biological matrices. According to SPME theory, analytes should be present in their neutral form in matrices; therefore, pH should be adjusted according to their pK values. Considering this fact, tricyclic antidepressant drugs were extracted from plasma samples at alkaline pH, i.e., 10 by using PDMS/DVB fibres (Cantú et al., 2006). The fact of suitability of SPME fibre according to polarity of analytes has been exploited for some polar drugs, such as pregabalin, which was converted into a less polar derivative by ethyl chloroformate derivatization directly in urine samples. The derivative thus formed was extracted by DI-SPME with mid-polar fibre (PDMS/DVB) for GC-MS analysis (Mudiam et al., 2012). Similarly, valproic acid (VPA) has been derivatized directly in plasma samples with isobutylchloroformate to produce VPA ethyl ester, which was relatively non-polar than VPA followed by its HS-SPME using non-polar fibre (i.e., PDMS) at 80 °C for 20 min (Deng et al., 2006).

2.2.8 Analysis of Volatile and Other Toxicants

Trichloroethylene metabolites, i.e., dichloroacetic acid, trichloroacetic acid, and trichloroethanol, has been analyzed in human plasma samples of exposed industrial workers by HS-SPME, coupled with GC-electron capture detector (GC-ECD). *In matrix* derivatization of analytes was performed directly on plasma samples with methyl chloroformate. Derivatized compounds were extracted by HS-SPME using PDMS fibre for 22 min. The method offered good sensitivity, with LODs in the range of 0.036–0.068 μg mL^{-1} (Mudiam et al., 2013). Ethyl alcohol has been analyzed in PM specimens, such as blood, urine, and vitreous humour by HS-SPME using PA fibre for 1 min at 60 °C. Samples were diluted with water, followed by the addition of ammonium sulphate to increase recoveries due to the salting out effect (De Martinis and Martin, 2002).

Cyanide, a short-acting powerful toxicant, has been determined in PM blood samples of fire victims. Cyanide has been converted into hydrogen cyanide by the addition of phosphoric acid, followed by HS-SPME for 10 min at 30 °C with carbowax/PDMS fibre. The method was found to be sensitive and offered detection limits of 0.006 μg mL^{-1}. It consumed less than 20 min for analysis. Under optimized conditions, cyanide was detected in PM blood samples at a concentration of 2 μg mL^{-1} (Frison et al., 2006). Halothane was determined in PM biological samples (blood, liver, kidney, brain, urine, and bile) in a case of double homicide. Biological samples, along with ammonium sulphate and sulphuric acid, were pre-heated for 15 min at 100 °C, followed by HS-SPME for another 15 min. The method offered linearity in the concentration range of 0.1–100 mg kg^{-1}, with a detection limit of 0.004 mg kg^{-1} for blood samples. The highest amount of halothane was detected in brain samples (91.5 and 94.4 mg kg^{-1}; Musshoff et al., 2000). A method was reported based on HS-SPME-GC-MS for analyzing strychnine, a toxic alkaloid in blood samples with a detection limit of 6.83 ng mL^{-1}. In this method, only 100 μL of blood was used for analysis. The blood sample was diluted with water and subjected for DI-SPME with carbowax/PDMS fibre for 20 min. The optimized method was applied to blood samples obtained from persons intoxicated with strychnine. Strychnine was detected in the range of 1.03–2.39 μg mL^{-1} (Barroso et al., 2005).

2.3 Conclusion

In recent years, SPME has found wide applications for the analysis of various drugs and poisons in biological specimens and PM matrices. The obvious advantages offered by SPME over conventional extraction techniques are its simplicity, low cost, ease of operation, high extraction efficiencies, complete elimination of toxic organic solvents, and availability of a wide range of fibres for almost all kinds of analytes. Additionally, configurations of automated SPME with analytical instruments are proving to be time and cost saving for forensic and clinical laboratories. Fortunately, now plenty of literature on applications of SPME are available that cover almost all analytes that are routinely tested in analytical toxicological laboratories. Therefore, analytical laboratories should consider this green, rapid, and sensitive sample preparation method for their routine analytical work, which can save them cost and time of analysis as well as protect their health and the environment from toxic organic solvents.

REFERENCE

Abdulra'uf, Lukman Bola, Wasiu Adebayo Hammed, and Guan Huat Tan. "SPME Fibers for the Analysis of Pesticide Residues in Fruits and Vegetables: A Review." *Critical Reviews in Analytical Chemistry* 42, no. 2 (2012): 152–61.

Álvarez, Iván, Ana María Bermejo, María Jesús Tabernero, Purificación Fernández, and Patricia López. "Determination of Cocaine and Cocaethylene in Plasma by Solid-Phase Microextraction and Gas Chromatography–Mass Spectrometry." *Journal of Chromatography B* 845, no. 1 (2007): 90–94.

Anzillotti, Luca, Erika Castrignanò, Sabina Strano Rossi, and Marcello Chiarotti. "Cannabinoids Determination in Oral Fluid by SPME–GC/MS and UHPLC–MS/MS and Its Application on Suspected Drivers." *Science & Justice* 54, no. 6 (2014): 421–6.

Anzillotti, Luca, Francesca Marezza, Luca Calò, Roberta Andreoli, Silvia Agazzi, Federica Bianchi, Maria Careri, and Rossana Cecchi. "Determination of Synthetic and Natural Cannabinoids in Oral Fluid by Solid-Phase Microextraction Coupled to Gas Chromatography/Mass Spectrometry: A Pilot Study." *Talanta* 201 (2019): 335–41.

Bagheri, Habib, Ali Es-haghi, Faezeh Khalilian, and Mohammad-Reza Rouini. "Determination of Fentanyl in Human Plasma by Head-Space Solid-Phase Microextraction and Gas Chromatography–Mass Spectrometry." *Journal of Pharmaceutical and Biomedical Analysis* 43, no. 5 (2007): 1763–68.

Barroso, Mario, E. Gallardo, Claudia Margalho, Sofia Avila, Estela P. Marques, Duarte Nuno Vieira, and Manuel López-Rivadulla. "Application of Solid Phase Microextraction to the Determination of Strychnine in Blood." *Journal of Chromatography B* 816, no. 1–2 (2005): 29–34.

Bermejo, Ana Maria, R. Seara, Ana Cyra dos Santos Lucas, María Jesús Tabernero, Purificacion Fernandez, and Remo Marsili. "Use of Solid-Phase Microextraction (SPME) for the Determination of Methadone and Its Main Metabolite, EDDP, in Plasma by Gas Chromatography-Mass Spectrometry." *Journal of Analytical Toxicology* 24, no. 1 (2000): 66–69.

Cantú, Marcelo Delmar, Daniel Rodrigo Toso, Cristina Alves Lacerda, Fernando Mauro Lanças, Emanuel Carrilho, and Maria Eugênia Costa Queiroz. "Optimization of Solid-Phase Microextraction Procedures for the Determination of Tricyclic Antidepressants and Anticonvulsants in Plasma Samples by Liquid Chromatography." *Analytical and Bioanalytical Chemistry* 386, no. 2 (2006): 256–63.

De Martinis, Bruno Spinosa, and Carmen Cinira Santos Martin. "Automated Headspace Solid-Phase Microextraction and Capillary Gas Chromatography Analysis of Ethanol in Postmortem Specimens." *Forensic Science International* 128, no. 3 (2002): 115–19.

Deng, Chunhui, Ning Li, Jie Ji, Bei Yang, Gengli Duan, and Xiangmin Zhang. "Development of Water-Phase Derivatization Followed by Solid-Phase Microextraction and Gas Chromatography/Mass Spectrometry for Fast Determination of Valproic Acid in Human Plasma." *Rapid Communications in Mass Spectrometry: An International Journal Devoted to the Rapid Dissemination of Up-to-the-Minute Research in Mass Spectrometry* 20, no. 8 (2006): 1281–87.

de Toledo, Fernanda Crossi Pereira, Mauricio Yonamine, Regina Lucia de Moraes Moreau, and Ovandir Alves Silva. "Determination of Cocaine, Benzoylecgonine and Cocaethylene in Human Hair by Solid-Phase Microextraction and Gas Chromatography–Mass Spectrometry." *Journal of Chromatography B* 798, no. 2 (2003): 361–5.

Dietz, Christian, Jon Sanz, and Carmen Cámara. "Recent Developments in Solid-Phase Microextraction Coatings and Related Techniques." *Journal of Chromatography A* 1103, no. 2 (2006): 183–92.

Dong, Liang, Yongzhe Piao, Xiao Zhang, Changxin Zhao, Yingmin Hou, and Zhongping Shi. "Analysis of Volatile Compounds from a Malting Process Using Headspace Solid-Phase Micro-extraction and GC–MS." *Food Research International* 51, no. 2 (2013): 783–9.

dos Santos Lucas, A. C., Ana Maria Bermejo, Purificacion Fernandez, and Maria Jesus Tabernero. "Solid-Phase Microextraction in the Determination of Methadone in Human Saliva by Gas Chromatography-Mass Spectrometry." *Journal of Analytical Toxicology* 24, no. 2 (2000): 93–96.

Emídio, Elissandro Soares, Vanessa de Menezes Prata, and Haroldo Silveira Dórea. "Validation of an Analytical Method for Analysis of Cannabinoids in Hair by Headspace Solid-Phase Microextraction and Gas Chromatography–Ion Trap Tandem Mass Spectrometry." *Analytica Chimica Acta* 670, no. 1–2 (2010): 63–71.

Flanagan, Robert J., Andrew Taylor, Ian D. Watson, and Robin Whelpton. *Fundamentals of Analytical Toxicology*, Vol. 455. Chichester: John Wiley & Sons, 2007.

Frison, Giampietro, Luciano Tedeschi, Sergio Maietti, and Santo Davide Ferrara. "Determination of Midazolam in Human Plasma by Solid-Phase Microextraction and Gas Chromatography/Mass Spectrometry." *Rapid Communications in Mass Spectrometry* 15, no. 24 (2001): 2497–501.

Frison, Giampietro, Flavio Zancanaro, Donata Favretto, and Santo Davide Ferrara. "An Improved Method for Cyanide Determination in Blood Using Solid-Phase Microextraction and Gas Chromatography/Mass Spectrometry." *Rapid Communications in Mass Spectrometry* 20, no. 19 (2006): 2932–38.

Fucci, Nadia, Nadia De Giovanni, and Marcello Chiarotti. "Simultaneous Detection of Some Drugs of Abuse in Saliva Samples by SPME Technique." *Forensic Science International* 134, no. 1 (2003): 40–5.

Hernandez, Felix, Elena Pitarch, Joaquin Beltran, and Francisco J. López. "Headspace Solid-Phase Microextraction in Combination with Gas Chromatography and Tandem Mass Spectrometry for the Determination of Organochlorine and Organophosphorus Pesticides in Whole Human Blood." *Journal of Chromatography B* 769, no. 1 (2002): 65–77.

Huang, Min-Kun, Chiareiy Liu, and Shang-Da Huang. "One Step and Highly Sensitive Headspace Solid-Phase Microextraction Sample Preparation Approach for the Analysis of Methamphetamine and Amphetamine in Human Urine." *Analyst* 127, no. 9 (2002): 1203–6.

Iwai, Masae, Hideki Hattori, Tetsuya Arinobu, Akira Ishii, Takeshi Kumazawa, Hiroki Noguchi, Hiroshi Noguchi, Osamu Suzuki, and Hiroshi Seno. "Simultaneous Determination of Barbiturates in Human Biological Fluids by Direct Immersion Solid-Phase Microextraction and Gas Chromatography–Mass Spectrometry." *Journal of Chromatography B* 806, no. 1 (2004): 65–73.

Jain, Rajeev, and Ritu Singh. "Applications of Dispersive Liquid–Liquid Micro-extraction in Forensic Toxicology." *TrAC Trends in Analytical Chemistry* 75 (2016a): 227–37.

Jain, Rajeev, and Ritu Singh. "Microextraction Techniques for Analysis of Cannabinoids." *TrAC Trends in Analytical Chemistry* 80 (2016b): 156–66.

Jha, Rakesh Roshan, Chetna Singh, Aditya B. Pant, and Devendra K. Patel. "Ionic Liquid Based Ultrasound Assisted Dispersive Liquid-liquid Micro-extraction for Simultaneous Determination of 15 Neurotransmitters in Rat Brain, Plasma and Cell Samples." *Analytica Chimica Acta* 1005 (2018a): 43–53.

Jha, Rakesh Roshan, Nivedita Singh, Rupender Kumari, and Devendra Kumar Patel. "Dispersion-Assisted Quick and Simultaneous Extraction of 30 Pesticides from Alcoholic and Non-alcoholic Drinks with the Aid of Experimental Design." *Journal of Separation Science* 41, no. 7 (2018b): 1625–34.

Jha, Rakesh Roshan, Nivedita Singh, Rupender Kumari, and Devendra Kumar Patel. "Ultrasound-Assisted Emulsification Microextraction Based on a Solidified Floating Organic Droplet for the Rapid Determination of 19 Antibiotics as Environmental Pollutants in Hospital Drainage and Gomti River Water." *Journal of Separation Science* 40, no. 13 (2017): 2694–702.

Kim, Miok, Na Rae Song, Jongki Hong, Jeongae Lee, and Heesoo Pyo. "Quantitative Analysis of Organochlorine Pesticides in Human Serum Using Headspace Solid-Phase Microextraction Coupled with Gas Chromatography–Mass Spectrometry." *Chemosphere* 92, no. 3 (2013): 279–85.

Kumari, Rupender, Devendra K. Patel, Smita Panchal, Rakesh R. Jha, G. N. V. Satyanarayana, Ankita Asati, Nasreen G. Ansari, Manoj K. Pathak, C. Kesavachandran, and Ramesh C. Murthy. "Fast Agitated Directly Suspended Droplet Microextraction Technique for the Rapid Analysis of Eighteen Organophosphorus Pesticides in Human Blood." *Journal of Chromatography A* 1377 (2015): 27–34.

Kumazawa, Takeshi, Hiroshi Seno, Kanako Watanabe-Suzuki, Hideki Hattori, Akira Ishii, Keizo Sato, and Osamu Suzuki. "Determination of Phenothiazines in Human Body Fluids by Solid-Phase Microextraction and Liquid Chromatography/Tandem Mass Spectrometry." *Journal of Mass Spectrometry* 35, no. 9 (2000): 1091–99.

Lord, Heather, and Janusz Pawliszyn. "Evolution of Solid-Phase Microextraction Technology." *Journal of Chromatography A* 885, no. 1–2 (2000): 153–93.

Moller, Monique, Katarina Aleksa, Paula Walasek, Tatyana Karaskov, and Gideon Koren. "Solid-phase microextraction for the detection of codeine, morphine and 6-monoacetylmorphine in human hair by gas chromatography–mass spectrometry." *Forensic Science International* 196, no. 1–3 (2010): 64–9.

Mudiam, Mohana Krishna Reddy, Abhishek Chauhan, Rajeev Jain, Ratnasekhar Ch, Ghizal Fatima, Ekta Malhotra, and R. C. Murthy. "Development, Validation and Comparison of Two Microextraction Techniques for the Rapid and Sensitive Determination of Pregabalin in Urine and Pharmaceutical Formulations After Ethyl Chloroformate Derivatization Followed by Gas Chromatography–Mass Spectrometric Analysis." *Journal of Pharmaceutical and Biomedical Analysis* 70 (2012): 310–19.

Mudiam, Mohana Krishna Reddy, Rajeev Jain, Meenu Varshney, Ratnasekhar Ch, Abhishek Chauhan, Sudhir Kumar Goyal, Haider A. Khan, and R. C. Murthy. "In Matrix Derivatization of Trichloroethylene Metabolites in Human Plasma with Methyl Chloroformate and Their Determination by Solid-Phase Microextraction–Gas Chromatography-Electron Capture Detector." *Journal of Chromatography B* 925 (2013): 63–69.

Musshoff, Frank, Heike Junker, and Burkhard Madea. "Rapid Analysis of Halothane in Biological Samples Using Headspace Solid-Phase Microextraction and Gas Chromatography-Mass Spectrometry—A Case of a Double Homicide." *Journal of Analytical Toxicology* 24, no. 5 (2000): 372–76.

Myung, Seung-Woon, Seungki Kim, Joon-Ho Park, Myungsoo Kim, Jong-Chul Lee, and Taek-Jae Kim. "Solid-Phase Microextraction for the Determination of Pethidine and Methadone in Human Urine Using Gas Chromatography with Nitrogen–Phosphorus Detection." *Analyst* 124, no. 9 (1999): 1283–6.

Okajima, Kazuo, Akira Namera, Mikio Yashiki, Ichiro Tsukue, and Tohru Kojima. "Highly Sensitive Analysis of Methamphetamine and Amphetamine in Human Whole Blood Using Headspace Solid-Phase Microextraction and Gas Chromatography–Mass Spectrometry." *Forensic Science International* 116, no. 1 (2001): 15–22.

Pragst, Fritz. "Application of Solid-Phase microextraction in analytical toxicology." *Analytical and Bioanalytical Chemistry* 388, no. 7 (2007): 1393–414.

Reubsaet, Karianne Johansen, Hans Ragnar Norli, Peter Hemmersbach, and Knut E. Rasmussen. "Determination of Benzodiazepines in Human Urine and Plasma with Solvent Modified Solid Phase Micro Extraction and Gas Chromatography; Rationalisation of Method Development Using Experimental Design Strategies." *Journal of Pharmaceutical and Biomedical Analysis* 18, no. 4–5 (1998): 667–80.

Riahi-Zanjani, Bamdad, Mahdi Balali-Mood, Ahmad Asoodeh, Zarrin Es'haghi, and Adel Ghorani-Azam. "Developing a New Sensitive Solid-Phase Microextraction Fiber Based on Carbon Nanotubes for Preconcentration of Morphine." *Applied Nanoscience* 8, no. 8 (2018): 2047–56.

Risticevic, Sanja, Vadoud H. Niri, Dajana Vuckovic, and Janusz Pawliszyn. "Recent Developments in Solid-Phase Microextraction." *Analytical and Bioanalytical Chemistry* 393, no. 3 (2009): 781–95.

Sha, Yunfei, Shunquing Shen, and Gengli Duan. "Rapid Determination of Tramadol in Human Plasma by Headspace Solid-Phase Microextraction and Capillary Gas Chromatography–Mass Spectrometry." *Journal of Pharmaceutical and Biomedical Analysis* 37, no. 1 (2005): 143–47.

Spietelun, Agata, Adam Kloskowski, Wojciech Chrzanowski, and Jacek Namieśnik. "Understanding Solid-Phase Microextraction: Key Factors Influencing the Extraction Process and Trends in Improving the Technique." *Chemical Reviews* 113, no. 3 (2013): 1667–85.

Spietelun, Agata, Michał Pilarczyk, Adam Kloskowski, and Jacek Namieśnik. "Current Trends in Solid-Phase Microextraction (SPME) Fibre Coatings." *Chemical Society Reviews* 39, no. 11 (2010): 4524–37.

Tsoukali, Heleni, Nikolas Raikos, Georgios Theodoridis, and Dimitrios Psaroulis. "Headspace Solid Phase Microextraction for the Gas Chromatographic Analysis of Methyl-parathion in Post-mortem Human Samples: Application in a Suicide Case by Intravenous Injection." *Forensic Science International* 143, no. 2–3 (2004): 127–32.

Ugland, Hege Grefslie, Mette Krogh, and Knut E. Rasmussen. "Automated Determination of 'Ecstasy' and Amphetamines in Urine by SPME and Capillary Gas Chromatography After Propylchloroformate Derivatisation." *Journal of Pharmaceutical and Biomedical Analysis* 19, no. 3–4 (1999): 463–75.

Walles, Markus, Wayne M. Mullett, and Janusz Pawliszyn. "Monitoring of Drugs and Metabolites in Whole Blood by Restricted-Access Solid-Phase Microextraction Coupled to Liquid Chromatography–Mass Spectrometry." *Journal of Chromatography A* 1025, no. 1 (2004): 85–92.

Watanabe, Tomohiko, Akira Namera, Mikio Yashiki, Yasumasa Iwasaki, and Tohru Kojima. "Simple Analysis of Local Anaesthetics in Human Blood Using Headspace Solid-Phase Microextraction and Gas Chromatography–Mass Spectrometry–Electron Impact Ionization Selected Ion Monitoring." *Journal of Chromatography B: Biomedical Sciences and Applications* 709, no. 2 (1998): 225–32.

Wille, Sarah M. R., and Willy E. E. Lambert. "Recent Developments in Extraction Procedures Relevant to Analytical Toxicology." *Analytical and Bioanalytical Chemistry* 388, no. 7 (2007): 1381–91.

Yonamine, Mauricio, and Alessandro Morais Saviano. "Determination of Cocaine and Cocaethylene in Urine by Solid-Phase Microextraction and Gas Chromatography–Mass Spectrometry." *Biomedical Chromatography* 20, no. 10 (2006): 1071–5.

Yonamine, Mauricio, Nadia Tawil, Regina Lucia de Moraes Moreau, and Ovandir Alves Silva. "Solid-Phase Micro-extraction–Gas Chromatography–Mass Spectrometry and Headspace-Gas Chromatography of Tetrahydrocannabinol, Amphetamine, Methamphetamine, Cocaine and Ethanol in Saliva Samples." *Journal of Chromatography B* 789, no. 1 (2003): 73–8.

Zhang, Zhouyao, Juergen Poerschmann, and Janusz Pawliszyn. "Direct Solid Phase Microextraction of Complex Aqueous Samples with Hollow Fibre Membrane Protection." *Analytical Communications* 33, no. 7 (1996): 219–21.

3

Applications of Micro Solid-Phase Extraction in Analytical Toxicology

Shahram Seidi[1] and Maryam Rezazadeh[2]
[1]*Department of Analytical Chemistry, Faculty of Chemistry,*
 K.N. Toosi University of Technology, Tehran, Iran
[2]*Chemical Analysis Laboratory, Geological Survey of Iran, Tehran, Iran*

CONTENTS

3.1 Introduction 23
3.2 Dispersive μSPE 24
3.3 Porous Membrane-Protected μSPE 25
3.4 Pipette Tip Micro Solid-Phase Extraction 37
3.5 Spin-Column Micro Solid-Phase Extraction (SC-μSPE) 43
3.6 Concluding Remarks and Future Trends 46
References 46

3.1 Introduction

Analytical toxicology is mainly concerned with qualitative and quantitative investigations of drugs and illegal compounds and their metabolites. The aim makes the treatment steps necessary for the samples regarding their complex matrices and low target concentrations in most cases. Sample preparation is of vital importance in forensic analysis since these kinds of matrices are often diverse and complicated. There are also some limitations due to small sample size, and considering the fact that forensic samples usually undergo rigorous legal scrutiny, making the selection of an appropriate custody preparation method is momentous. On the other hand, applying standardized analysis methods is difficult in such cases as each forensic sample is unique and requires exclusive studies. Thus, modern analysis approaches may be useful in analytical toxicology since the conditions are still under consideration.

Among different known sample preparation methods, liquid-based and solid-based microextraction techniques are operational for analytical toxicology considering the limited availability of such samples, instead of traditional liquid-liquid extraction or solid-phase extraction. Solid-based pre-treatment methods are widely used for extraction of different types of analytes in various matrices as simple and relatively selective approaches. In this class of extraction techniques, a solid sorbent is used in order to isolate the analyte of interest from a given sample matrix. Depending on the sorbent types and amounts, the extraction could be classified into three main groups, including bulk solid-phase extraction (SPE), micro solid-phase extraction (μSPE), and solid-phase microextraction (SPME). Miniaturized designs of the extraction methods involve the advantages of small sample solution requirements, saving time and money, and being efficient, in some cases. μSPE could be an alternative to the disadvantages of SPE while offering unique benefits such as exhaustive or near-exhaustive recoveries in comparison with equilibrium-based SPME (Seidi et al. 2019).

Different μSPE modes are available based on various designs for sorbent introductions into the sample solution, including dispersive μSPE (D-μSPE), membrane-protected μSPE (MP- μSPE), pipette

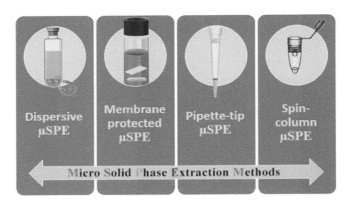

FIGURE 3.1 Different types of micro solid-phase extraction methods.

tip μSPE (PT-μSPE), and spin-column μSPE (SC-μSPE). These methods are schematically presented in Figure 3.1.

This chapter discusses the present best practices and developments for analysis of forensic samples with a focus on those using the diverse modes of the μSPE technique. Although it is very challenging to cover all published works, all efforts have been made to include as many papers as possible. More details are available in the references.

3.2 Dispersive μSPE

Since sorbent-analyte interactions are limited in the SPE approach via sample flow rate, dispersive SPE has been introduced as an alternative. Thus, the close contact between the dispersed sorbent particles and the sample solution tremendously enhances the extraction kinetic and the overall process efficiency as a result (Anastassiades et al. 2003). The process includes dispersion of optimized amounts of the sorbent into the sample solution. Analytes were then isolated in a clean eluent via sorbent collection throughout the sample and its washing afterward. Extraction selectivity could also be raised by sorbent modification, eluent composition, and controlling the extraction conditions (Chisvert et al. 2019).

The technique is also called dispersive micro solid-phase extraction (D-μSPE) when limited amounts of the sorbent (a few milligrams) are used, or it is referred to as dispersive solid-phase microextraction in some cases. Also, it could be known as magnetic solid-phase extraction if the sorbents have some magnetic properties (Chisvert et al. 2019).

Dispersion of the sorbent could be performed via an auxiliary energy, such as ultrasound (Aghaie and Hadjmohammadi 2016; Dil et al. 2016; Krawczyk and Stanisz 2016; Krawczyk-Coda and Stanisz 2017) and vortex (Ojeda and Rojas 2018; Galán-Cano et al. 2013; Cai et al. 2017). These external energies could also enhance the analytes' mass transfer and the extraction efficiency. Ultrasound radiation is a stronger auxiliary energy than the mechanical vortex agitation, and it may positively affect the extractability by increasing the analytes' diffusion, reducing the sorbent particle size and raising its contact surface. Ultrasonication could also influence the extraction kinetic and extremely decrease the extraction time (Chisvert et al. 2019). However, uncontrolled radiation may diminish the extraction recovery duo, increasing the temperature.

Vortex is simpler and lighter auxiliary energy that is widely used to enhance the extractability regarding its cost and availability. This technique offers a mechanical agitation for mass transfer reinforcement without the temperature increase problems faced by the ultrasound-assisted D-μSPE. Also, the back-and-forth movement of a glass syringe plunger was used for facilitation of sorbent dispersion into the sample solution, called air-assisted D-μSPE (Rajabi et al. 2016).

Chemical solvents or chemical reactions may be used for sorbent dispersion or in situ formation of a dispersed sorbent. Jamali et al. chose benzophenone as the solid sorbent since it can be solved in some water-miscible organic solvents, such as acetonitrile and methanol. As benzophenone solvent entered the sample solution, tiny solid sorbent particles were formed throughout the aqueous sample (Jamali et al. 2013). In another work, in situ formation of carbon dioxide bubbles was used for sorbent dispersion in the sample solution, called effervescence-assisted D-µSPE (Lasarte-Aragonés et al. 2011).

D-µSPE has a relatively fast kinetic, in comparison with other µSPE modes, regarding its large sorbent-sample contact surface. Although the dispersion strategy is the method bottle neck, the sorbent characteristics indicate the method selectivity, which influences its application for analysis of complicated forensic matrices. Various sorbents introduced for employment in this method include common micromaterials, nanostructured materials and nanoparticles, metal-organic frameworks, layered double hydroxides, molecular-imprinted polymers, hybrid materials, etc. A web search may result in many papers using this technique for analysis of different analytes in forensic samples since it is a simple and fast extraction method with controllable selectivity and pre-concentration. Table 3.1 presents a summary of the recent works on D-µSPE for analysis of biological samples. In all works, sorbents are finally recovered by means of filtration, centrifugation, and applying a magnetic field in cases where magnetic sorbents are used. A typical D-µSPE procedure is shown in Figure 3.2.

Another feature of D-µSPE that makes this method more interesting is its hyphenation with other sample preparation methods. Different goals are pursued from these combinations. For example, low sample clean-up is one of the main problems in solid-liquid extraction methods, such as microwave-assisted extraction (MAE), accelerated solvent extraction (ASE), and ultrasound-assisted extraction (USAE). Moreover, due to the large volume of the extracts, a solvent evaporation step is often used to enhance the pre-concentration factor and limit of detection (LOD) in these methods, which is time-consuming. To overcome these issues, an extra sample preparation method is often applied. Liquid-liquid extraction (LLE) and SPE are the common approaches used to this aim. However, these methods have some drawbacks, too, such as being time-consuming, having sorbent blockage, and using large amounts of hazardous organic solvents. Applying D-µSPE eliminates or reduces these drawbacks and thus has found considerable attention among researchers as a further sample clean-up procedure.

Besides, D-µSPE has been combined with liquid-phase microextraction methods, such as dispersive liquid-liquid microextraction (DLLME) and ultrasound-assisted emulsification microextraction (USAEME). In this case, D-µSPE can be performed as both pre- and post-hyphenated extraction methods. In the pre-hyphenated approach, the eluent of the D-µSPE step (e.g., methanol, acetonitrile) is mixed with a microliter volume of a water-immiscible organic solvent, and the mixture is rapidly injected into a low volume of an aqueous solution (pH adjustment is required for the ionized analytes). Then, the resulted cloudy solution is collected by different approaches, such as centrifugation, withdrawn by a microsyringe, and injected to the analytical instrument for further analysis. This combination is useful, especially for the biological samples, due to enhancement of the sample clean-up and pre-concentration factor.

In the post-hyphenated approaches, D-µSPE is applied to overcome the centrifugation challenge in DLLME and USAEME. By addition of the sorbent into the cloudy solution, the fine droplets of dispersed water-immiscible organic solvent are adsorbed on the surface of the dispersed sorbent, mainly a magnetic sorbent. Then, the sorbent is separated from the sample solution, eluted with a microliter volume of a suitable organic solvent, and injected to the analytical equipment. This strategy has created facilities for automation of dispersive liquid-phase microextraction methods.

3.3 Porous Membrane-Protected µSPE

As can be deduced from the method name, it is based on applying a piece of porous flat sheet membrane in which a small bag-shaped pocket is formed and filled with a few milligrams of a special sorbent. This method was first reported by Basheer et al., in 2006 (Basheer et al. 2006). The most useable pocket configuration is made by heat-sealing the three other edges of a folded sheet membrane (Figure 3.3a). A similar configuration (Figure 3.3b) can also be created by heat-sealing two edges so that the square membrane bends in the direction of one of its diameters and the two matching edges are closed by heat,

TABLE 3.1

Applicability of D-μSPE method for quantitative analysis of different analytes in biological samples

Sorbent	Instrument	Analyte(s)	Matrix	LOD	RSD%	Ref.
Magnetic molecularly imprinted polymer ($Fe_3O_4@SiO_2$-MIP)	HPLC-UV	Melatonin	Urine and plasma	0.046 ng mL^{-1}	<6.1%	Dil et al. (2021)
C_{18}+ primary secondary amine	LC-MS-MS	Multi-pesticide residues	Fish and shrimp	<5 μg kg^{-1}	<20%	Shin et al. (2021)
Molecularly imprinted polymer		Propranolol	Bovine serum	0.002 μM	2.3%–3.7%	Tu et al. (2021)
Pectin/Fe_3O_4/GO	ICP/OES	Pb, Cd, Hg, Co, Ni	Fish	0.01–0.21 μg g^{-1}		Bozorgzadeh et al. (2021)
Mannitol capped magnetic nanoparticles	Fluorescence	Sparfloxacin, Orbifloxacin	Milk	5.0 × 10^{-10} M (SPX) 4.0 × 10^{-10} M (ORX)	>1.3	Ali et al. (2021)
Dummy magnetic molecularly imprinted polymers	UPLC-MS-MS	Sarcosine	Urine	0.02 nM	2.6%–11.5%	Chen et al. (2021)
Reticulated imine-based triazine-cored covalent organic framework	GC-MS-MS	Polybrominated diphenyl ethers (PBDEs)	Fish and milk	0.03–0.13 ng L^{-1}	>4.7	Liu et al. (2021)
SCX, PAX, COOH, PCX	LC-MS	Fluoroquinolones	Swine serum and urine	0.02–0.03 μg L^{-1}	2.1%–8.2%	Wang et al. (2021)
Magnetic graphene oxide-Sodium dodecyl sulfate	HPLC-DAD	Metoprolol, atenolol, propranolol	Plasma and urine	10 ng mL^{-1} (metoprolol) 0.8 ng mL^{-1} (atenolol) 2 ng mL^{-1} (propranolol)	2.96%	Ameri Akhtiar Abadi et al. (2021)
Acetic acid-functionalized Fe_2O_3 nanoparticles modified by (3-amino-propyl)-tri-ethoxy silane	HPLC/FL	Letrozole	Plasma	23 ng mL^{-1}	<15%	Shaban et al. (2021)
C18 + primary secondary amine	RPLC-PDA	Nine biogenic amines (BAs)	Fish-shrimp-shellfish	0.08–0.25 mg kg^{-1}	0.44%–6.83%	Wang et al. (2021)
Mixed-mode cation exchange sorbent	HPLC-MS	Veterinary drugs	Egg	0.03–0.33 μg kg^{-1}	1.2%–9.1%	Wang et al. (2021)
Magnetic multiwalled carbon nanotubes/ Fe_3O_4 @poly(2-aminopyrimidine)	HPLC-PDA	Phenolphthalein	Urine	0.01 μg L^{-1}	2.7%–3.4%	Jalilian et al. (2021)
Poly(ionic liquid)	UPLC-DAD	Oligonucleotides	Serum	0.30–0.40 μM	2.31%–3.09%	Nuckowski et al. (2021)

Sorbent	Technique	Analyte	Matrix	LOD/Concentration	RSD	Reference
NH_2-MIL-53(Al)@Chitosan whit coated Fe_3O_4	GC	Diazinon Ethion	Mice liver	0.07, 0.04 µg L^{-1}	≤6.45%	Mohammadi et al. (2021)
Graphene oxide (GO)	LC-ESI-MS/MS	145 insecticides	Guttation fluid	—	<20%	Hrynko et al. (2021)
C18 sorbent	GC-MS	multiPesticides	Fish	0.001–0.029 µg mL^{-1}	≤20%	Mandal et al. (2021)
ZSM-5 zeolite/Fe_2O_3	Square-wave anodic stripping voltammetry	Cd	Urine	0.5–1.0 µg L^{-1}	<14	Baile et al. (2020)
Graphene oxide modified with sodium hydroxide	ICP-AES	Pb, Cd, Ba, Zn, Cu, and Ni	Poultry-pork-beef	0.01 and 0.21 µg g^{-1}	1.9%	Manousi et al. (2020)
	HPLC-UV	1-naphthol and 2-naphthol	Urine	0.3 µg L^{-1} and 0.5 µg L^{-1}	3.1%–9.0%	Omidi et al. (2020)
Nickel metal organic modified-Al_2O_3 nanoparticles	HPLC-UV	Atorvastatin	Plasma	0.05 ng mL^{-1}	<5%	Bahrani et al. (2020)
Fe_3O_4@SiO_2	GC-MS	Valproic acid, phenobarbital, levetiracetam, pregabalin	Urine	<0.11 ng mL^{-1}	4.7%	Mohammadi et al. (2020)
PSA, MgSO4, C18EC, Z-Sep	GC-MS	Synthetic musks and organophosphorus pesticides	Human adipose tissue	4–9 ng g^{-1} 1–7 ng g^{-1}	<14%	Sousa et al. (2020)
Vinyl-functionalized COU-2 mesoporous carbon	HPLC-UV	Azole antifungal drugs	Urine and plasma	0.4–1.6 µg L^{-1}	7.5%–13.4%	Yahaya et al. (2020)
Molecularly imprinted polymer nanoparticles	HPLC-UV	Albendazole sulfoxide	Urine and plasma	0.074 ng mL^{-1}	2.2%–4.4%	Alipanahpour Dil et al. (2020)
Mesoporous silica sorbent	HPLC-UV	Lamotrigine and carbamazepine	Urine and plasma	0.02 ng mL^{-1}	4.4%–7.9%	Behbahani et al. (2020)
Graphene oxide	HPLC-FD	Ochratoxin	Chicken liver	0.02 ng mL^{-1}	<3.0%	Cui et al. (2020)
Molecularly imprinted polymer	LC-MS	Aflatoxins	Fish	0.29–0.61 µg kg^{-1}	<12%	Jayasinghe et al. (2020)
Tetracycline-grafted polyacrylamide polymer	HPLC-DAD	Vitamin A and E	Milk and egg yolk	5.71 ng mL^{-1} (vitamin A) 14.28 ng mL^{-1} (Vitamine E)	3.25%(vitamin A) 2.85% (Vitamine E)	Köseoğlu et al. (2020)
Magnetic restricted-access carbon nanotubes	GC-MS	Organophosphates (chlorpyrifos, malathion, disulfoton, pirimiphos)	Milk	0.36–0.95 µg L^{-1}	10.47%–19.85%	Campos do Lago et al. (2020)

(Continued)

TABLE 3.1 (Continued)
Applicability of D-μSPE method for quantitative analysis of different analytes in biological samples

Sorbent	Instrument	Analyte(s)	Matrix	LOD	RSD%	Ref.
Magnetic molecularly imprinted polymer	HPLC-DAD	Valsartan and atorvastatin	Urine	0.1 μg L^{-1} (Valsartan) 0.2 μg L^{-1} (Atorvastatin)	<4%	Abbasi et al. (2020)
Magnetic attapulgite/polypyrrole/ Fe_3O_4	HPLC-DAD	Five pyrethroids	Honey	0.21–0.34 μg L^{-1}	81.42%	Yang et al. (2020)
Dummy magnetic molecularly imprinted polymer	UHPLC-MS/MS	Globotriaosylsphingosine	Plasma	0.01nM	<5.2%	Hu et al. (2020)
Fe_3O_4-SiO_2-NH_2@UiO-66	HPLC	Muconicacid	Urine	0.001 μg mL^{-1}	3.7%–4.5%	Rahimpoor et al. (2020)
Fe_3O_4@multiwalled carbon nanotubes	UHPLC-HRMS		Urine	<0.03μg L^{-1}	<12%	Arroyo-Manzanares et al. (2020)
Nano graphene oxide polypyrolle composite	HPLC-UV	Methamphetamine	Urine	9 ng mL^{-1}	<5.51%	Jabbari et al. (2020)
Zeolite imidazole framework@ hydroxyapatite composite	HPLC-VWD	Benzodiazepines	Urine	0.7–1.4 ng mL^{-1}	3.0%–10.3%	Li et al. (2020)
Molecularly imprinted polymer – metal organic framework	UPLC-PDA	Tetracyclines	Chicken meat	<0.6 ng g^{-1}	<4.7%	Ma et al. (2020)
Graphene oxide	IMS	Ethambutol	Plasma, saliva, breastmilk, and artificial tears	0.4 μg L^{-1}	1.6%	Shafiee et al. (2020)
Molybdenum disulfide	CE	Ibuprofen	Urine	0.025 μg mL^{-1}	<2.3%	Naghdi et al. (2020)
Fe_3O_4@Cu-Fe layered double hydroxides	GC-FID	Tramadol	Biological samples	≤2.4 μg L^{-1}	<8.1%	Ezoddin et al. (2019)
Fe_3O_4@TiO_2	HPLC-UV	Toluene and xylene biomarkers	Urine	≤1.0 μg L^{-1}	<7.0%	Omidi et al. (2019)
Magnetic graphene oxide/ polypyrrole	UV–Vis	Methotrexate	Human urine samples	10 ng mL^{-1}	≤10.2	Hamidi et al. (2019)
Sulfur-doped tin oxide nanoparticles loaded on activated carbon	HPLC-UV	Glibenclamide(GB)	Urine	≤0.16 mg L^{-1}	1.15%–6.84%	Eilami et al. (2019)
Benzophenone	HPLC-UV	Diclofenac	Human serum	0.47 μg L^{-1}	2.1%	Nakhaei et al. (2019)
Magnetic graphene oxide	EAAS	Cobalt ion	Saliva and urine	0.023 μg L^{-1}	3.8	AlKinani et al. (2019)

Sorbent	Technique	Analyte	Matrix	LOD	RSD	Reference
MgSO4 + modified biochar of Cocos nucifera husk	GC	Phthalate esters	Breastmilk, urine	0.012–0.020 µg L^{-1}	<20%	Adenuga et al. (2020)
Multiwalled carbon nanotubes / Fe$_3$O$_4$@poly(2-aminopyrimidine)	HPLC-DAD	Acidic, basic, and amphoteric 4 drugs	Urine and plasma	≤3.5 µg L^{-1}	1.4%–10.5%	Jalilian et al. (2018)
Polydopamine-coated Fe$_3$O$_4$ nanoparticles with multi-walled carbon nanotubes	HPLC	Antiepileptic drugs	Biological matrices samples	≤3.1 µg L^{-1}	<8.2%	Zhang et al. (2018a)
PCX	CE	Glycopyrrolate stereoisomers	Rat plasma	2.0 µg L^{-1}	<13%	Liu et al. (2018)
Magnetic-nylon 6 composite	HPLC-UV	Bisphenol A	Milk	3.05 ng L^{-1}	9.1%	Reyes-Garcés et al. (2018)
Polypyrrole-sodium dodecylbenzenesulfonate/zinc oxide	HPLC-VWD	Metoprolol, propranolol, and carvedilol	Urine and plasma	≤1.5 µg L^{-1}	<6.3%	Hemmati et al. (2017)
Cu@SnS/SnO nanoparticles	HPLC-UV	Atorvastatin	Urine and plasma	0.0608 µg L^{-1}	4.75%–1.10%	Dastkhoon et al. (2016)
Fe$_3$O$_4$@graphene	OAMTLS	Fluoxetine	Urine	≤1.2 µg L^{-1}	2.1%	Kazemi et al. (2016)
Carbon fibers	GC/MS	Chlorophenols	Urine	≤0.9 ng mL^{-1}	<13%	García-Valverde et al. (2016)
Oxidized multiwall carbon nanotubes	F-AAS, ET-AAS	Pb(II)	Fish	≤0.26 µg L^{-1}	<2.9%	Feist (2016)
G-SiO2 hybrid	ETAAS	Pb (II), Cd (II), and Cr (III)	Biological samples	≤12.5 µg L^{-1}	3.1%–3.8 %	Ghazaghi et al. (2016)
CTAB-coated Fe$_3$O$_4$@caprylic acid nanoparticles	HPLC-DAD	Estrogens	Pork	≤0.033 µg L^{-1}	1.87%–2.92%	Wang et al. (2016)
COU-2	HPLC-UV	Penicillins	Milk	≤3.3 ng mL^{-1}	6.2%–8.8 %	Yahaya et al. (2015)
MIL-101(Cr)	HPLC-DAD	Hormones	Urine and hormones	≤0.91 4 µg L^{-1}	<6.1%	Zhai et al. (2014)
Fe$_3$O$_4$/polyaniline-polypyrrole	UPLC-UV	Lorazepam and nitrazepam	Urine and plasma	≤2.0 µg L^{-1}	<7.8%	Asgharinezhad et al. (2014)
Coated titanium dioxide nanotube and bared titanium dioxide nanotube	LC	Naproxen and ketoprofen	Saliva and urine	≤110 ng mL^{-1}	<8.5% (ketoprofen) <6.6% (naproxen)	García-Valverde et al. (2014)
Multiwall carbon nanotubes	X-ray fluorescence spectrometry	Selenium (IV)	Mineral water and biological samples	0.06 ng mL^{-1}	3.2	Skorek et al. (2012)

(Continued)

TABLE 3.1 (Continued)
Applicability of D-μSPE method for quantitative analysis of different analytes in biological samples

Sorbent	Instrument	Analyte(s)	Matrix	LOD	RSD%	Ref.
Layered double hydroxide-coated magnetic nanoparticles	HPLC-UV	Cholesterol-lowering drugs	Urine and plasma	0.3–0.5 ng mL^{-1}	≤7.8	Arghavani-Beydokhti et al. (2017)
MIL-101(Cr)	HPLC-UV	Pantoprazole	Rat plasma	–	≤12.8	Cai et al. (2019)
Modified ZSM-5 zeolite/Fe2O3 composite	Inductively coupled plasma optical emission spectrometry	Cd(II), Hg(II), Pb(II)	Urine samples	Cd 0.15–0.2 μg L^{-1} Hg 0.42–0.73 μg L^{-1} Pb 0.23–0.39 μg L^{-1}	≥5	Baile et al. (2018)
Magnetic ZnFe2O4 nanotubes	ICP-MS	Co(II), Ni(II), Mn(II), Cd(II)	Bovine liver or human hair	Co(II) 0.10 pg mL^{-1} Ni(II) 3.7 pg mL^{-1} Mn(II) 1.2 pg mL^{-1} Cd(II) 0.09 pg mL^{-1}	≥4.3	Chen et al. (2019)
Melamine-phytate supermolecular aggregate	HPLC-UV	Tyrosine kinase inhibitors	Biological sample	0.12–0.2 μg L^{-1}	≤7	Adlnasab et al. (2018)
Multiwall carbon nanotubes –magnetic molecularly imprinted polymer	HPLC-UV	Sotalol	Biological fluids	0.31 ng mL^{-1}	4.50	Ansari and Masoum (2018)
MCM-41@NH2	GFAAS	Pb	Water and urine	0.05 μgL^{-1}	≤9.2	Sobhi et al. (2019)
Fe$_3$O$_4$@SiO2@polythiophene	GFAAS	Cd(II)	Water, urine and serum	0.8 ng L^{-1}	6	Behbahani et al. (2018)
Magnetic p-Phenylenediamine functionalized reduced graphene oxide Quantum Dots@ Ni nanocomposites	HPLC-DAD	Duloxetine, venlafaxine, and Atomoxetine	Human urine and real water	1.1 ng mL^{-1}	≥4.6	Ghorbani et al. (2018)
Magnetic carbon nanotubes	GC/MS	2-aminothiazoline-4-carboxylic acid	Biological samples	Synthetic urine 15 ng mL^{-1} Bovine blood 30 ng mL^{-1}	≥2.9	Li et al. (2019)
Honey@magnetic carbon nanotubes	HPLC-UV	Sunitinib	Biological samples	1.58 ng mL^{-1}	3.15	Hooshmand and Es'haghi (2018)
Fe$_3$O$_4$/NiO@SiO2-OP$_2$O$_3$H	HPLC-DAD	Meloxicam and Piroxicam	Real water, biological and milk samples	Meloxicam 0.2 ng mL^{-1} Piroxicam 0.3 ng mL^{-1}	4.1	Ghorbani et al. (2019)

μSPE Applications in Analytical Toxicology

Sorbent	Technique	Analyte	Sample	LOD/Range	RSD (%)	Reference
Fe_3O_4/GO nanocomposites	UHPLC-MS/MS	Neurosteroids	Rat blood	0.06–0.12 pg mL^{-1}	≤8.5	Xu et al. (2019)
ZSM-5/Fe_2O_3	SWASV	Pb(II)	Urine samples	1.0–2.0 μg L^{-1}	–	Fernández et al. (2020)
ZSM-5 zeolite-based composite decorated with iron oxide magnetic nanoparticles and modified with hexadecyltrimethylammonium bromide surfactant	LC-DAD	NSAIDs	Water and urine samples	0.5–3.0 μg L^{-1}	≤5	Baile et al. (2019)
Octadecyl-bonded silica and graphitized carbon black	HPLC-PDA	Flavor enhancers	Seafood products	0.06–0.15 μg g^{-1}	≤5.5	Ma et al. (2020)
Carboxylated ZIF-8	HPLC-UV	Methamphetamine	Urine samples	10 ng mL^{-1}	4.50	Taghvimi et al. (2019)
Molecularly imprinted polymers	HPLC	Sialic acid	Rabbit serum	0.025 mg mL^{-1}	≤2.8	Huang et al. (2019)
Mesoporous poly (melamin-formaldehyde)	LC–MS/MS	Busulfan	Plasma samples	–	≤12.14	Jahed et al. (2020)
Fe_3O_4@SiO_2@N3	HPLC	Amitriptyline, Nortriptyline	Pharmaceutical wastewater and urine	0.05 ng mL^{-1} 0.03 ng mL^{-1}	1.12.04	Fahimirad et al. (2019)
Reduced graphene oxide Fe_3O_4 nanocomposite	UHPLC-DAD	Apixaban, Dabigatran, Rivaroxaban	Human plasma	0.003 μg mL^{-1}	0.81%–8.97%	Ferrone et al. (2020)
Magnetic molecularly imprinted polymer	HPLC-UV	Valsartan	Water, urine, and plasma sample	0.0004–0.0012 μg mL^{-1}	≤5.5	Kaabipour et al. (2020)
MIL-53(Al)	UPLC-MS/MS	Estrogens and Glucocorticoids	Water and urine samples	≤1.0 μg L^{-1}	≤10.0	Gao et al. (2018)

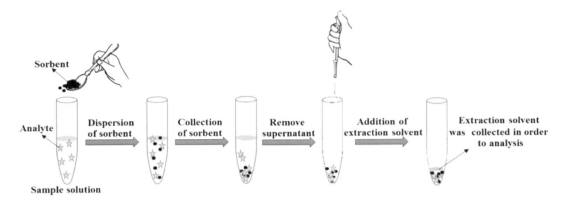

FIGURE 3.2 Schematic presentation of a typical D-μSPE procedure (Moyakao et al. 2018).

leading to a triangular pocket device (Rozaini et al. 2017). However, recently, a new and more feasible configuration (Figure 3.3c) has been introduced by Sánchez-González et al., similar to what is used to prepare a filter paper (Sánchez-González et al. 2015). As a result, just one end of the proposed configuration needs to heat-seal. It should be noticed that the type of the membrane used in MP-μSPE has a great deal of importance, and it should have special properties such as good chemical resistance in the sample solution and a different organic solvent, suitable flexibility, and proper heat-seal ability. Polypropylene is the most utilized membrane for the MP-μSPE purpose. Different membrane-protected micro solid-phase extraction (MP-μSPE) configurations and the preparation approaches are schematically shown in Figure 3.3 (Sánchez-González et al. 2015; Basheer et al. 2007).

To carry out a MP-μSPE procedure, the prepared pocket should be cleaned and conditioned before use. To this aim, the pocket is dipped into a suitable organic solvent and sonicated for a specified time and then kept in the same organic solvent for further MP-μSPE experiments. During use, the fiber is first air-dried, located into the sample solution, and agitated for a specified period of time. Finally, the pocket is withdrawn from the sample solution, washed with ultrapure water, dried using a Kleenex, and put into a small vial for desorption of the extracted analytes using a suitable eluent. To improve the desorption efficiency, sonication is often applied. The schematic presentation of a typical MP-μSPE procedure is shown in Figure 3.4 (Sajid et al. 2016).

One of the probable limitations in the conventional configuration of MP-μSPE is destruction of the heat-sealed edges of the prepared pocket due to contact with organic solvents, such as dichloromethane (Sánchez-González et al. 2015). Due to the possibility of preparing the filter-paper-like configuration with long length (Figure 3.3c), heat-sealing can be done in its upper part that is not in contact with the solvent (Sánchez-González et al. 2015).

Compared to the conventional SPE, the main supremacies of MP-μSPE are (Sajid 2017): easy handling; cost-effectiveness; low usage of sorbent amount and hazardous organic solvents; acceptable robustness; high sample clean-up due to applying a porous membrane eliminating the required sample treatment step of the complex matrices and making it attractive for the biological fluids; decreasing the sorbent surface contamination in complex matrices and thus improving the adsorption efficiency and sorbent reusability; eliminating the sorbent blockage or back-pressure in cartridge-based SPE; eliminating the sorbent collection in dispersive SPE; providing higher pre-concentration factors due to desorption possibility with lower eluent volumes, and shortening the total extraction time.

Considering the advantages of MP-μSPE, several developments have been reported in this research field of interest that can mainly be classified as three categories, including setup modification, application of new sorbent types, and hyphenation with the other extraction or microextraction methods.

The main factor in performing a successful extraction in all SPE methods is choosing the suitable adsorbent, so this factor can be called the heart of the method. The sorbent should have some properties, such as a large surface area and high adsorption capacity, adsorption selectivity, good chemical stability, good reusability, and fast adsorption kinetic. Besides these features, the sorbent should also have low tendency to stick to the membrane surface because it interferes with the effective

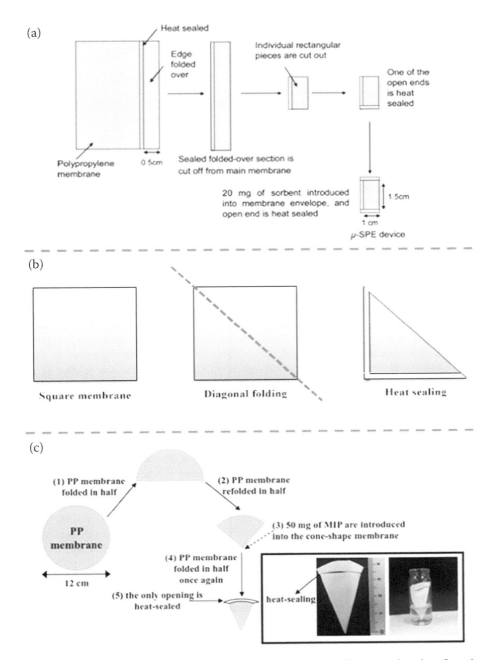

FIGURE 3.3 Rectangular pocket shaped (a), triangular pocket shaped (b), and filter paper shaped configurations (c) of MP-μSPE. Reproduced with permission from Basheer et al. (2007) and Sajid et al. (2016).

heat-sealing of the fabricated pocket membrane and opens it during extraction or desorption with a solvent (Lim et al. 2013).

So far, different sorbent materials have been applied and reported in MP-μSPE. These sorbents include commercial materials, such as C2, C8, C18, activated carbon, carbograph, Haye-Sep A, Haye-Sep B, and natural sorbents (e.g. seed powder of *Moringa oleifera*); polymeric sorbents, such as molecularly imprinted polymers (MIP); and synthesized nanomaterials, such as single-wall carbon nanotubes (SWCNTs), multiwall carbon nanotubes (MWCNTs), graphene and graphene oxide, carbon

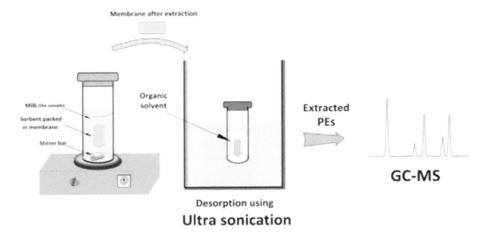

FIGURE 3.4 Schematic presentation of a typical MP-µSPE procedure. Reproduced with permission from Sajid (2017).

fibers, mesoporous silica-based materials (e.g., SBA-15), layered double hydroxides (LDHs), zeolites, and metal organic frameworks. These nanomaterials have been used as bared, surface modified, and composite sorbents. Each of these sorbents can provide special type(s) of interaction(s) with the target analytes, such as hydrophobic interaction, π-π interaction, electrostatic interaction, hydrogen bonding, etc. As a result, when there is a wide range of compounds with different polarities in a sample, a combination of two or more adsorbents with different interaction mechanisms can be used to provide an effective extraction. For example, a combination of Haye-SepA and C18 has been reported by Basheer et al. for more effective extraction of persistent organic pollutants in human ovarian cancer tissues (Basheer et al. 2008). The recent MP-µSPE works based on different sorbent materials for preconcentration of different analytes in biological samples are summarized in Table 3.2.

Another aspect of MP-µSPE is devoted to setup development or its hyphenation with the other sample preparation methods to eliminate or compensate its shortcomings. These aspects are classified based on literature (Sajid 2017) and depicted in Figure 3.5. As discussed above, MP-µSPE can be performed using both pocket-shaped (rectangular and triangular) and filter-paper configurations (Basheer et al. 2006; Rozaini et al. 2017; Sánchez-González et al. 2015). As can be seen in Figure 3.5, setup developments are focused on agitation approaches, including vortex-assisted MP-µSPE (VA-MP-µSPE) (Guo and Lee 2013), stir bar supported MP-µSPE (Sajid et al. 2017), handheld battery operated stirring MP-µSPE (Abidin et al. 2014), and magnetic sorbents-based MP-µSPE (Naing et al. 2016).

The results showed that vortex agitation is more efficient than common magnetic stirring, leading to increasing the mass transfer rate and decreasing the extraction time (Guo and Lee 2013). In stir bar supported MP-µSPE, a stir bar beside the sorbent is located in the pocket membrane (Sajid et al. 2017; Sajid and Basheer 2016). This setup provides better immersion of the pocket membrane, more efficient agitation via motion and rotation, and consequently improved extraction recovery to save the analysis time (Sajid 2017). The handheld battery operated stirring MP-µSPE makes this method suitable for onsite sampling (Abidin et al. 2014). Applying a magnetic sorbent for performing MP-µSPE not only adsorbs the analyte but also promotes the agitation of the pocket membrane, and eliminates the need to use a stir bar, which simplifies the method (Sajid and Basheer 2016).

One of the interesting aspects of the MP-µSPE is its hyphenation with other exhaustive (e.g., MAE, ASE) (Kanimozhi et al. 2011; Sajid et al. 2015; Jiao et al. 2015) or equilibrium-based sample preparation methods (e.g., DLLME, USAEME, SBSE) (Cai et al. 2019; Guo and Lee 2013; Tsai et al. 2009; Ge and Lee 2012; Mao et al. 2016). The main advantages of hyphenated methods are reducing the use of hazardous organic solvents; decreasing the required time for solvent evaporation; decreasing the extraction time; and improving the pre-concentration factors, extraction efficiency, sample clean-up, and limit of detection.

In MAE-MP-µSPE, besides the extraction solvent and solid sample, the pocket device is also located in the extraction vessel; thus, the µSPE procedure is performed simultaneously (Kanimozhi et al. 2011;

TABLE 3.2
Applicability of MP-μSPE method for quantitative analysis of different analytes in biological samples

Sorbent	Instrument	Analyte	Matrix	LOD	Rsd%	Ref.
Haye-Sep A/C18	GC-MS	Organochlorine pesticides	Ovarian cancer tissues	0.002–0.009 ng/g	0.14–12.7	(Basheer et al. 2008)
C2	GC-MS	Estrogens	Ovarian cyst fluid	9–22 μg/L	4–11	(Kanimozhi et al. 2011)
Molecularly imprinted polymerbeads/particles	LC-MS/MSLC-MS/MS	Cocaine and its metabolites	Human urine, human plasma	0.16–1.7 ng/L 0.061–0.87 ng/L	3–8≤10	(Sánchez-González 2015; Sánchez-González et al. 2016)
Haye-Sep A	HPLC-UV	Parabens	Ovarian cancer tissues	0.005–0.0244 ng/g	0.09–2.81	(Sajid et al. 2015)
Molecularly imprinted polymer	LC-MS/MS	Cannabinoids	Human plasma, urine	0.11–0.15 ng/L 0.14–0.17 ng/L	≤8	(Sánchez-González et al. 2017)
Molecularly imprinted polymer	LC-MS/MS	Hyperoside and isoquercitrin	Rat plasma	0.92 and 0.89 μg/mL	≤8.4	(Yin et al. 2012)
Surfactant template MCM141	LC-MS/MS	Perfluorinated carboxylic acids	Plasma samples	21.23–65.07 ng/L	≤15.22	(Lashgari and Lee 2016)
C18	GC-MS	Polychlorinated biphenyl congeners	Serum samples	0.003–0.047 ng/mL	3.1–7.2	(Sajid and Basheer 2016)
Layered double hydroxide/graphene hybrid	GC-MS	Organochlorine pesticides	Urine samples	0.22–1.38 ng/mL	2.7–9.5	(Sajid et al. 2017)
Molecularly imprinted polymer	LC-MS/MS	Synthetic cannabinoids	Urine samples	0.032–0.75 ng/mL	≤8	(Sánchez-González et al. 2018)
Molecularly imprinted polymer	LC-MS/MS	aflatoxins	Cultured fish	0.29–0.61 μg/Kg	≤19	(Jayasinghe et al. 2020)
PS/SiO2@ Polydopamine	LC-MS	OH- Monohydroxy polycyclic aromatic hydrocarbons(2-OHN, 3-OHPhe, 1-OHP)	Urine	0.002–0.021 ng/mL	2.0–9.1	(Chen and Xu 2020)

(*Continued*)

TABLE 3.2 (Continued)
Applicability of MP-μSPE method for quantitative analysis of different analytes in biological samples

Sorbent	Instrument	Analyte	Matrix	LOD	Rsd%	Ref.
Molecularly imprinted polymer crushed monolith	HPLC-FLD	Ochratoxin A	Coffee, grape juice, and urine	0.02–0.06 ng/g	≤2.3	(Lee et al. 2012)
Poly methacrylic acid ethylene glycol dimethacrylate	HPLC-UV	Sulfonamides	Milk and chicken muscle samples	380–620 ng/L	2.7–15.2	(Huang et al. 2012)
b-cyclodextrin	LC-MS/MS	Steroids	Urine	250–500 ng/L	2.1–4.9	(Manaf et al. 2018)
Polymeric reversed-phase sorbent HLB, silica-based sorbent C18, and multiwalled carbon nanotubes	GC-IMS	Aromatic hydrocarbons and aldehydes	Saliva	0.18–0.31 mg/L 0.38–0.49 mg/L	8.4–21.2	(Criado-García 2016)
Cetyltrimethylammonium bromide-templated MCM-41	LC-MS/MS	Perfluorinated carboxylic acids	Fish fillet	0.97–2.70 ng/g	5.4–13.5	(Lashgari and Lee 2014)
Copper (II) isonicotinate	HPLC-UV	Tetracycline antibiotics	Pork, chicken, and clam meat	7.4–16.3 ng/g	5.7–9.3	(Jiao et al. 2015)

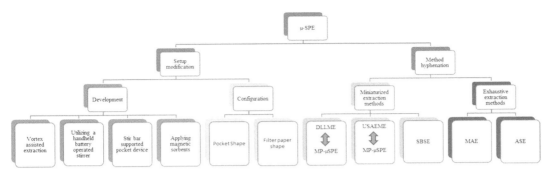

FIGURE 3.5 Setup development of MP-μSPE and its hyphenation with the other sample preparation methods (Lim et al. 2013).

Sajid et al. 2015). In addition to other benefits mentioned above, this work decreases analysis steps and helps to prove higher recoveries and selectivity. Improving the extraction efficiency is due to the continuous and simultaneous adsorption of the analytes extracted from the sample matrix into the extraction solvent, which prevents solvent saturation (Sajid 2017). For MAE of non-polar analytes, extraction selectivity can be enhanced using a non-polar solvent, which decreases the extraction of polar interferences. Although non-polar solvents do not absorb microwaves, this problem is solved by absorbing the waves by the sorbent filled inside the pocket device as well as the sample matrix (Wang et al. 2013). This issue is one of the interesting advantages of simultaneous locating of the MP-μSPE pocket device inside the MAE vessel.

As can be seen in the literature, the sorbent materials have also been dispersed into an extract or used as the packed material in a cartridge for further pre-concentration and clean-up (Huang et al. 2013; Yang 2011). Compared to the dispersive approach, sorbent filled into the pocket membrane has the main advantages, including easy collection, low sorbent contamination with interferences, and more reusability. Also, despite the SPE cartridges, the sorbent blockage is eliminated by the pocket device.

MP-μSPE has also been hyphenated with some microextraction methods, including DLLME, USAEME, and SBSE (Cai et al. 2019; Tsai et al. 2009; Ge and Lee 2012; Mao et al. 2016). As shown in Figure 3.5, MP-μSPE can be performed before and after DLLME and USAEME procedures. If MP-μSPE is first performed and then a DLLME or USAEME procedure is used, the main aims are eliminating the solvent evaporation step of the desorption eluent to reach the higher pre-concentration factors, save the analysis time, and improve the clean-up. On the other hand, applying MP-μSPE after a DLLME or USAEME procedure is mainly due to eliminating the centrifugation challenge in DLLME or USAEME and decreasing the required long extraction time (> 30 min) in MP-μSPE (Sajid 2017). The recent applications of MP-μSPE for pre-concentration of different analytes in biological samples are summarized in Table 3.2.

3.4 Pipette Tip Micro Solid-Phase Extraction

PT-μSPE is the miniaturized form of the conventional SPE introduced by William Brewer (University of South Carolina, USA) (Brewer 2003). The main goal of this technique is to reduce the amounts of sample and hazardous organic solvents as much as practicable in a highly efficient SPE process. Micropipettes utilized in PT-μSPE techniques are made of polypropylene, polyethylene, polytetrafluoroethylene, and polyolefin with a distal shape of one end and a conical shape of the other (Bordin et al. 2016). Desired sorbent is packed between two pieces of filter, which can be common frits or ungreased cotton. To eliminate the air bubbles inside the packed sorbent, the pipette tip should be sonicated precisely. Prior to an actual extraction process, the sorbent should be washed by an appropriate solvent-like deionized water to eliminate contaminants (Seidi et al. 2019).

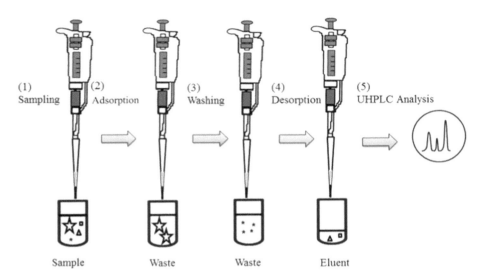

FIGURE 3.6 Schematic presentation of a PT-µSPE procedure. Reproduced with permission from Xie et al. (2009) with a brief modification.

In a PT-µSPE procedure, the target sample is aspirated into the pipette tip, while during this step the sample is passed through the packed sorbent, and analytes of interest are extracted by the sorbent. To make sure the highest possible extraction efficiencies are achieved, this step should be replicated several times. Afterward, the extracts can be desorbed into a desorption solvent (like methanol, acetonitrile, as well as acidic or basic solutions). Just like the previous step, several desorption cycles should be carried out by the same desorption solvent to make sure that the extracts are completely desorbed out of the pipette tip (Buszewski and Szultka 2012; Hasegawa et al. 2007). A typical PT-µSPE procedure is shown in Figure 3.6.

PT-µSPE can bring about extraction efficiencies comparable with the conventional SPE procedures in which large SPE disks or cartridges are utilized. Interestingly, PT-µSPE does not have the practical challenges of the conventional SPE procedures. For instance, his technique does not require vacuum pumps for passing the sample or extraction solvent through the sorbent. Instead, sample or extraction solvent can be easily passed through the sorbent with the aid of a pipettor. In addition, this simple strategy can facilitate the automation of an SPE procedure (Kumazawa et al. 2010). On top of this, the overall amounts of material consumption comprising the target sample, extraction solvents, and sorbents are dramatically lower compared to a conventional SPE procedure (Pereira et al. 2013). Furthermore, PT-µSPE is a well-suited technique for the extraction of target analytes from very low sample volumes.

PT-µSPE can be performed as other features: (1) locating the conventional SPME fibers into the micropipette tips (Xie et al. 2009); and (2) eliminating the upper frit of the sorbent in the pipette tip and a combination of the advantages of both dispersive µSPE and LLE (Fred D. Foster et al.). Moreover, PT-µSPE can be applied in combination with other extraction methods to decrease the drawbacks of enhancing the sample clean-up and extraction efficiency. For example, a PT-µSPE procedure was applied after accelerated solvent extraction for the efficient enrichment and analysis of atrazine and its degradation products in Chinese yam (Wu et al. 2021).

Along with the mentioned advantages of PT-µSPE, this technique suffers a few drawbacks. The first and most important limitation of this technique is that PT-µSPE is a low-throughput extraction technique due to the low amounts of utilized sorbents. It should be noted that utilization of higher amounts of sorbent cannot tackle this problem, as the higher amounts of sorbent results in higher pipette tip pressure. The second limitation of this technique is that the packed sorbent can be easily clogged during the extraction of samples with various amounts of contaminants. This problem usually occurs during the extraction of analytes from biological samples with hundreds of thousands of biomolecules along with the target analytes (Seidi et al. 2019). The recent applications of PT-µSPE for quantitative analysis of different analytes in biological samples are summarized in Table 3.3.

TABLE 3.3

Applicability of the PT-μSPE method for quantitative analysis of different analytes in biological samples

Sorbent	Instrument	Analyte	Matrix	LOD	RSD%	Ref.
Graphene oxide/ZnCr layered double hydroxide	GFAAS	Pb(II)	Hair	0.2 ng mL^{-1}, 0.1 μg g^{-1}	4.5	(Mirzaee et al. 2021)
Restricted access monolithic tip	HPLC-UV	Magnolol, Honokiol	Rat plasma	25.18 ng mL^{-1}, 25.05 ng mL^{-1}	6.49, 6.39	(Zhou et al. 2020), (Zhou et al. 2020)
Schiff base network-1	HPLC-PDA	Sulfathiazole, Sulfamerazine, Sulfamethazine, Sulfamonomethoxine	Milk and honey	0.13 ng mL^{-1}, 0.19 ng mL^{-1}, 0.21 ng mL^{-1}, 0.25 ng mL^{-1}	0.6, 1.0, 0.7, 1.1	(Zhang 2020), (Zhang et al. 2020), (Zhang et al. 2020), (Zhang et al. 2020)
Polymeric ionic liquid –molecularly imprinted/graphene oxide	HPLC-UV	Clorprenaline, Clenbuterol	Swine urine	0.56 ng mL^{-1}, 0.20 ng mL^{-1}	<3.8, <4.8	(Yuan et al. 2020a), (Yuan et al. 2020a)
Deep eutectic solvent functionalized graphene oxide composite	HPLC-DAD	Hippuric acid, Methylhippuric acid	Urine	1.66 - 2.89 ng mL^{-1}, 1.66 - 2.89 ng mL^{-1}	≤ 5.2, ≤ 5.2	(Yuan et al. 2020), (Yuan et al. 2020)
Polyacrylonitrile@covalent organic framework-Sichuan University	HPLC-PDA	Oxytetracycline, Tetracycline, Chlortetracycline	Duck and grass carp	0.6 ng ml^{-1}, 0.6 ng mL^{-1}, 3 ng mL^{-1}	4.57, 8.46, 5.97	(Wang et al., 2020), (Wang et al., 2020), (Wang et al., 2020)
Modified divinyl benzene	GC-MS/MS	Amphetamine, Methamphetamine, 3,4-Methylenedioxyamphetamine, 3,4-Methylenedioxymethamphetamine	Urine	8.25 ng mL^{-1}, 5.88 ng mL^{-1}, 2.52 ng mL^{-1}, 4.56 ng mL^{-1}	<8.6, <6.0, <9.5, <7.2	(Shi et al., 2020), (Shi et al., 2020), (Shi et al., 2020)
Poly styrene-codivinyl benzene + C8Poly styrene-codivinyl benzene + C18	UHPLC-MS/MS	Tyrosol, Farnesol	Vaginal fluid	0.16 ng mL^{-1}, 0.16 ng mL^{-1}	<11.0, <7.0	(Pilařová et al., 2020), (Pilařová et al., 2020)
Ionic liquid-thiol-graphene oxide composite	HPLC-UV	Fipronil	Egg	4.76 μg kg^{-1}	≤6.3	(Li et al., 2020)
Molecularly imprinted polymer	Spectrophotometry	Ciprofloxacin	Plasma	1.50 μg L^{-1}	≤4.9	(Hashemi et al., 2020)
Polyacrylonitrile/molecularly imprinted polymer -53(Fe) nanofibers	HPLC-DAD	Oxazepam, Oxazepam, Nitrazepam, Nitrazepam	Urine, Plasma	1.5 ng mL^{-1}, 1.5 ng mL^{-1}, 2.5 ng mL^{-1}, 2.5 ng mL^{-1}	≤5.1, ≤5.6, ≤8.8, ≤9.2	(Amini 2020), (Amini et al., 2020), (Amini et al., 2020), (Amini et al., 2020)

(Continued)

TABLE 3.3 (Continued)

Applicability of the PT-μSPE method for quantitative analysis of different analytes in biological samples

Sorbent	Instrument	Analyte	Matrix	LOD	RSD%	Ref.
Molecularly imprinted polymer	HPLC-DAD	(R)-(+)-Albendazole sulfoxide(S)-(−)-Albendazole sulfoxide	Urine	–	78.2 ± 0.2, 69.7 ± 1.7	(Anacleto et al., 2019), (Anacleto et al., 2019)
Carboxyl cotton chelator-zirconium-based metal–organic framework	HPLC-UV	Gallic acid, Protocatechuic acid	Rabbit serum	0.024 μg mL^{-1}, 0.023 μg mL^{-1}	2.42, 3.89	(Bao et al., 2019), (Bao et al., 2019)
Covalent organic framework on the surface of functionalized metal organic framework	HPLC-VWD	Sulfadiazine, Sulfamerazine, Sulfamethazine, Sulfamonomethoxine, Sulfamethoxazole, Sulfadimethoxine	Milk	10 ng mL^{-1}, 2 ng mL^{-1}, 1 ng mL^{-1}, 1 ng mL^{-1}, 5 ng mL^{-1}, 2 ng mL^{-1}	≤4.7, ≤5.3, ≤6.6, ≤5.6, ≤5.3, ≤6.2	(Chen et al., 2019), (Chen et al., 2019), (Chen et al., 2019), (Chen et al., 2019), (Chen et al., 2019), (Chen et al., 2019)
Molecularly imprinted polymer/ZnO	Spectrophotometry	Nicotine	Human plasma	0.33 μg L^{-1}	5.3	(Hashemi and Keykha, 2019)
Carbon nanotubes	Spectrofluorometry	Thiamine	Human plasma	0.16 μg L^{-1}	2.3	(Hashemi et al., 2019)
Tantalum metal organic framework	HPLC-UV	Nicotine, Nicotine	Saliva, Urine	0.7 μg L^{-1}, 0.7 μg L^{-1}	<2.7, <4.7	(Reza and Kahkha, 2019), (Reza and Kahkha, 2019)
2-(hexyloxy) naphthalenesulfate doped polyaniline polymer	HPLC/UV-Vis	Sulfathiazole, Sulfapyridine, Sulfadiazine	Honey and milk	10 ng mL^{-1}, 16.5 ng mL^{-1}, 9.5 ng mL^{-1}	4.5, 2.5, 4.9	(Sadeghi and Olieaei, 2019), (Sadeghi and Olieaei, 2019), (Sadeghi and Olieaei, 2019)
Methacrylic acid-co-ethylene glycol dimethacrylate	UHPLC-MS/MS	Drugs	Oral fluid	0.03–0.6 μg L^{-1}	<8.5	(Sorribes-Soriano et al. 2019)
Graphene/multi-walled carbon nanotubes	HPLC-FLD	17β-estradiol	Milk	0.7 ng mL^{-1}	≤5.5	(Yuan et al., 2019)
Hydrophilic interaction liquid chromatography (HILIC) powder based on amide-modified silica	LC-MS	Glyphosate, Aminomethylphosphonic acid	Urine	3 μg L^{-1}, 1 μg L^{-1}	≤9.3, ≤7.7	(Chen et al., 2019), (Chen et al., 2019)

Sorbent	Technique	Analytes	Matrix	LOD	Recovery (%)	Reference
Styrene divinyl benzene	LC-MS/MS	Norepinephrine, Epinephrine, Dopamine, Normetanephrine, Metanephrine	Urine	0.18 ng mL^{-1}, 0.22 ng mL^{-1}, 0.15 ng mL^{-1}, 0.003 ng mL^{-1}, 0.03 ng mL^{-1}	<5, <6, <4, <5, <5	(Mastrianni et al., 2019), (Mastrianni et al., 2019), (Mastrianni et al., 2019), (Mastrianni et al., 2019), (Mastrianni et al., 2019)
SNW-1@PAN nanofiber	HPLC-PDA	Sulfadiazine, Sulfamerazine, Sulfamethazine, Sulfamonomethoxine, Sulfamethoxazole	Pork and chicken	1.7 ng mL^{-1}, 1.7 ng mL^{-1}, 2.0 ng mL^{-1}, 2.4 ng mL^{-1}, 2.7 ng mL^{-1}	7.5, 6.3, 8.5, 5.4, 6.8	(Yan et al., 2018), (Yan et al., 2018), (Yan et al., 2018), (Yan et al., 2018), (Yan et al., 2018)
Ionic liquid-graphene oxide	GC-MS	PAHs	Blood	0.002–0.004 µg L^{-1}	<9.39	(Zhang et al., 2018b)
Molecularly imprinted polymer	GC-MS	Bisphenol A	Urine	–	<13.6	(Brigante et al., 2017)
Carbon nanotube-ZnS	LC-UV	Mefenamic acid	Urine	0.075 µg L^{-1}	<4.20	(Kakhka et al., 2016)
Molecularly imprinted polymer	LC-DAD	(−)-(2S,4R) ketoconazole (+)-(2R,4S) ketoconazole	Urine	–	<5.14, <5.14	(da Silva et al., 2016), (da Silva et al., 2016)
Carbon nanotube-polystyrene divinylbenzene	LC-UV	Magnoflorine, Epiberberin, Jatrorrhizine, Palmatine, Epimedin A, Epimedin B, Epimedin C, Icariin	Biological samples	0.16 µg L^{-1}, 0.68 µg L^{-1}, 0.45 µg L^{-1}, 0.61 µg L^{-1}, 0.75 µg L^{-1}, 0.46 µg L^{-1}, 0.76 µg L^{-1}, 0.27 µg L^{-1}	<4.88, <2.89, <4.33, <3.83, <4.58, <4.67, <4.50, <4.61	(Wang et al., 2017), (Wang et al., 2017), (Wang et al., 2017), (Wang et al., 2017), (Wang et al., 2017), (Wang et al., 2017), (Wang et al., 2017), (Wang et al., 2017)
C8, C18, Silica gel, Polystyrene	LC-DAD	Flavonoids	Rat serum	–	<5.62	(Wang et al., 2015)
Molecularly imprinted polymer	LC-DAD	Prednisolone	Urine	0.085 µg L^{-1}	<3.10	(Wang et al., 2016)
Molecularly imprinted polymer	LC-FLD	Estradiol	Milk	6.00 ng L^{-1}	<3.10	(Wang et al., 2016)
Graphene oxide	LC-UV	Sulfadimidine, Sulfachloropyridazine, Sulfamonomethoxine, Sulfachloropyrazine	Bovine milk	0.004 µg g^{-1}, 0.008 µg g^{-1}, 0.006 µg g^{-1}, 0.012 µg g^{-1}	6.30, –, 3.70, <5.90	(Sun et al., 2014), (Sun et al., 2014), (Sun et al., 2014), (Sun et al., 2014)

(Continued)

TABLE 3.3 (Continued)

Applicability of the PT-μSPE method for quantitative analysis of different analytes in biological samples

Sorbent	Instrument	Analyte	Matrix	LOD	RSD%	Ref.
Polyaniline-styrene divinylbenzene	LC-FLD	Fluoxetine, Norfluoxetine	Plasma	—	<9.70, <12.1	(Chaves et al., 2015), (Chaves et al., 2015)
Molecularly imprinted polymer	LC-DAD	Ciprofloxacin, Enrofloxacino, Marbofloxacino, Norfloxacin	Urine	18.0 ng mL^{-1}, 18.0 ng mL^{-1}, 18.0 ng mL^{-1}, 18.0 ng mL^{-1}	<3.55, <1.45, <2.50, <2.74	(de Oliveira et al., 2016), (de Oliveira et al., 2016), (de Oliveira et al., 2016), (de Oliveira et al., 2016)
Graphene oxide	LC-UV	Indomethacin, Acemetacin	Urine	0.027 μg mL^{-1}, 0.026 μg mL^{-1}	<6.40, <5.00	(Yuan et al., 2016), (Yuan et al., 2016)
Molecularly imprinted polymer	LC-UV	Met-enkephalin, Leu-enkephalin	Cerebrospinal fluid	0.08 nM, 0.05 nM	<5.90, <5.50	(Li and Li, 2015), (Li and Li, 2015)
Cigarette filter	LC-DAD	Ketoconazole	Urine	6.25 ng mL^{-1}	<8.80	(Andrade et al., 2015)
Carbon nanotube–polymethacrylate	LC-UV	Mianserine, Desipramine, Amitriptyline, Trimipramine	Urine	13.0 μg L^{-1}, 9.00 μg L^{-1}, 15.0 μg L^{-1}, 5.00 μg L^{-1}	<7.00, <4.00, <14.0, <10.0	(Fresco-Cala et al., 2018), (Fresco-Cala et al., 2018), (Fresco-Cala et al., 2018), (Fresco-Cala 2018)
Graphene oxide-polypyrrole	LC-UV	Sulfathiazole, Sulfapyridine, Sulfamethizole, Sulfadoxine, Sulfisoxazole, Sulfamethoxazole, Sulfadimethoxine	Honey and milk	1.50 μg mL^{-1}, 1.04 μg mL^{-1}, 1.42 μg mL^{-1}, 1.04 μg mL^{-1}, 1.12 μg mL^{-1}, 1.19 μg mL^{-1}, 1.21 μg mL^{-1}	<9.10, <10.7, <10.4, <9.30, <10.5, <11.2, <10.8	(Qi et a., 2018), (Qi et al. 2018), (Qi et al. 2018), (Qi et al. 2018), (Qi et al.,2018), (Qi et al. 2018)
Molecularly imprinted polymer	LC-UV	Ofloxacin, Pefloxacin, Norfloxacin, Ciprofloxacin, Enrofloxacin	Egg	1.07 μg kg^{-1}, 0.68 μg kg^{-1}, 0.53 μg kg^{-1}, 0.79 μg kg^{-1}, 0.65 μg kg^{-1}	<4.60, <4.40, <3.70, <3.60, <4.80	(Liu et al. 2013), (Liu et al. 2013), (Liu et al. 2013), (Liu et al. 2013), (Liu et al. 2013)
Au@Lectin	LC-UV	Galactosylated protein	E. coli	—	<8.00	(Alwael et al. 2011)

3.5 Spin-Column Micro Solid-Phase Extraction (SC-μSPE)

Spin column micro solid-phase extraction (SC-μSPE) is one of the most utilized and popular configurations of μSPE techniques. This technique differs from the other micro solid-phase extraction methods based on the method exploited for passing the sample through the solid extraction phase (Hansen and Pedersen-Bjergaard 2019). In SC-μSPE, a few milligrams of the desired solid sorbent is packed between two frit pieces to generate a column tip, which is subsequently inserted into a centrifuge microtube. Then, the sample solution containing the target analytes is loaded in the top of column tip followed by successive movement of sample through the adsorbent by the aid of a spinning rotator applying a centrifugal force. In this method, flow rate of sample solution across the solid phase can easily be manipulated by programming of centrifuge speed, to attain efficient analytes isolation and prevent the prolonged extractions. However, in most cases, repeated cycles of sample aspiration and dispensing through the extraction phase are required to give the analyte enough time for interaction with the solid phase and maintain a favorable extraction efficiency (Seidi et al. 2019). SC-μSPE is schematically shown in Figure 3.7.

SC-μSPE can be accounted as an alternative to PT-μSPE, which eliminates the manual handling of the sample solution during the extraction procedure and provides much better precision and capacity for analyte entrapment compared to PT-μSPE. On the other side, as the repeated sample aspirating/dispensing cycles are needed to be performed manually and coupling of the centrifuge systems with analytical instruments is coming with several challenges, there is no report in the literature describing an automated SC-μSPE procedure. The latter issue imposes severe restriction on utilization of this method in routine laboratory tasks. Additionally, as another challenge associated with this technique, the possibility of column clogging where complicated sample matrices are in use should not be rolled out (Seidi et al. 2019).

The developments in SC-μSPE have been mainly focused on the utilized sorbent and implementation of SC-μSPE procedures for extraction of a wide variety of analytes from diverse sample matrices. The type of solid phase sorbent can be considered as the most effective parameter on the efficiency of SC-μSPE, since the utilized extraction phase controls the back pressure in the spin column, the fouling of adsorbent surface during the extraction, the selectivity of extraction, and the capacity for quantitative analytes capture.

So far, a great variety of adsorbents have been exploited in SC-μSPE procedures, which have demonstrated to have a remarkable effect on the extraction efficiency of target analytes. These utilized solid extraction phases are ranging from commercially available sorbents, including C_{18}, Styrene-divinylbenzene-reverse phase sulfonated, cation exchanger (Svačinová et al. 2012), and metal oxides (La Barbera 2018), to synthetic adsorbents, such as monolithic silica rods with different modifications (Namera et al. 2008; Namera et al. 2012; Namera et al. 2011), metal organic frameworks (Esrafili et al. 2020), as well as the electrospun nanofibers of polyamide-graphene oxide-polypyrrole (Seidi et al. 2019), polyacrylonitrile/Ni-metal-organic framework 74 (Amini et al. 2020), and polyacrylonitrile/metal-organic framework of MIL-53(Fe) (Amini et al. 2020).

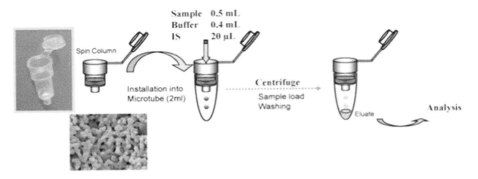

FIGURE 3.7 Schematic presentation of a SC-μSPE procedure. Reproduced with permission from (Alwael et al. 2011).

TABLE 3.4

Applicability of the SC-μSPE method for quantitative analysis of different analytes in biological samples

Sorbent	Instrument	Analyte	Matrix	LOD	RSD%	Ref.
Polyacrylonitrile/Ni-metal-organic framework-74	HPLC-DAD	Atenolol, Captopril	Plasma, Urine	0.13 ng mL^{-1}, 0.15 ng mL^{-1}	≤7.8, ≤6.9	(Amini 2020)
Poly glycidyl methacrylate-co-1-allyl-3-methylimidazolium chloride-co-ethylene glycol dimethacrylate monolith	HPLC-UV	Carteolol, Acebutolol, Oxprenolol, Alprenolol, Propranolol	Urine	2.0 μg·L^{-1}, 1.4 μg·L^{-1}, 40.0 μg·L^{-1}, 40.0 μg·L^{-1}, 20.0 μg·L^{-1}	5.4, 4.2, 5.1, 3.3, 6.5	(Mompó-roselló et al. 2020)
Poly glycidyl methacrylate-co-ethylene glycol dimethacrylate monolith metal organic framework functionalized	HPLC-DAD	Ketoprofen, Ibuprofen, Flurbiprofen	Urine	0.2 μg·L^{-1}, 6.2 μg·L^{-1}, 0.3 μg·L^{-1}	0.8, 1.2, 1.8	(Giesbers et al. 2019)
Polyamide-graphene oxide-polypyrrole nanofibers	HPLC/UV	Methyl paraben, Ethyl paraben, Propyl paraben	Milk	3.0 ng mL^{-1}, 5.0 ng mL^{-1}, 7.0 ng mL^{-1}	≤8.6, ≤6.4, ≤6.8	(Seidi et al. 2019)
Monolithic silica bonded to C18 and a cation-exchange phase	HPLC-MS	Pindolol, Carazolol, Bisoprolol, Propranolol, Carvedilol	Serum	2 ng mL^{-1}, 2 ng mL^{-1}, 2 ng mL^{-1}, 2 ng mL^{-1}, 2 ng mL^{-1}	≤8.2, ≤2.9, ≤3.5, ≤4.6, ≤3.2	(Namera et al. 2018)
Poly glycidyl methacrylate-co-ethylene glycol dimethacrylate with oxidized single-walled carbon nanohorns	HPLC-DAD	Naproxen, Fenbufen, Flurbiprofen, Ibuprofen	Urine	0.1 μg·L^{-1}, 0.5 μg·L^{-1}, 0.5 μg·L^{-1}, 10 μg·L^{-1}	4.0, 11.8, 5.4, 5.8	(Fresco-Cala et al. 2017)

μSPE Applications in Analytical Toxicology

Method	Technique	Analytes	Sample	LOD	RSD (%)	Reference
Titanium (IV) immobilized metal ion affinity chromatography	LC-MS/MS	Phosphopeptides	Biological samples	–	1.91	(Yao 2017; Seidi 2019; Abbasi 2020)
Monolithic silica disk-packed spin column with phenylboronate moieties	HPLC-UV	3,4-dihydroxyphenylglycol, Dopamine, Epinephrine, 3,4-dihydroxyphenylalanine, Norepinephrine	Mouse urine	3 nM, 4 nM, 5 nM, 8 nM, 10 nM	≤8.7, ≤6.3, ≤9.2, ≤8.5, ≤7.7	(Kanamori et al. 2015)
Monolithic silica bonded to propyl benzenesulfonic acid	HPLC-FLD	N^G-Monomethyl-L-arginineN^G, N^G-dimethyl-L-arginineN^G, $N^{G'}$-dimethyl-L-arginine L-Arginine	Plasma	3.75 fmol, 7.5 fmol, 3.75 fmol, 9.0 fmol	≤4.78, ≤4.34, ≤4.50, ≤4.78	(Nonaka et al. 2014)
TopTip C18 spin tips	LC/MS-MS	Methamphetamine, Amphetamine	Plasma	0.5 ng mL^{-1}, 0.5 ng mL^{-1}	≤3.9, ≤4.4	(Marumo et al. 2013)
Mixed-mode monolithic silica	GC-MS	Arsenite, Arsenate, Methylarsenate	Urine	1 ng mL^{-1}, 1 ng mL^{-1}, 1 ng mL^{-1}	≤6.60, ≤10.5, ≤8.6	(Namera et al. 2012)
Mixed-mode monolithic silica	GC-MS	Phenacetine, Secobarbital, Phenobarbital, Lidocaine, Amitriptyline Imipramine, Diazepam, Chlorpromazine, Clotiazepam	Urine	5 ng mL^{-1}, 5 ng mL^{-1}, 1 ng mL^{-1}, 1 ng mL^{-1}, 5 ng mL^{-1}, 1 ng mL^{-1}, 25 ng mL^{-1}, 5.0 ng mL^{-1}	≤11.5, ≤10.5, ≤8.6, ≤5.5, ≤14.7, ≤14.6, ≤3.1, ≤10.9, ≤5.1	(Namera et al. 2011)

Apart from the developments in type of solid phase adsorbents, there are a few reports in the literature dealing with improvement of the SC-μSPE performance through exploitation of multi-stage SC-μSPE and mix-mode SC-μSPE. In multi-stage SC-μSPE, a layer-by-layer packing of different sorbents is utilized to provide much higher clean-up and selectivity by removing the interferences of different types (Svačinová et al. 2012). In mix-mode SC-μSPE, different modifications are applied onto the adsorbent phase to enable simultaneous or sequential extraction of diverse analytes by a single spin column (Namera et al. 2012; Namera et al. 2011).

To draw partial conclusions, although the research into SC-μSPE has been mainly dealing with the developments in the type of extraction phase, more developments for SC-μSPE are needed to alleviate some of the typical drawbacks of this method such as manual repeated aspirating/dispensing cycles of sample solution through exploration of new configurations and device designs capable of automated recycling of sample solution through the solid phase. The recent applications of SC-μSPE for quantitative analysis of different analytes in biological samples are summarized in Table 3.4.

3.6 Concluding Remarks and Future Trends

This chapter focused on the most well-known μSPE formats, which are useful in forensic sample analysis. Among these discussed extraction techniques, MP-μSPE offers the most sample clean-up since the process consists of analytes migration across the polymeric membrane and their adsorption on the solid sorbent afterward. Thus, it could be useful for analysis of complicated forensic samples. However, the fastest extraction mechanism belongs to D-μSPE, which benefited from the wide contact surface of nanoscaled sorbents. SC-μSPE has a simple setup, sorbent collection, and desorption approach. On the other hand, PT-μSPE is the most suitable μSPE mode for automation since a sample flow and an eluent flow may be applied for presenting the extraction process. μSPE could perform an appropriate sample cleanup, which reduces the matrix effect; regarding its dual extraction mechanism, it includes analytes adsorption on the solid sorbent and their further elution. The approach may consist of some washing steps for diminishing the interference effects. Also, miniaturized μSPE scale decreases the required sample size and eluent volume and may have a positive effect on extraction recovery and its rapidness. One of the main imaginable futures for all extraction methods is the automation possibility, which may be conducted in some μSPE modes in the close future. However, regarding the matrix of interest in this chapter, introducing some novel, selective, available, and inexpensive sorbents is the most important aim of solid-based extraction techniques. Due to the unique potential ability of each μSPE mode, the design of some lab-on-chip devices could be expected as the round goals. Finally, the combination of different μSPE methods with other extraction and microextraction methods is another interesting aspect that can provide considerable advantages, such as more sample cleanup, higher pre-concentration factors, and reduced analysis time.

REFERENCES

Abbasi, Shahryar, Seyed A. Haeri, Ali Naghipour, and Sami Sajjadifar. "Enrichment of Cardiovascular Drugs Using Rhamnolipid Bioaggregates after Dispersive Solid Phase Extraction Based Water Compatible Magnetic Molecularly Imprinted Biopolymers." *Microchemical Journal* 157, February (2020): 104874. 10.1016/j.microc.2020.104874

Abidin, Nurul Nabilah Zainal, Mohd Marsin Sanagi, Wan Aini Wan Ibrahim, Salasiah Endud, Dyia Syaleyana Md Shukri. "Portable Micro-Solid Phase Extraction for the Determination of Polycyclic Aromatic Hydrocarbons in Water Samples." *Analytical Methods*, 6 (2014): 5512. 10.1039/c4ay00474d

Adenuga, Adeniyi A., Olawole Ayinuola, Ebunoluwa A. Adejuyigbe, and Aderemi O. Ogunfowokan. "Biomonitoring of Phthalate Esters in Breast-Milk and Urine Samples as Biomarkers for Neonates' Exposure, Using Modified Quechers Method with Agricultural Biochar as Dispersive Solid-Phase Extraction Absorbent." *Microchemical Journal* 152 (2020): 104277. 10.1016/j.microc.2019.104277

Adlnasab, Laleh, Maryam Ezoddin, Rouh A. Shojaei, and Fezzeh Aryanasab. "Ultrasonic-Assisted Dispersive Micro Solid-Phase Extraction Based on Melamine-Phytate Supermolecular Aggregate as a Novel

Bio-Inspired Magnetic Sorbent for Preconcentration of Anticancer Drugs in Biological Samples Prior to HPLC-UV Analysis." *Journal of Chromatography B: Analytical Technologies in the Biomedical and Life Sciences* 1095, August (2018): 226–34. 10.1016/j.jchromb.2018.08.001

Aghaie, Ali B. G., and Mohammad R. Hadjmohammadi. "Fe3O4@p-Naphtholbenzein as a Novel Nano-Sorbent for Highly Effective Removal and Recovery of Berberine: Response Surface Methodology for Optimization of Ultrasound Assisted Dispersive Magnetic Solid Phase Extraction." *Talanta* 156–157 (2016): 18–28. 10.1016/j.talanta.2016.04.034

Ali, Hassan Refat H., Ahmad I. Hassan, Yasser F. Hassan, and Mohamed M. El-Wekil. "Mannitol Capped Magnetic Dispersive Micro-Solid-Phase Extraction of Polar Drugs Sparfloxacin and Orbifloxacin from Milk and Water Samples Followed by Selective Fluorescence Sensing Using Boron-Doped Carbon Quantum Dots." *Journal of Environmental Chemical Engineering* 9 (2021): 105078. 10.1016/j.jece.2021.105078

Alipanahpour Dil, Ebrahim, Arash Asfaram, and Hamedreza Javadian. "A New Approach for Microextraction of Trace Albendazole Sulfoxide Drug from the Samples of Human Plasma and Urine, and Water by the Molecularly Imprinted Polymer Nanoparticles Combined with HPLC." *Journal of Chromatography B: Analytical Technologies in the Biomedical and Life Sciences* 1158 (2020): 122249. 10.1016/j.jchromb.2020.122249

AlKinani, Alaa, Mohammad Eftekhari, and Mohammad Gheibi. "Ligandless Dispersive Solid Phase Extraction of Cobalt Ion Using Magnetic Graphene Oxide as an Adsorbent Followed by Its Determination with Electrothermal Atomic Absorption Spectrometry." *International Journal of Environmental Analytical Chemistry* 101, 2021, 1 (2019): 17–34. 10.1080/03067319.2019.1659254

Alwael, Hassan, Damian Connolly, Paul Clarke, Roisin Thompson, Brendan Twamley, Brendan O'Connor, and Brett Paull. "Pipette-Tip Selective Extraction of Glycoproteins with Lectin Modified Gold Nano-Particles on a Polymer Monolithic Phase." *Analyst* 136, no. 12 (2011): 2619–28.

Ameri Akhtiar Abadi, Mohammad, Mahboubeh Masrournia, Mohamad R. Abedi. "Simultaneous Extraction and Preconcentration of Three Beta (β)-Blockers in Biological Samples with an Efficient Magnetic Dispersive Micro-Solid Phase Extraction Procedure Employing in Situ Sorbent Modification." *Microchemical Journal* 163 (2021): 105937. 10.1016/j.microc.2021.105937

Amini, Shima, Homeira Ebrahimzadeh, Shahram Seidi, and Niloofar Jalilian. "Polyacrylonitrile/MIL-53 (Fe) Electrospun Nanofiber for Pipette-Tip Micro Solid Phase Extraction of Nitrazepam and Oxazepam Followed by HPLC Analysis." *Microchimica Acta* 187, no. 2 (2020a): 1–10.

Amini, Shima, Homeira Ebrahimzdeh, Shahram Seidi, and Niloofar Jalilian. "Preparation of Electrospun Polyacrylonitrile/Ni-MOF-74 Nanofibers for Extraction of Atenolol and Captopril Prior to HPLC-DAD." *Microchimica Acta* 187, no. 9 (2020b): 1–12.

Anacleto, Sara da Silva, Hanna L. De Oliveira, Anny Talita, Tienne Aparecida, Marcella Matos Cordeiro Borges, Ricky Cássio, Arnaldo C. Pereira, Keyller B. Borges; D. D. C. Naturais; et al. "Assessment of the Performance of Solid Phase Extraction Based on Pipette Tip Employing a Hybrid Molecularly Imprinted Polymer as an Adsorbent for Enantioselective Determination of Albendazole Sulfoxide." *Journal of Chromatographic Science* 57 (2019), 671–8. 10.1093/chromsci/bmz036.

Anastassiades, Michelangelo, Steven J. Lehotay, Darinka Štajnbaher, and Frank J. Schenck. "Fast and Easy Multiresidue Method Employing Acetonitrile Extraction/Partitioning and 'Dispersive Solid-Phase Extraction' for the Determination of Pesticide Residues in Produce." *Journal of AOAC International* 86 (2003): 412–31. 10.1093/jaoac/86.2.412

Andrade, Roseane. T., Ricky C. S. da Silva, Arnaldo C. Pereira, and Keyller B. Borges. "Self-Assembly Pipette Tip-Based Cigarette Filters for Micro-Solid Phase Extraction of Ketoconazole Cis-Enantiomers in Urine Samples Followed by High-Performance Liquid Chromatography/Diode Array Detection." *Analytical Methods* 7, no. 17 (2015): 7270–9.

Ansari, Saeedeh, and Saeed Masoum. "A Multi-Walled Carbon Nanotube-Based Magnetic Molecularly Imprinted Polymer as a Highly Selective Sorbent for Ultrasonic-Assisted Dispersive Solid-Phase Microextraction of Sotalol in Biological Fluids." *Analyst* 143, no. 12 (2018): 2862–75. 10.1039/c7an02077e

Arghavani-Beydokhti, Somayeh, Maryam Rajabi, and Alireza Asghari. "Combination of Magnetic Dispersive Micro Solid-Phase Extraction and Supramolecular Solvent-Based Microextraction Followed by High-Performance Liquid Chromatography for Determination of Trace Amounts of Cholesterol-Lowering

Drugs in Complicated Matrices." *Analytical and Bioanalytical Chemistry* 409, no. 18 (2017): 4395–407. 10.1007/s00216-017-0383-x

Arroyo-Manzanares, Natalia, Rosa Peñalver-Soler, Natalia Campillo, and Pilar Viñas. "Dispersive Solid-Phase Extraction Using Magnetic Carbon Nanotube Composite for the Determination of Emergent Mycotoxins in Urine Samples." *Toxins (Basel)* 12, no. 1 (2020): 1–12. 10.3390/toxins12010051

Asgharinezhad, Ali A., Homeira Ebrahimzadeh, Fatemeh Mirbabaei, Narges Mollazadeh, and Nafiseh Shekari. "Dispersive Micro-Solid-Phase Extraction of Benzodiazepines from Biological Fluids Based on Polyaniline/Magnetic Nanoparticles Composite." *Analytica Chimica Acta* 844 (2014): 80–9. 10.1016/j.aca.2014.06.007

Bahrani, Sonia, Mehrorang Ghaedi, Arash Asfaram, Mohammad J. K. Mansoorkhani, and Hamedreza Javadian. "Rapid Ultrasound-Assisted Microextraction of Atorvastatin in the Sample of Blood Plasma by Nickel Metal Organic Modified with Alumina Nanoparticles." *Journal of Separation Science* 43, no. 24 (2020): 4469–79. 10.1002/jssc.202000660

Baile, Paola, Lorena Vidal, and Antonio Canals. "A Modified Zeolite/Iron Oxide Composite as a Sorbent for Magnetic Dispersive Solid-Phase Extraction for the Preconcentration of Nonsteroidal Anti-Inflammatory Drugs in Water and Urine Samples." *Journal of Chromatography A* 1603 (2019): 33–43. 10.1016/j.chroma.2019.06.039

Baile, Paola, Lorena Vidal, and Antonio Canals. "Magnetic Dispersive Solid-Phase Extraction Using ZSM-5 Zeolite/Fe2O3 Composite Coupled with Screen-Printed Electrodes Based Electrochemical Detector for Determination of Cadmium in Urine Samples." *Talanta* 220, March (2020): 121394. 10.1016/j.talanta.2020.121394

Baile, Paola, Lorena Vidal, Miguel Á. Aguirre, and Antonio Canals. "A Modified ZSM-5 Zeolite/Fe2O3 Composite as a Sorbent for Magnetic Dispersive Solid-Phase Microextraction of Cadmium, Mercury and Lead from Urine Samples Prior to Inductively Coupled Plasma Optical Emission Spectrometry." *Journal of Analytical Atomic Spectrometry* 33, no. 5 (2018): 856–66. 10.1039/c7ja00366h

Bao, Tao, Ying Su, Nan Zhang, Yan Gao, and Sicen Wang. "Hydrophilic Carboxyl Cotton for in Situ Growth of UiO-66 and Its Application as Adsorbents." *Industrial & Engineering Chemistry Research* 58 (2019): 20331–9. 10.1021/acs.iecr.9b05172.

Basheer, Chanbasha, Anass Ali Alnedhary, B. S. Madhava Rao, Suresh Valliyaveettil, and Hian Kee Lee. "Development and Application of Porous Membrane-Protected Carbon Nanotube Micro-Solid-Phase Extraction Combined with Gas Chromatography/Mass Spectrometry." *Analytical Chemistry* 78 (2006): 2853–8. 10.1021/ac060240i

Basheer, Chanbasha, Guang Chong Han, Ming Hii Toh, and Kee Lee Hian. "Application of Porous Membrane-Protected Micro-Solid-Phase Extraction Combined with HPLC for the Analysis of Acidic Drugs in Wastewater." *Analytical Chemistry* 79 (2007): 6845–50. 10.1021/ac070372r

Basheer, Chanbasha, Kothandaraman Narasimhan, Meihui Yin, Changqing Zhao, Mahesh Choolani, and Hian Kee Lee. "Application of Micro-Solid-Phase Extraction for the Determination of Persistent Organic Pollutants in Tissue Samples." *Journal of Chromatography A* 1186, no. 1–2 (2008): 358–64. 10.1016/j.chroma.2007.10.015

Behbahani, Mohammad, Ali Veisi, Fariborz Omidi, Aminreza Noghrehabadi, Ali Esrafili, and Mohammad H. Ebrahimi. "Application of a Dispersive Micro-Solid-Phase Extraction Method for Pre-Concentration and Ultra-Trace Determination of Cadmium Ions in Water and Biological Samples." *Applied Organometallic Chemistry* 32, no. 3 (2018): 1–10. 10.1002/aoc.4134

Behbahani, Mohammad, Saman Bagheri, and Mostafa M. Amini. "Developing an Ultrasonic-Assisted d-μ-SPE Method Using Amine-Modified Hierarchical Lotus Leaf-like Mesoporous Silica Sorbent for the Extraction and Trace Detection of Lamotrigine and Carbamazepine in Biological Samples." *Microchemical Journal* 158, July (2020): 105268. 10.1016/j.microc.2020.105268

Bordin, Dayanne C. M., Marcela N. R. Alves, Eduardo G. De Campos, and Bruno S. De Martinis. "Disposable Pipette Tips Extraction: Fundamentals, Applications and State of the Art." *Journal of Separation Science* 39 (2016): 1168–72. 10.1002/jssc.201500932

Bozorgzadeh, Elahe, Ardalan Pasdaran, and Heshmatollah Ebrahimi-Najafabadi. "Determination of Toxic Heavy Metals in Fish Samples Using Dispersive Micro Solid Phase Extraction Combined with Inductively Coupled Plasma Optical Emission Spectroscopy." *Food Chemistry* 346 (2021): 128916. 10.1016/j.foodchem.2020.128916

Brewer, William E. Disposable Pipette Extraction. Google Patents , May 2003.

Brigante, Tamires A. V., Luis F. C. Miranda, Israel D. de Souza, Vinicius R. Acquaro Junior, and Maria E. C. Queiroz. "Pipette Tip Dummy Molecularly Imprinted Solid-Phase Extraction of Bisphenol A from Urine Samples and Analysis by Gas Chromatography Coupled to Mass Spectrometry." *Journal of Chromatography B* 1067 (2017): 25–33. 10.1016/j.jchromb.2017.09.038

Buszewski, Boguslaw, and Malgorzata Szultka. "Past, Present, and Future of Solid Phase Extraction: A Review." *Critical Reviews in Analytical Chemistry* 42, no. 3 (2012): 198–213.

Cai, QianQian, Lianjun Zhang, Pan Zhao, Xiaowen Lun, Wei Li, Yi Guo, and XiaoHong Hou. "A Joint Experimental-Computational Investigation: Metal Organic Framework as a Vortex Assisted Dispersive Micro-Solid-Phase Extraction Sorbent Coupled with UPLC-MS/MS for the Simultaneous Determination of Amphenicols and Their Metabolite in Aquaculture Water." *Microchemical Journal* 130 (2017): 263–70. 10.1016/j.microc.2016.09.014

Cai, Qianqian, Tianyu Zhao, Lianjun Zhang, Pan Zhao, Yajie Zhu, Haiyan Xu, and Xiaohong Hou. "A New Strategy for Extraction and Depuration of Pantoprazole in Rat Plasma: Vortex Assisted Dispersive Micro-Solid-Phase Extraction Employing Metal Organic Framework MIL-101(Cr) as Sorbent Followed by Dispersive Liquid–Liquid Microextraction Based on Solidification of a Floating Organic Droplet." *Journal of Pharmaceutical and Biomedical Analysis* 172 (2019): 86–93. 10.1016/j.jpba.2019.04.016

Campos do Lago, Ayla, Marcello H. da Silva Cavalcanti, Mariana A. Rosa, Alberto T. Silveira, Cesar R. Teixeira Tarley, Eduardo C. Figueiredo. "Magnetic Restricted-Access Carbon Nanotubes for Dispersive Solid Phase Extraction of Organophosphates Pesticides from Bovine Milk Samples." *Analytica Chimica Acta* 1102 (2020): 11–23. 10.1016/j.aca.2019.12.039

Chaves, Andrea R., Bruno H. F. Moura, Juciene A. Caris, Denilson Rabelo, and Maria E. C. Queiroz. "The Development of a New Disposable Pipette Extraction Phase Based on Polyaniline Composites for the Determination of Levels of Antidepressants in Plasma Samples." *Journal of Chromatography A* 1399 (2015): 1–7.

Chen, Dawei, Hong Miao, Yunfeng Zhao, and Yongning Wu SC. *Journal of Chromatography A* 1587 (2019): 73–8. 10.1016/j.chroma.2018.11.030

Chen, Dan, and Hui Xu. "Electrospun Core-Shell Nanofibers as an Adsorbent for on-Line Micro-Solid Phase Extraction of Monohydroxy Derivatives of Polycyclic Aromatic Hydrocarbons from Human Urine, and Their Quantitation by LC-MS." *Microchimica Acta* 187, no. 1 (2020): 57. 10.1007/s00604-019-4007-3

Chen, Shi-En, Jingwen Hu, Ping Yan, Jing Sun, Wenhui Jia, Shuyun Zhu, Xian-En Zhao, and Huwei Liu. "12-Plex UHPLC-MS/MS Analysis of Sarcosine in Human Urine Using Integrated Principle of Multiplex Tags Chemical Isotope Labeling and Selective Imprint Enriching." *Talanta*, 224 (2021): 121788. 10.1016/j.talanta.2020.121788

Chen, Shizhong, Juntao Yan, Jianfen Li, and Dengbo Lu. "Dispersive Micro-Solid Phase Extraction Using Magnetic ZnFe 2 O 4 Nanotubes as Adsorbent for Preconcentration of Co(II), Ni(II), Mn(II) and Cd(II) Followed by ICP-MS Determination." *Microchemical Journal* 147, January (2019): 232–8. 10.1016/j.microc.2019.02.066

Chen, Zhipeng, Chen Yu, Jiangbo Xi, Sheng Tang, Tao Bao, and Juan Zhang. "A Hybrid Material Prepared by Controlled Growth of a Covalent Organic Framework on Amino-Modified MIL-68 for Pipette Tip Solid-Phase Extraction of Sulfonamides Prior to Their Determination by HPLC." *Microchimica Acta* 186, no. 6 (2019): 1–11. 10.1007/s00604-019-3513-7

Chisvert, Alberto, Soledad Cárdenas, and Rafael Lucena. "Dispersive Micro-Solid Phase Extraction." *TrAC – Trends in Analytical Chemistry* 112 (2019): 226–33. 10.1016/j.trac.2018.12.005

Criado-García, Laura, and Lourdes Arce. "Extraction of Toxic Compounds from Saliva by Magnetic-Stirring-Assisted Micro-Solid-Phase Extraction Step Followed by Headspace-Gas Chromatography-Ion Mobility Spectrometry." *Analytical and Bioanalytical Chemistry* 408, no. 24 (2016): 6813–22. 10.1007/s00216-016-9808-1

Cui, Yixuan, Haiyan Ma, Di Liu, Mengjiao Li, Ruisen Hao, Junmei Li, and Ye Jiang. "Graphene Oxide Adsorbent-Based Dispersive Solid Phase Extraction Coupled with Multi-Pretreatment Clean-up for Analysis of Trace Ochratoxin A in Chicken Liver." *Chromatographia* 83, no. 10 (2020): 1307–14. 10.1007/s10337-020-03942-8

da Silva, Ricky C. S., Valdir Mano, Arnaldo C. Pereira, Eduardo C. de Figueiredo, and Keyller B. Borges. "Development of Pipette Tip-Based on Molecularly Imprinted Polymer Micro-Solid Phase Extraction for Selective Enantioselective Determination of (−)-(2 S, 4 R) and (+)-(2 R, 4 S) Ketoconazole in Human Urine Samples Prior to HPLC-DAD." *Analytical Methods* 8, no. 20 (2016): 4075–85.

Dastkhoon, Mehdi, Mehrorang Ghaedi, Arash Asfaram, Maryam Arabi, Abbas Ostovan, and Alireza Goudarzi. "Cu@SnS/SnO2 Nanoparticles as Novel Sorbent for Dispersive Micro Solid Phase Extraction of Atorvastatin in Human Plasma and Urine Samples by High-Performance Liquid Chromatography with UV Detection: Application of Central Composite Design (CCD)." *Ultrasonics Sonochemistry* 36 (2016): 42–9. 10.1016/j.ultsonch.2016.10.030

de Oliveira, Hanna L., Sara da Silva Anacleto, Anny T. M. da Silva, Arnaldo C. Pereira, Warley de Souza Borges, Eduardo C. Figueiredo, and Keyller B. Borges. "Molecularly Imprinted Pipette-Tip Solid Phase Extraction for Selective Determination of Fluoroquinolones in Human Urine Using HPLC-DAD." *Journal of Chromatography B: Analytical Technologies in the Biomedical and Life Sciences* 1033–1034 (2016): 27–39. 10.1016/j.jchromb.2016.08.008

Dil, Ebrahim A., Mehrorang Ghaedi, Arash Asfaram, Fatemeh Mehrabi, Ali A. Bazrafshan, and Abdol M. Ghaedi. "Trace Determination of Safranin O Dye Using Ultrasound Assisted Dispersive Solid-Phase Micro Extraction: Artificial Neural Network-Genetic Algorithm and Response Surface Methodology." *Ultrasonics Sonochemistry* 33 (2016): 129–40. 10.1016/j.ultsonch.2016.04.031

Dil, Ebrahim A., Amir H. Doustimotlagh, Hamedreza Javadian, Arash Asfaram, and Mehrorang Ghaedi. "Nano-Sized Fe3O4@SiO2-Molecular Imprinted Polymer as a Sorbent for Dispersive Solid-Phase Microextraction of Melatonin in the Methanolic Extract of Portulaca Oleracea, Biological, and Water Samples." *Talanta* 221, July 2020 (2021): 121620. 10.1016/j.talanta.2020.121620

Eilami, Owrang, Arash Asfaram, Mehrorang Ghaedi, and Alireza Goudarzi. "Effective Determination of Trace Residues of Glibenclamide in Urine Samples Using Dispersive Micro Solid-Phase Extraction and Its Final Detection by Chromatographic Analysis." *Analytical Methods* 11, no. 5 (2019): 627–34. 10.1039/c8ay01965g

Esrafili, Ali, Mahnaz Ghambarian, Mohammad Tajik, and Mahroo Baharfar. "Spin-Column Micro-Solid Phase Extraction of Chlorophenols Using MFU-4l Metal-Organic Framework." *Microchimica Acta* 187, no. 1 (2020): 39.

Ezoddin, Maryam, Laleh Adlnasab, Akram Afshari Kaveh, Mohammad A. Karimi, Behnaz Mahjoob. "Development of Air-Assisted Dispersive Micro-Solid-Phase Extraction-Based Supramolecular Solvent-Mediated Fe3O4@Cu–Fe–LDH for the Determination of Tramadol in Biological Samples." *Biomedical Chromatography* 33, no. 9 (2019): 1–11. 10.1002/bmc.4572

Fahimirad, Bahareh, Maryam Rajabi, and Ali Elhampour. "A Rapid and Simple Extraction of Anti-Depressant Drugs by Effervescent Salt-Assisted Dispersive Magnetic Micro Solid-Phase Extraction Method Using New Adsorbent Fe 3 O 4 @SiO 2 @N 3." *Analytica Chimica Acta* 1047 (2019): 275–84. 10.1016/j.aca.2018.10.028

Feist, Barbara. "Selective Dispersive Micro Solid-Phase Extraction Using Oxidized Multiwalled Carbon Nanotubes Modified with 1,10-Phenanthroline for Preconcentration of Lead Ions." *Food Chemistry* 209 (2016): 37–42. 10.1016/j.foodchem.2016.04.015

Fernández, Elena, Lorena Vidal, Joaquin Silvestre-Albero, and Antonio Canals. "Magnetic Dispersive Solid-Phase Extraction Using a Zeolite-Based Composite for Direct Electrochemical Determination of Lead (II) in Urine Using Screen-Printed Electrodes." *Microchimica Acta* 187, no. 1 (2020): 1–10. 10.1007/s00604-019-4062-9

Ferrone, Vincenzo, Sabrina Todaro, Maura Carlucci, Antonella Fontana, Alessia Ventrella, Giuseppe Carlucci, and Edoardo Milanetti. "Optimization by Response Surface Methodology of a Dispersive Magnetic Solid Phase Extraction Exploiting Magnetic Graphene Nanocomposite Coupled with UHPLC-PDA for Simultaneous Determination of New Oral Anticoagulants (NAOs) in Human Plasma." *Journal of Pharmaceutical and Biomedical Analysis* 179, November (2020): 112992. 10.1016/j.jpba.2019.112992

Foster, Fred D., John R. Stuff, Edward A. Pfannkoch and William E. Brewer. Determination of Pain Management Drugs Using Automated Disposable Pipette Extraction and LC-MS/MS. GERSTEL AppNote AN-2011-06.

Fresco-Cala, Beatriz, Óscar Mompó-Roselló, Ernesto F. Simó-Alfonso, Soledad Cárdenas, and Jose M. Herrero-Martínez. "Carbon Nanotube-Modified Monolithic Polymethacrylate Pipette Tips for (Micro) Solid-Phase Extraction of Antidepressants from Urine Samples. *Microchimica Acta* 185, no. 2 (2018): 127. 10.1007/s00604-017-2659-4

Fresco-Cala, Beatriz, Soledad Cárdenas, and Jose M. Herrero-Martínez. "Preparation of Porous Methacrylate Monoliths with Oxidized Single-Walled Carbon Nanohorns for the Extraction of Nonsteroidal

Anti-Inflammatory Drugs from Urine Samples." *Microchimica Acta* 184 (2017): 1863–71. 10.1007/s00604-017-2203-6

Galán-Cano, Francisco, Rafael Lucena, Soledad Cárdenas, and Miguel Valcárcel. "Dispersive Micro-Solid Phase Extraction with Ionic Liquid-Modified Silica for the Determination of Organophosphate Pesticides in Water by Ultra Performance Liquid Chromatography." *Microchemical Journal* 106 (2013): 311–7. 10.1016/j.microc.2012.08.016

Gao, Guihua, Sijia Li, Shuo Li, Yudan Wang, Pan Zhao, Xiangyu Zhang, and Xiaohong Hou. "A Combination of Computational–experimental Study on Metal-Organic Frameworks MIL-53(Al) as Sorbent for Simultaneous Determination of Estrogens and Glucocorticoids in Water and Urine Samples by Dispersive Micro-Solid-Phase Extraction Coupled to UPLC-MS/MS." *Talanta* 180 (2018): 358–67. 10.1016/j.talanta.2017.12.071

García-Valverde, Maria Teresa, Rafael Lucena, Francisco Galán-Cano, Soledad Cárdenas, and Miguel A. Valcárcel. "Carbon Coated Titanium Dioxide Nanotubes: Synthesis, Characterization and Potential Application as Sorbents in Dispersive Micro Solid Phase Extraction." *Journal of Chromatography A* 1343 (2014): 26–32. 10.1016/j.chroma.2014.03.062

García-Valverde, Maria Teresa, Rafael Lucena, Soledad Cárdenas, and Miguel A. Valcárcel. "In-Syringe Dispersive Micro-Solid Phase Extraction Using Carbon Fibres for the Determination of Chlorophenols in Human Urine by Gas Chromatography/Mass Spectrometry." *Journal of Chromatography A* 1464 (2016): 42–9. 10.1016/j.chroma.2016.08.036

Ge, Dandan, and Hian K. Lee. "Sonication-Assisted Emulsification Microextraction Combined with Vortex-Assisted Porous Membrane-Protected Micro-Solid-Phase Extraction Using Mixed Zeolitic Imidazolate Frameworks 8 as Sorbent." *Journal of Chromatography A* 1263 (2012): 1–6. 10.1016/j.chroma.2012.09.016

Ghazaghi, Mehri, Hassan Z. Mousavi, Hamid Shirkhanloo, and Alimorad Rashidi. "Ultrasound Assisted Dispersive Micro Solid-Phase Extraction of Four Tyrosine Kinase Inhibitors from Serum and Cerebrospinal Fluid by Using Magnetic Nanoparticles Coated with Nickel-Doped Silica as an Adsorbent." *Microchimica Acta* 183, no. 10 (2016): 2779–89. 10.1007/s00604-016-1927-z

Ghorbani, Mahdi, Mahmoud Chamsaz, Mohsen Aghamohammadhasan, and Alireza Shams. "Ultrasonic Assisted Magnetic Dispersive Solid Phase Microextraction for Pre Concentration of Serotonin–Norepinephrine Reuptake Inhibitor Drugs." *Analytical Biochemistry* 551 (2018): 7–18. 10.1016/j.ab.2018.05.003

Ghorbani, Mahdi, Mohsen Aghamohammadhasan, Alireza Shams, Farzaneh Tajfirooz, Reza Pourhassan, Seyyede R. Bana Khosravi, Elahe Karimi, and Ali Jampour. "Ultrasonic Assisted Magnetic Dispersive Solid Phase Microextraction for Preconcentration of Two Nonsteroidal Anti-Inflammatory Drugs in Real Water, Biological and Milk Samples Employing an Experimental Design." *Microchemical Journal* 145 (2019): 1026–35. 10.1016/j.microc.2018.12.019

Giesbers, Myrthe, Enrique J. Carrasco-correa, Ernesto F. Simó-alfonso, and Jose M. Herrero-martínez. "Hybrid Monoliths with Metal-Organic Frameworks in Spin Columns for Extraction of Non-Steroidal Drugs Prior to Their Quantitation by Reversed-Phase HPLC." *Microchimica Acta* 101 (2019): 1–9.

Guo, Liang, and Hian Kee Lee. "Vortex-Assisted Micro-Solid-Phase Extraction Followed by Low-Density Solvent Based Dispersive Liquid-Liquid Microextraction for the Fast and Efficient Determination of Phthalate Esters in River Water Samples." *Journal of Chromatography A* 1300 (2013): 24–30. 10.1016/j.chroma.2013.01.030

Hamidi, Samin, Ayda Azami, and Elnaz Mehdizadeh Aghdam. "A Novel Mixed Hemimicelles Dispersive Micro-Solid Phase Extraction Using Ionic Liquid Functionalized Magnetic Graphene Oxide/Polypyrrole for Extraction and Pre-Concentration of Methotrexate from Urine Samples Followed by the Spectrophotometric Method." *Clinica Chimica Acta* 488, June 2018 (2019): 179–88. 10.1016/j.cca.2018.11.006

Hansen, Frederik A., and Stig Pedersen-Bjergaard. "Emerging Extraction Strategies in Analytical Chemistry." *Analytical Chemistry* 92, no. 1 (2019): 2–15.

Hasegawa, Chika, Takeshi Kumazawa, Xiao-Pen Lee, Akemi Marumo, Natsuko Shinmen, Hiroshi Seno, and Keizo Sato. "Pipette Tip Solid-Phase Extraction and Gas Chromatography–Mass Spectrometry for the Determination of Methamphetamine and Amphetamine in Human Whole Blood." *Analytical and Bioanalytical Chemistry* 389, no. 2 (2007): 563–70.

Hashemi, Sayyed H., and Fateme Keykha. "Application of the Response Surface Methodology in the Optimization of Modi Fi Ed Molecularly Imprinted Polymer Based Pipette-Tip Micro-Solid Determination of Nicotine in Seawater and Human." *Analytical Methods* (2019): 5405–12. 10.1039/c9ay01496a

Hashemi, Sayyed H., Morteza Ziyaadini, Massoud Kaykhaii, Ahmad Jamali Keikha, Nasrin Naruie. "Separation and Determination of Ciprofloxacin in Seawater, Human Blood Plasma and Tablet Samples Using Molecularly Imprinted Polymer Pipette-Tip Solid Phase Extraction and Its Optimization by Response Surface Methodology." *Journal of Separation Science* 43, no. 2 (2020): 505–13. 10.1002/jssc.201900923

Hashemi, Sayyed Hossein, Hossein Yahyavi, Massoud Kaykhaii, and Mohammad Hashemi. "Spectrofluorometrical Determination of Vitamin B 1 in Different Matrices Using Box-Behnken Designed Pipette Tip Solid Phase Extraction by a Carbon Nanotube Sorbent." *ChemistrySelect* 4, no. 11 (2019): 3052–7. 10.1002/slct.201803731.

Hemmati, Maryam, Maryam Rajabi, and Alireza Asghari. "Ultrasound-Promoted Dispersive Micro Solid-Phase Extraction of Trace Anti-Hypertensive Drugs from Biological Matrices Using a Sonochemically Synthesized Conductive Polymer Nanocomposite." *Ultrasonics Sonochemistry* 39 (2017): 12–24. 10.1016/j.ultsonch.2017.03.024

Hooshmand, Sara, and Zarrin Es'haghi. "Hydrophilic Modified Magnetic Multi-Walled Carbon Nanotube for Dispersive Solid/Liquid Phase Microextraction of Sunitinib in Human Samples." *Analytical Biochemistry* 542, August 2017 (2018): 76–83. 10.1016/j.ab.2017.11.019

Hrynko, Izabela, Bozena Łozowicka, and Piotr Kaczyński. "Development of Precise Micro Analytical Tool to Identify Potential Insecticide Hazards to Bees in Guttation Fluid Using LC–ESI–MS/MS." *Chemosphere* 263 (2021): 128143. 10.1016/j.chemosphere.2020.128143

Hu, Jingwen, Shuyun Zhu, Shi-En Chen, Ruisheng Liu, Jing Sun, Xian-En Zhao, and Huwei Liu. "Multiplexed Derivatization Strategy-Based Dummy Molecularly Imprinted Polymers as Sorbents for Magnetic Dispersive Solid Phase Extraction of Globotriaosylsphingosine Prior to UHPLC-MS/MS Quantitation." *Microchimica Acta* 187 (2020): 373. 10.1007/s00604-020-04341-4

Huang, Jiangeng, Juanjuan Liu, Cong Zhang, Jiaojiao Wei, Li Mei, Shan Yu, Gao Li, and Li Xu. "Determination of Sulfonamides in Food Samples by Membrane-Protected Micro-Solid Phase Extraction Coupled with High Performance Liquid Chromatography." *Journal of Chromatography A* 1219 (2012): 66–74. 10.1016/j.chroma.2011.11.026

Huang, Man C., Hsin C. Chen, Ssu C. Fu, and Wang H. Ding. "Determination of Volatile N-Nitrosamines in Meat Products by Microwave-Assisted Extraction Coupled with Dispersive Micro Solid-Phase Extraction and Gas Chromatography-Chemical Ionisation Mass Spectrometry." *Food Chemistry* 138 (2013): 227–33. 10.1016/j.foodchem.2012.09.119

Huang, Wei, Xingyu Hou, Yukui Tong, and Miaomiao Tian. "Determination of Sialic Acid in Serum Samples by Dispersive Solid-Phase Extraction Based on Boronate-Affinity Magnetic Hollow Molecularly Imprinted Polymer Sorbent." *RSC Advances* 9, no. 10 (2019): 5394–401. 10.1039/c9ra00511k

Jabbari, Neghin R., Arezou Taghvimi, Siavoush Dastmalchi, and Yousef Javadzadeh. "Dispersive Solid-Phase Extraction Adsorbent of Methamphetamine Using in-Situ Synthesized Carbon-Based Conductive Polypyrrole Nanocomposite: Focus on Clinical Applications in Human Urine." *Journal of Separation Science* 43 (2020): 606–13. 10.1002/jssc.201900773

Jahed, Fatemeh S., Samin Hamidi, Saba Ghaffary, and Babak Nejati. "Dispersive Micro Solid Phase Extraction of Busulfan from Plasma Samples Using Novel Mesoporous Sorbent Prior to Determination by HPLC-MS/MS." *Journal of Chromatography B: Analytical Technologies in the Biomedical and Life Sciences* 1145, February (2020): 122091. 10.1016/j.jchromb.2020.122091

Jalilian, Niloofar, Homeira Ebrahimzadeh, and Ali A. Asgharinezhad. "Determination of Acidic, Basic and Amphoteric Drugs in Biological Fluids and Wastewater after Their Simultaneous Dispersive Micro-Solid Phase Extraction Using Multiwalled Carbon Nanotubes/Magnetite Nanoparticles@poly (2-Aminopyrimidine) Composite." *Microchemical Journal* 143 (2018): 337–49. 10.1016/j.microc.2018.08.037

Jalilian, Niloofar, Homeira Ebrahimzadeh, Ali A. Asgharinezhad, and Parisa Khodayari. "Magnetic Molecularly Imprinted Polymer for the Selective Dispersive Micro Solid Phase Extraction of Phenolphthalein in Urine Samples and Herbal Slimming Capsules Prior to HPLC-PDA Analysis." *Microchemical Journal* 160, PB (2021): 105712. 10.1016/j.microc.2020.105712

Jamali, Mohammad Reza, Ahmad Firouzjah, and Reyhaneh Rahnama. "Solvent-Assisted Dispersive Solid Phase Extraction." *Talanta* 116 (2013): 454–9. 10.1016/j.talanta.2013.07.023

Jayasinghe, G. D. Thilini Madurangika, Raquel Domínguez-González, Pilar Bermejo-Barrera, and Antonio Moreda-Piñeiro. "Miniaturized Vortex Assisted-Dispersive Molecularly Imprinted Polymer Micro-Solid Phase Extraction and HPLC-MS/MS for Assessing Trace Aflatoxins in Cultured Fish." *Analytical Methods* 12, no. 35 (2020): 4351–62. 10.1039/d0ay01259a

Jiao, Zhe, Dan Zhu, and Weixuan Yao. "Combination of Accelerated Solvent Extraction and Micro-Solid-Phase Extraction for Determination of Trace Antibiotics in Food Samples." *Food Analytical Methods* 8 (2015): 2163–8. 10.1007/s12161-015-0105-y

Kaabipour, Maryam, Saeid Khodadoust, and Fatemeh Zeraatpisheh. "Preparation of Magnetic Molecularly Imprinted Polymer for Dispersive Solid-Phase Extraction of Valsartan and Its Determination by High-Performance Liquid Chromatography: Box-Behnken Design." *Journal of Separation Science* 43, no. 5 (2020): 912–9. 10.1002/jssc.201901058

Kahkha, Mohammad R. R., Massoud Kaykhaii, Mahdi S. Afarani, and Zahra Sepehri. "Determination of Mefenamic Acid in Urine and Pharmaceutical Samples by HPLC after Pipette-Tip Solid Phase Microextraction Using Zinc Sulfide Modified Carbon Nanotubes." *Analytical Methods* 8, no. 30 (2016): 5978–83.

Kanamori, Takahiro, Muneki Isokawa, Takashi Funatsu, and Makoto Tsunoda. "Development of Analytical Method for Catechol Compounds in Mouse Urine Using Hydrophilic Interaction Liquid Chromatography with Fluorescence Detection." *Journal of Chromatography B* 985 (2015): 142–8.

Kanimozhi, Sivarajan, Chanbasha Basheer, Kothandaraman Narasimhan, Lin Liu; Stephen Koh, Feng Xue, Mahesh Choolani, and Hian Kee Lee. "Application of Porous Membrane Protected Micro-Solid-Phase-Extraction Combined with Gas Chromatography-Mass Spectrometry for the Determination of Estrogens in Ovarian Cyst Fluid Samples." *Analytica Chimica Acta* 687, no. 1 (2011), 56–60. 10.1016/j.aca.2010.12.007.

Kazemi, Elahe, Ali M. Haji Shabani, Shayessteh Dadfarnia, Amir Abbasi, Mohammad R. Rashidian Vaziri, and Abbas Behjat. "Development of a Novel Mixed Hemimicelles Dispersive Micro Solid Phase Extraction Using 1-Hexadecyl-3-Methylimidazolium Bromide Coated Magnetic Graphene for the Separation and Preconcentration of Fluoxetine in Different Matrices Before Its Determination by Fiber Optic Linear Array Spectrophotometry and Mode-Mismatched Thermal Lens Spectroscopy." *Analytica Chimica Acta* 905 (2016): 85–92. 10.1016/j.aca.2015.12.012

Köseoğlu, Kadir, Halil Ibrahim Ulusoy, Erkan Yilmaz, and Mustafa Soylak. "Simple and Sensitive Determination of Vitamin A and E in the Milk and Egg Yolk Samples by Using Dispersive Solid Phase Extraction with Newly Synthesized Polymeric Material." *Journal of Food Composition and Analysis* 90, March (2020): 103482. 10.1016/j.jfca.2020.103482

Krawczyk, Magdalena, and Ewa Stanisz. "Ultrasound-Assisted Dispersive Micro Solid-Phase Extraction with Nano-TiO2 as Adsorbent for the Determination of Mercury Species." *Talanta* 161 (2016): 384–91. 10.1016/j.talanta.2016.08.071

Krawczyk-Coda, Magdalena, and Ewa Stanisz. "Determination of Fluorine in Herbs and Water Samples by Molecular Absorption Spectrometry After Preconcentration on Nano-TiO2 Using Ultrasound-Assisted Dispersive Micro Solid Phase Extraction." *Analytical and Bioanalytical Chemistry* 409 (2017): 6439–49. 10.1007/s00216-017-0589-y

Kumazawa, Takeshi, Chika Hasegawa, Xiao-Pen Lee, and Keizo Sato. "New and Unique Methods of Solid-Phase Extraction for Use before Instrumental Analysis of Xenobiotics in Human Specimens." *Forensic Toxicology* 28, no. 2 (2010): 61–8.

La Barbera, Giorgia, Anna L. Capriotti, Chiara Cavaliere, Francesca Ferraris, Michele Laus, Susy Piovesana, Katia Sparnacci, and Aldo Laganà. "Development of an Enrichment Method for Endogenous Phosphopeptide Characterization in Human Serum." *Analytical and Bioanalytical Chemistry* 410, no. 3 (2018): 1177–1185. 10.1007/s00216-017-0822-8

Lasarte-Aragonés, Guillermo, Rafael Lucena, Soledad Cárdenas, and Miguel Valcárcel. "Effervescence-Assisted Dispersive Micro-Solid Phase Extraction." *Journal of Chromatography A* 1218 (2011): 9128–34. 10.1016/j.chroma.2011.10.042

Lashgari, Maryam, and Hian K. Lee. "Determination of Perfluorinated Carboxylic Acids in Fish Fillet by Micro-Solid Phase Extraction, Followed by Liquid Chromatography-Triple Quadrupole Mass Spectrometry." *Journal of Chromatography A* 1369 (2014): 26–32. 10.1016/j.chroma.2014.09.082

Lashgari, Maryam, and Hian K. Lee. "Micro-Solid Phase Extraction of Perfluorinated Carboxylic Acids from Human Plasma." *Journal of Chromatography A* 1432 (2016): 7–16. 10.1016/j.chroma.2016.01.005

Lee, Tien P., Bahruddin Saad, Wejdan S. Khayoon, and Baharuddin Salleh. "Molecularly Imprinted Polymer as Sorbent in Micro-Solid Phase Extraction of Ochratoxin A in Coffee, Grape Juice and Urine." *Talanta* 88 (2012): 129–35. 10.1016/j.talanta.2011.10.021

Li, Hua, and Dan Li. "Preparation of a Pipette Tip-Based Molecularly Imprinted Solid-Phase Microextraction Monolith by Epitope Approach and Its Application for Determination of Enkephalins in Human Cerebrospinal Fluid." *Journal of Pharmaceutical and Biomedical Analysis* 115 (2015): 330–8. 10.1016/j.jpba.2015.07.033

Li, Mengyuan, Chunliu Yang, Hongyuan Yan, Yehong Han, and Dandan Han. "An Integrated Solid Phase Extraction with Ionic Liquid-Thiol-Graphene Oxide as Adsorbent for Rapid Isolation of Fipronil Residual in Chicken Eggs." *Journal of Chromatography A* 1631 (2020): 461568. 10.1016/j.chroma.2020.461568

Li, Sun Yi, Ilona Petrikovics, and Jorn (Chi C.) Yu. "Development of Magnetic Carbon Nanotubes for Dispersive Micro Solid Phase Extraction of the Cyanide Metabolite, 2-Aminothiazoline-4-Carboxylic Acid, in Biological Samples." *Journal of Chromatography B: Analytical Technologies in the Biomedical and Life Sciences* 1109, December 2018 (2019): 67–75. 10.1016/j.jchromb.2019.01.020

Li, Zi Ling, Zi Yang Zhang, Teng Wen Zhao, Chun Yan Meng, Qian Ying Zhang, and Man Man Wang. "In-Situ Fabrication of Zeolite Imidazole Framework@hydroxyapatite Composite for Dispersive Solid-Phase Extraction of Benzodiazepines and Their Determination with High-Performance Liquid Chromatography-VWD Detection." *Microchimica Acta* 187, no. 9 (2020): 540. 10.1007/s00604-020-04517-y

Lim, Tze Han, Lingna Hu, Cong Yang, Chaobin He, and Hian Kee Lee. "Membrane Assisted Micro-Solid Phase Extraction of Pharmaceuticals with Amino and Urea-Grafted Silica Gel." *Journal of Chromatography A* 1316 (2013): 8–14. 10.1016/j.chroma.2013.09.034

Liu, Lu, Xiao-Xing Wang, Xia Wang, Gui-Ju Xu, Yan-Fang Zhao, Ming Lin Wang, Jing Ming Lin, Ru Song Zhao, and Yongning Wu. "Triazine-Cored Covalent Organic Framework for Ultrasensitive Detection of Polybrominated Diphenyl Ethers from Real Samples: Experimental and DFT Study." *Journal of Hazardous Materials* 403, July 2020 (2021): 123917. 10.1016/j.jhazmat.2020.123917

Liu, Suting, Hongyuan Yan, Mingyu Wang, and Lihui Wang. "Water-Compatible Molecularly Imprinted Microspheres in Pipette Tip Solid-Phase Extraction for Simultaneous Determination of Five Fluoroquinolones in Eggs." *Journal of Agricultural and Food Chemistry* 61, no. 49 (2013): 11974–80. 10.1021/jf403759t

Liu, Yongjing, Lishuang Yu, Hua Zhang, and Dawei Chen. "Dispersive Micro-Solid-Phase Extraction Combined with Online Preconcentration by Capillary Electrophoresis for the Determination of Glycopyrrolate Stereoisomers in Rat Plasma." *Journal of Separation Science* 41, no. 6 (2018): 1395–404. 10.1002/jssc.201700753

Ma, Ning, Cheng Feng, Ping Qu, Ge Wang, Jing Liu, Ju X. Liu, and Jian P. Wang. "Determination of Tetracyclines in Chicken by Dispersive Solid Phase Microextraction Based on Metal-Organic Frameworks/Molecularly Imprinted Nano-Polymer and Ultra Performance Liquid Chromatography." *Food Analytical Methods* 13, no. 5 (2020): 1211–9. 10.1007/s12161-020-01744-0

Ma, Yun Jiao, An Qi Bi, Xiao Yuan Wang, Lei Qin, Ming Du, Liang Dong, Xian Bing Xu. "Dispersive Solid-Phase Extraction and Dispersive Liquid–Liquid Microextraction for the Determination of Flavor Enhancers in Ready-to-Eat Seafood by HPLC-PDA." *Food Chemistry* 309 (2020): 125753. 10.1016/j.foodchem.2019.125753

Manaf, Normaliza A., Bahruddin Saad, Mohamed H. Mohamed, Lee D. Wilson, and Aishah A. Latiff. "Cyclodextrin Based Polymer Sorbents for Micro-Solid Phase Extraction Followed by Liquid Chromatography Tandem Mass Spectrometry in Determination of Endogenous Steroids." *Journal of Chromatography A* 1543 (2018): 23–33. 10.1016/j.chroma.2018.02.032

Mandal, Swagata, Rajlakshmi Poi, Sudip Bhattacharyya, Inul Ansary, Subrata D. Roy, Dipak K. Hazra, and Rajib Karmakar. "Multiclass Multipesticide Residue Analysis in Fish Matrix by a Modified QuEChERS Method Using Gas Chromatography with Mass Spectrometric Determination." *Journal of AOAC International* 103, no. 1 (2021): 62–7. 10.5740/jaoacint.19-0205

Manousi, Natalia, Eleni Deliyanni, and George Zachariadis. "Multi-Element Determination of Toxic and Nutrient Elements by ICP-AES after Dispersive Solid-Phase Extraction with Modified Graphene Oxide." *Applied Science* 10, no. 23 (2020): 1–15. 10.3390/app10238722

Mao, Xiangju, Man He, Beibei Chen, and Bin Hu. "Membrane Protected C18 Coated Stir Bar Sorptive Extraction Combined with High Performance Liquid Chromatography-Ultraviolet Detection for the Determination of Non-Steroidal Anti-Inflammatory Drugs in Water Samples." *Journal of Chromatography A* 1472 (2016): 27–34. 10.1016/j.chroma.2016.10.051

Mastrianni, Kaylee R., William E. Kemnitzer, and Kevin W. P. Miller. "A Novel, Automated Dispersive Pipette Extraction Technology Greatly Simplifies Catecholamine Sample Preparation for Downstream LC-MS / MS Analysis." *Translating Life Sciences Innovation* 24, no. 24 (2019): 117–23. 10.1177/2472630318792659.

Mirzaee, Mahsa T., Shahram Seidi, and Reza Alizadeh. "Pipette-Tip SPE Based on Graphene/ZnCr LDH for Pb(II) Analysis in Hair Samples Followed by GFAAS." *Analytical Biochemistry* 612, March 2020 (2021): 113949. 10.1016/j.ab.2020.113949

Mohammadi, Parisa, Mahboubeh Masrournia, Zarrin Es'haghi, and Mehdi Pordel. "Determination of Four Antiepileptic Drugs with Solvent Assisted Dispersive Solid Phase Microextraction – Gas Chromatography–Mass Spectrometry in Human Urine Samples." *Microchemical Journal* 159, September (2020): 105542. 10.1016/j.microc.2020.105542

Mohammadi, Parisa, Mahdi Ghorbani, Parinaz Mohammadi, Majid Keshavarzi, Ayoob Rastegar, Mohsen Aghamohammadhassan, and Ava Saghafi. "Dispersive Micro Solid-Phase Extraction with Gas Chromatography for Determination of Diazinon and Ethion Residues in Biological, Vegetables and Cereal Grain Samples, Employing D-Optimal Mixture Design." *Microchemical Journal* 160 (2021): 105680. 10.1016/j.microc.2020.105680

Mompó-roselló, Oscar, Ana Ribera-castelló, Ernesto F. Simó-alfonso, Maria J. Ruiz-angel, Maria C. García-alvarez-coque, and Jose M. Herrero-martínez. "Extraction of β-Blockers from Urine with a Polymeric Monolith Modified with 1-Allyl-3-Methylimidazolium Chloride in Spin Column Format." *Talanta* 214, February (2020): 120860. 10.1016/j.talanta.2020.120860

Moyakao, Khwankaew, Yanawath Santaladchaiyakit, Supalax Srijaranai, and Jitlada Vichapong. "Preconcentration of Trace Neonicotinoid Insecticide Residues Using Vortex-Assisted Dispersive Micro Solid-Phase Extraction with Montmorillonite as an Efficient Sorbent." *Molecules* 23 (2018): 883. 10.3390/molecules23040883

Naghdi, Elahe, Ali R. Fakhari, and Jahan B. Ghasemi. "Enantioseparation and Quantitative Determination of Ibuprofen Using Vancomycin-Mediated Capillary Electrophoresis Combined with Molybdenum Disulfide-Assisted Dispersive Solid-Phase Extraction: Optimization Using Experimental Design." *Journal of the Iranian Chemical Society* 17, 6 (2020): 1467–77. 10.1007/s13738-020-01874-6

Naing, Nyi Nyi, Sam Fong Yau Li, and Hian Kee Lee. "Magnetic Micro-Solid-Phase-Extraction of Polycyclic Aromatic Hydrocarbons in Water." *Journal of Chromatography A* 1440 (2016): 23–30. 10.1016/j.chroma.2016.02.046

Nakhaei, Jamshid M., Mohammad R. Jamali, Shabnam Sohrabnezhad, and Reyhaneh Rahnama. "Solvent-Assisted Dispersive Solid Phase Extraction of Diclofenac from Human Serum and Pharmaceutical Tablets Quantified by High-Performance Liquid Chromatography." *Microchemical Journal* 152 (2019): 104260. 10.1016/j.microc.2019.104260

Namera, Akira, Akihiro Nakamoto, Manami Nishida, Takeshi Saito, Izumi Kishiyama, Shota Miyazaki, Midori Yahata, Mikio Yashiki, and Masataka Nagao. "Extraction of Amphetamines and Methylenedioxyamphetamines from Urine Using a Monolithic Silica Disk-Packed Spin Column and High-Performance Liquid Chromatography–Diode Array Detection." *Journal of Chromatography A* 1208, no. 1–2 (2008): 71–5.

Namera, Akira, Akito Takeuchi, Takeshi Saito, Shota Miyazaki, Hiroshi Oikawa, Tatsuro Saruwatari, and Masataka Nagao. "Sequential Extraction of Inorganic Arsenic Compounds and Methyl Arsenate in Human Urine Using Mixed-mode Monolithic Silica Spin Column Coupled with Gas Chromatography-mass Spectrometry." *Journal of Separation Science* 35, no. 18 (2012): 2506–13.

Namera, Akira, Shinobu Yamamoto, Takeshi Saito, Shota Miyazaki, Hiroshi Oikawa, Akihiro Nakamoto, and Masataka Nagao. "Simultaneous Extraction of Acidic and Basic Drugs from Urine Using Mixed-mode Monolithic Silica Spin Column Bonded with Octadecyl and Cation-exchange Group." *Journal of Separation Science* 34, no. 16–17 (2011): 2232–9.

Namera, Akira, Takeshi Saito, Yoshimoto Seki, Taro Mizutani, Kazuhiro Murata, and Masataka Nagao. "High-Throughput MonoSpin Extraction for Quantification of Cardiovascular Drugs in Serum Coupled to High-Performance Liquid Chromatography – Mass Spectrometry." *Acta Chromatographica* 31 (2018): 1–5. 10.1556/1326.2018.00493

Nonaka, Satoko, Masae Sekine, Makoto Tsunoda, Yuji Ozeki, Kumiko Fujii, Kazufumi Akiyama, Kazutaka Shimoda, Takemitsu Furuchi, Masumi Katane, Yasuaki Saitoh, and Hiroshi Homma. "Simultaneous Determination of NG-monomethyl-L-Arginine, NG, NG-Dimethyl-L-Arginine, NG, NG-Dimethyl-L-Arginine, and L-Arginine Using Monolithic Silica Disk-Packed Spin Columns and a Monolithic Silica Column." *Journal of Separation Science* 37 (2014): 2087–94. 10.1002/jssc.201400240

Nuckowski, Łukasz, Ewa Zalesińska, Krzysztof Dzieszkowski, Zbigniew Rafiński, and Sylwia Studzińska. "Poly(Ionic Liquid)s as New Adsorbents in Dispersive Micro-Solid-Phase Extraction of Unmodified and Modified Oligonucleotides." *Talanta* 221, March 2020 (2021): 121662. 10.1016/j.talanta.2020.121662

Ojeda, Catalina B., and F. Sanchez Rojas. "Vortex-Assisted Liquid–Liquid Microextraction (VALLME): The Latest Applications." *Chromatographia* 77 (2018): 745–54. 10.1007/s10337-017-3403-2

Omidi, Fariborz, Fatemeh Dehghani, and Seyed Jamaleddin Shahtaheri. "N-Doped Mesoporous Carbon as a New Sorbent for Ultrasonic-Assisted Dispersive Micro-Solid-Phase Extraction of 1-Naphthol and 2-Naphthol, the Biomarkers of Exposure to Naphthalene, from Urine Samples." *Journal of Chromatography B: Analytical Technologies in the Biomedical and Life Sciences* 1160, June (2020): 122353. 10.1016/j.jchromb.2020.122353

Omidi, Fariborz, Mohammad Behbahani, Monireh Khadem, Farideh Golbabaei, and Seyed J. Shahtaheri. "Application of a New Sample Preparation Method Based on Surfactant-Assisted Dispersive Micro Solid Phase Extraction Coupled with Ultrasonic Power for Easy and Fast Simultaneous Preconcentration of Toluene and Xylene Biomarkers from Human Urine Samples." *Journal of the Iranian Chemical Society* 16, no. 6 (2019): 1131–8. 10.1007/s13738-018-01588-w

Pereira, Jorge, João Gonçalves, Vera Alves, and José S. Câmara. "Microextraction Using Packed Sorbent as an Effective and High-Throughput Sample Extraction Technique: Recent Applications and Future Trends." *Sample Preparation* 1, no. 2013 (2013): 38–53.

Pilařová, Veronika, Hana Kočová Vlčková, Ondrej Jung, Michele Protti, Vladimir Buchta, Laura Mercolini, Frantisek Svec, and Lucie Nováková. "Unambiguous Determination of Farnesol and Tyrosol in Vaginal Fluid Using Fast and Sensitive UHPLC-MS/MS Method." *Analytical and Bioanalytical Chemistry* 412, no. 24 (2020): 6529–41. 10.1007/s00216-020-02699-1

Qi, Mengyu, Chunyan Tu, Zhaoqian Li, Weiping Wang, Jianrong Chen, and Ai-Jun Wang. "Determination of Sulfonamide Residues in Honey and Milk by HPLC Coupled with Novel Graphene Oxide/Polypyrrole Foam Material-Pipette Tip Solid Phase Extraction." *Food Analytical Methods* 11, no. 10 (2018): 2885–96. 10.1007/s12161-018-1271-5.

Rahimpoor, Razzagh, Abdulrahman Bahrami, Davood Nematollahi, Farshid Ghorbani Shahna, and Maryam Farhadian. "Sensitive Determination of Urinary Muconic Acid Using Magnetic Dispersive-Solid-Phase Extraction by Magnetic Amino-Functionalised UiO-66." *International Journal of Environmental Analytical Chemistry* 00, no. 00 (2020): 1–14. 10.1080/03067319.2020.1727460

Rajabi, Maryam, Ahmad G. Moghadam, Behruz Barfi, and Alireza Asghari. "Air-Assisted Dispersive Micro-Solid Phase Extraction of Polycyclic Aromatic Hydrocarbons Using a Magnetic Graphitic Carbon Nitride Nanocomposite." *Microchimica Acta* 183 (2016): 1449–58. 10.1007/s00604-016-1780-0

Reyes-Garcés, Nathaly, Emanuela Gionfriddo, German A. Gómez-Ríos, Md. Nazmul Alam, Ezel Boyacl, Barbara Bojko, Varoon Singh, Jonathan Grandy, and Janusz Pawliszyn. "Advances in Solid Phase Microextraction and Perspective on Future Directions." *Analytical Chemistry* 90, no. 1 (2018): 302–60. 10.1021/acs.analchem.7b04502

Rozaini, Muhammad N. H., Noorfatimah Yahaya, Bahruddin Saad, Sazlinda Kamaruzaman, and Nor Suhaila Mohamad Hanapi. "Rapid Ultrasound Assisted Emulsification Micro-Solid Phase Extraction Based on Molecularly Imprinted Polymer for HPLC-DAD Determination of Bisphenol A in Aqueous Matrices." *Talanta* 171 (2017): 242–9. 10.1016/j.talanta.2017.05.006

Sadeghi, Susan, and Samieh Olieaei. "Nanostructured Polyaniline Based Pipette Tip Solid Phase Extraction Coupled with High-Performance Liquid Chromatography for the Selective Determination of Trace Levels of Three Sulfonamides in Honey and Milk Samples with the Aid of Experimental Design Methodology." *Microchemical Journal* 146 (2019): 974–85. 10.1016/j.microc.2019.02.020

Sajid, Muhammad. "Porous Membrane Protected Micro-Solid-Phase Extraction: A Review of Features, Advancements and Applications." *Analytica Chimica Acta* 965 (2017): 36–53. 10.1016/j.aca.2017.02.023

Sajid, Muhammad, and Chanbasha Basheer. "Stir-Bar Supported Micro-Solid-Phase Extraction for the Determination of Polychlorinated Biphenyl Congeners in Serum Samples. *Journal of Chromatography A* 1455 (2016): 37–44. 10.1016/j.chroma.2016.05.084

Sajid, Muhammad, Chanbasha Basheer, Abdulnaser Alsharaa, Kothandaraman Narasimhan, Abdelbaset Buhmeida, Mohammed Al Qahtani, Mahmoud S. Al-Ahwal. "Development of Natural Sorbent Based Micro-Solid-Phase Extraction for Determination of Phthalate Esters in Milk Samples." *Analytica Chimica Acta* 924 (2016): 35–44. 10.1016/j.aca.2016.04.016

Sajid, Muhammad, Chanbasha Basheer, Kothandaraman Narasimhan, Mahesh Choolani, and Hian K. Lee. "Application of Microwave-Assisted Micro-Solid-Phase Extraction for Determination of Parabens in Human Ovarian Cancer Tissues." *Journal of Chromatography B: Analytical Technologies in the Biomedical and Life Sciences* 1000 (2015): 192–8. 10.1016/j.jchromb.2015.07.020

Sajid, Muhammad, Chanbasha Basheer, Muhammad Daud, Abdulnaser Alsharaa. "Evaluation of Layered Double Hydroxide/Graphene Hybrid as a Sorbent in Membrane-Protected Stir-Bar Supported Micro-Solid-Phase Extraction for Determination of Organochlorine Pesticides in Urine Samples." *Journal of Chromatography A* 1489 (2017): 1–8. 10.1016/j.chroma.2017.01.089

Sánchez-González, Juan, Maria J. Tabernero, Ana M. Bermejo, Pilar Bermejo-Barrera, and Antonio Moreda-Piñeiro. "Porous Membrane-Protected Molecularly Imprinted Polymer Micro-Solid-Phase Extraction for Analysis of Urinary Cocaine and Its Metabolites Using Liquid Chromatography - Tandem Mass Spectrometry." *Analytica Chimica Acta* 898 (2015): 50–9. 10.1016/j.aca.2015.10.002

Sánchez-González, Juan, Rocio Salgueiro-Fernández, Pamela Cabarcos, Ana M. Bermejo, Pilar Bermejo-Barrera, and Antonio Moreda-Piñeiro. "Cannabinoids Assessment in Plasma and Urine by High Performance Liquid Chromatography–Tandem Mass Spectrometry after Molecularly Imprinted Polymer Microsolid-Phase Extraction." *Analytical and Bioanalytical Chemistry* 409, no. 5 (2017): 1207–20. 10.1007/s00216-016-0046-3

Sánchez-González, Juan, Sara García-Carballal, Pamela Cabarcos, Maria J. Tabernero; Pilar Bermejo-Barrera, and Antonio Moreda-Piñeiro. "Determination of Cocaine and Its Metabolites in Plasma by Porous Membrane-Protected Molecularly Imprinted Polymer Micro-Solid-Phase Extraction and Liquid Chromatography-Tandem Mass Spectrometry." *Journal of Chromatography A* 1451 (2016): 15–22. 10.1016/j.chroma.2016.05.003

Sánchez-González, Juan, Sara Odoardi, Ana M. Bermejo, Pilar Bermejo-Barrera, Francesco S. Romolo, Antonio Moreda-Piñeiro, Sabina Strano-Rossi. "Development of a Micro-Solid-Phase Extraction Molecularly Imprinted Polymer Technique for Synthetic Cannabinoids Assessment in Urine Followed by Liquid Chromatography–Tandem Mass Spectrometry." *Journal of Chromatography A* 1550 (2018): 8–20. 10.1016/j.chroma.2018.03.049

Seidi, Shahram, Elnaz Sadat Karimi, Ahmad Rouhollahi, Mahroo Baharfar, Maryam Shanehsaz, and Mohammad Tajik. "Synthesis and Characterization of Polyamide-Graphene Oxide-Polypyrrole Electrospun Nanofibers for Spin-Column Micro Solid Phase Extraction of Parabens in Milk Samples." *Journal of Chromatography A* 1599 (2019): 25–34. 10.1016/j.chroma.2019.04.014.

Seidi, Shahram, Mohammad Tajik, Mahroo Baharfar, and Maryam Rezazadeh. "Micro Solid-Phase Extraction (Pipette Tip and Spin Column) and Thin Film Solid-Phase Microextraction: Miniaturized Concepts for Chromatographic Analysis." *TrAC - Trends in Analytical Chemistry* 118 (2019): 810–27. 10.1016/j.trac.2019.06.036

Shaban, Mina, Saba Ghaffary, Jalal Hanaee, Ayda Karbakhshzadeh, and Somaieh Soltani. "Synthesis and Characterization of New Surface Modified Magnetic Nanoparticles and Application for the Extraction of Letrozole from Human Plasma and Analysis with HPLC-Fluorescence." *Journal of Pharmaceutical and Biomedical Analysis* 193 (2021): 113659. 10.1016/j.jpba.2020.113659

Shafiee, Ali, Behzad Aibaghi, and Xu Zhang. "Determination of Ethambutol in Biological Samples Using Graphene Oxide Based Dispersive Solid-Phase Microextraction Followed by Ion Mobility Spectrometry." *International Journal of Ion Mobility Spectrometry* 23, no. 1 (2020): 19–27. 10.1007/s12127-019-00253-z

Shi, Jia Wei, Jing Feng Zhou, Xiong He, and Yun Zhang. "Rapid Analysis of Four Amphetamines in Urine by Self-Made Pipette-Tip Solid-Phase Extraction Followed by GC-MS/MS." *Journal of Chromatographic Science* 58, no. 6 (2020): 569–75. 10.1093/chromsci/bmaa018

Shin, Dasom, Joohye Kim, and Hui-Seung Kang. "Simultaneous Determination of Multi-Pesticide Residues in Fish and Shrimp Using Dispersive-Solid Phase Extraction with Liquid Chromatography–Tandem Mass Spectrometry." *Food Control* 120, March 2020 (2021): 107552. 10.1016/j.foodcont.2020.107552

Simultaneous Extraction of Acidic and Basic Drugs from Urine Using Mixed-Mode Monolithic Silica Spin Column Bonded with Octadecyl and Cation-Exchange Group. 2011, 2232–2239. 10.1002/jssc.201100165.

Skorek, Robert, Edyta Turek, Beata Zawisza, Eva Marguí, Ignasi Queralt, Marek Stempin, Piotr Kucharski, and Rafal Sitko. "Determination of Selenium by X-Ray Fluorescence Spectrometry Using Dispersive Solid-Phase Microextraction with Multiwalled Carbon Nanotubes as Solid Sorbent." *Journal of Analytical Atomic Spectrometry* 27, no. 10 (2012): 1688–93. 10.1039/c2ja30179b

Sobhi, Hamid R., Alireza Mohammadzadeh, Mohammad Behbahani, and Ali Esrafili. "Implementation of an Ultrasonic Assisted Dispersive μ-Solid Phase Extraction Method for Trace Analysis of Lead in Aqueous and Urine Samples." *Microchemical Journal* 146, January (2019): 782–8. 10.1016/j.microc.2019.02.008

Sousa, Sara, Diogo Pestana, Gil Faria, Fernando Vasconcelos, Cristina Delerue-Matos, Conceição Calhau, and Valentina F. Domingues. "Method Development for the Determination of Synthetic Musks and Organophosphorus Pesticides in Human Adipose Tissue." *Journal of Pharmaceutical and Biomedical Analysis* 191, no. xxxx (2020): 113598. 10.1016/j.jpba.2020.113598

Sun, Ning, Yehong Han, Hongyuan Yan, and Yanxue Song. "A Self-Assembly Pipette Tip Graphene Solid-Phase Extraction Coupled with Liquid Chromatography for the Determination of Three Sulfonamides in Environmental Water." *Analytica Chimica Acta* 810 (2014): 25–31.

Svačinová, Jana, Ondrej Novák, Lenka Plačková, Rene Lenobel, Josef Holík, Miroslav Strnad, and Karel Doležal. "A New Approach for Cytokinin Isolation from Arabidopsis Tissues Using Miniaturized Purification: Pipette Tip Solid-Phase Extraction." *Plant Methods* 8, no. 1 (2012): 17.

Taghvimi, Arezou, Ahad B. Tabrizi, Siavoush Dastmalchi, and Yousef Javadzadeh. "Metal Organic Framework Based Carbon Porous as an Efficient Dispersive Solid Phase Extraction Adsorbent for Analysis of Methamphetamine from Urine Matrix." *Journal of Chromatography B: Analytical Technologies in the Biomedical and Life Sciences* 1109, August 2018 (2019): 149–54. 10.1016/j.jchromb.2019.02.005

Marumo, Akemi, Takeshi Kumazawa, Xiao-Pen Lee, Chika Hasegawa, andKeizo Sato. Spin tip solid-phase extraction and HILIC-MS-MS for quantitative determination of methamphetamine and amphetamine in human plasma." *Journal of Liquid Chromatography & Related Technologies* 37, no. 3 (2013): 420–32.

Tsai Wen H., Hung Y. Chuang, Ho H. Chen, Joh J. Huang, Hwi C. Chen, Shou H. Cheng, and Tzou C. Huang. "Application of Dispersive Liquid-Liquid Microextraction and Dispersive Micro-Solid-Phase Extraction for the Determination of Quinolones in Swine Muscle by High-Performance Liquid Chromatography with Diode-Array Detection." *Analytica Chimica Acta*, 656 (2009): 56–62. 10.1016/j.aca.2009.10.008

Tu, Xiaozheng, Xiaohui Shi, Man Zhao, and Huiqi Zhang. "Molecularly Imprinted Dispersive Solid-Phase Microextraction Sorbents for Direct and Selective Drug Capture from the Undiluted Bovine Serum." *Talanta* 226 (2021): 122142. 10.1016/j.talanta.2021.122142

Sorribes-Soriano, Aitor, A. Valencia, Francesc Albert Esteve-Turrillas, Sergio Armenta, and José Manuel Herrero-Martínez. "Development of Pipette Tip-Based Poly (Methacrylic Acid- Co -Ethylene Glycol Dimethacrylate) Monolith for the Extraction of Drugs of Abuse from Oral Fluid Samples." *Talanta* 205 (2019): 120158. 10.1016/j.talanta.2019.120158

Wang, Chengfei, Xiaowei Li, Fugen Yu, Yingyu Wang, Dongyang Ye, Xue Hu, Lan Zhou, Jingjing Du, and Xi Xia. "Multi-Class Analysis of Veterinary Drugs in Eggs Using Dispersive-Solid Phase Extraction and Ultra-High Performance Liquid Chromatography-Tandem Mass Spectrometry." *Food Chemistry* 334, July 2020 (2021): 127598. 10.1016/j.foodchem.2020.127598

Wang, Jingyu, Zhidong Liu, and Yinghong Qu. "Ultrasound-Assisted Dispersive Solid-Phase Extraction Combined with Reversed-Phase High-Performance Liquid Chromatography-Photodiode Array Detection for the Determination of Nine Biogenic Amines in Canned Seafood." *Journal of Chromatography A* 1636 (2021): 461768. 10.1016/j.chroma.2020.461768

Wang, Juan, Zhiyan Chen, Zhiming Li, and Yaling Yang. "Magnetic Nanoparticles Based Dispersive Micro-Solid-Phase Extraction as a Novel Technique for the Determination of Estrogens in Pork Samples." *Food Chemistry* 204 (2016): 135–40. 10.1016/j.foodchem.2016.02.016

Wang, Lu, Hongyuan Yan, Chunliu Yang, Zan Li, and Fengxia Qiao. "Synthesis of Mimic Molecularly Imprinted Ordered Mesoporous Silica Adsorbent by Thermally Reversible Semicovalent Approach for Pipette-Tip Solid-Phase Extraction-Liquid Chromatography Fluorescence Determination of Estradiol in Milk." *Journal of Chromatography A* 1456 (2016): 58–67.

Wang, Nani, Xiaowen Huang, Xuping Wang, Yang Zhang, Renjie Wu, and Dan Shou. "Pipette Tip Solid-Phase Extraction and High-Performance Liquid Chromatography for the Determination of Flavonoids from Epimedii Herba in Rat Serum and Application of the Technique to Pharmacokinetic Studies." *Journal of Chromatography B* 990 (2015): 64–72.

Wang, Nani, Hailiang Xin, Qiaoyan Zhang, Yiping Jiang, Xuping Wang, Dan Shou, and Luping Qin. "Carbon Nanotube-Polymer Composite for Effervescent Pipette Tip Solid Phase Microextraction of Alkaloids and Flavonoids from Epimedii Herba in Biological Samples." *Talanta* 162 (2017): 10–8. 10.1016/j.talanta.2016.09.059

Wang, Ronglin, Canru Li, Qianlian Li, Sunxian Zhang, Feng lv, and Zhiming Yan. "Electrospinning Fabrication of Covalent Organic Framework Composite Nanofibers for Pipette Tip Solid Phase Extraction of Tetracycline Antibiotics in Grass Carp and Duck." *Journal of Chromatography A* 1622 (2020): 461098. 10.1016/j.chroma.2020.461098

Wang, Rui, Si Li, Dawei Chen, Yunfeng Zhao, Yongning Wu, and Kemin Qi. "Selective Extraction and Enhanced-Sensitivity Detection of Fluoroquinolones in Swine Body Fluids by Liquid Chromatography–High Resolution Mass Spectrometry: Application in Long-Term Monitoring in Livestock." *Food Chemistry* 341, February 2020 (2021): 128269. 10.1016/j.foodchem.2020.128269

Wang, Ziming, Xin Zhao, Xu Xu, Lijie Wu, Rui Su, Yajing Zhao, Chengfei Jiang, Hanqi Zhang, Qiang Ma, Chunmei Lu, and Deming Dong. "An Absorbing Microwave Micro-Solid-Phase Extraction Device Used in Non-Polar Solvent Microwave-Assisted Extraction for the Determination of Organophosphorus Pesticides." *Analytica Chimica Acta* 760 (2013): 60–8. 10.1016/j.aca.2012.11.031

Wu, Xingqiang, Shigang Shen, Hongyuan Yan, Yanan Yuan, and Xi Chen. "Efficient Enrichment and Analysis of Atrazine and Its Degradation Products in Chinese Yam Using Accelerated Solvent Extraction and Pipette Tip Solid-Phase Extraction Followed by UPLC–DAD." *Food Chemistry*, 337 (2021): 127752. 10.1016/j.foodchem.2020.127752

Xie, W., Wayne M. Mullett, Cynthia M. Miller-Stein, and Janusz Pawliszyn. "Automation of In-Tip Solid-Phase Microextraction in 96-Well Format for the Determination of a Model Drug Compound in Human Plasma by Liquid Chromatography with Tandem Mass Spectrometric Detection." *Journal of Chromatography B* 877, no. 4 (2009): 415–20.

Xu, Yanqiu, Luping Sun, Xin Wang, Shuyun Zhu, Jinmao You, Xian En Zhao, Yu Bai, and Huwei Liu. "Integration of Stable Isotope Labeling Derivatization and Magnetic Dispersive Solid Phase Extraction for Measurement of Neurosteroids by in Vivo Microdialysis and UHPLC-MS/MS." *Talanta* 199 (2019): 97–106. 10.1016/j.talanta.2019.02.011

Yahaya, Noorfatimah, Mohd M. Sanagi, Takahito Mitome, Norikazu Nishiyama, Wan Aini Wan Ibrahim, and Hadi Nur. "Dispersive Micro-Solid Phase Extraction Combined with High-Performance Liquid Chromatography for the Determination of Three Penicillins in Milk Samples." *Food Analytical Methods* 8, no. 5 (2015): 1079–87. 10.1007/s12161-014-9991-7

Yahaya, Noorfatimah, Sazlinda Kamaruzaman, Mohd Marsin Sanagi, Wan Aini Wan Ibrahim, Takahito Mitome, Norikazu Nishiyama, Hadi Nur, Zainab Abdul Ghaffar, Mohd Yusmaidie Aziz, and Hafizuddin Mohamed Fauzi. "Vinyl-Functionalized Mesoporous Carbon for Dispersive Micro-Solid Phase Extraction of Azole Antifungal Agents from Aqueous Matrices." *Separation Science and Technology* 55, no. 17 (2020): 3102–12. 10.1080/01496395.2019.1675699

Yan, Zhiming, Biqing Hu, Qianlian Li, Sunxian Zhang, Jie Pang, and Chunhua Wu "Facile Synthesis of Covalent Organic Framework Incorporated Electrospun Nanofiber and Application to Pipette Tip Solid Phase Extraction of Sulfonamides in Meat Samples." *Journal of Chromatography A* 1584 (2018): 33–41. 10.1016/j.chroma.2018.11.039.

Yang, Ru Z., Jin H. Wang, Ming L. Wang, Rong Zhang, Xiao Y. Lu, and Wei H. Liu Dispersive Solid-Phase Extraction Cleanup Combined with Accelerated Solvent Extraction for the Determination of Carbamate Pesticide Residues in Radix Glycyrrhizae Samples by UPLC-MS-MS." *Journal of Chromatographic Science* 49 (2011): 702–8. 10.1093/chrsci/49.9.702

Yang, Xiaoling, Yiduo Mi, Fang Liu, Jing Li, Haixiang Gao, Sanbing Zhang, Wenfeng Zhou, and Runhua Lu. "Preparation of Magnetic Attapulgite/Polypyrrole Nanocomposites for Magnetic Effervescence-Assisted Dispersive Solid-Phase Extraction of Pyrethroids from Honey Samples." *Journal of Separation Science* 43, 12 (2020): 2419–28. 10.1002/jssc.202000049

Yao, Yating, Jing Dong, Mingming Dong, Fangjie Liu, Yan Wang, Jiawei Mao, Mingliang Ye, and Hanfa Zou. "An Immobilized Titanium (IV) Ion Affinity Chromatography Adsorbent for Solid Phase Extraction of Phosphopeptides for Phosphoproteome Analysis.". *Journal of Chromatography A* 1498 (2017): 22–28. 10.1016/j.chroma.2017.03.026.

Yin, Xiao Ying, Yong Ming Luo, Jian Jiang Fu, You Quan Zhong, and Qing Shan Liu. "Determination of Hyperoside and Isoquercitrin in Rat Plasma by Membrane-Protected Micro-Solid-Phase Extraction with High-Performance Liquid Chromatography." *Journal of Separation Science* 35, no. 3 (2012): 384–91. 10.1002/jssc.201100867

Yuan, Yanan, Yehong Han, Chunliu Yang, Dandan Han, and Hongyuan Yan. "Deep Eutectic Solvent Functionalized Graphene Oxide Composite Adsorbent for Miniaturized Pipette-Tip Solid-Phase Extraction of Toluene and Xylene Exposure Biomarkers in Urine Prior to Their Determination with HPLC-UV." *Microchimica Acta* 187, no. 7 (2020a): 1–9. 10.1007/s00604-020-04370-z

Yuan, Yanan, Hailiang Nie, Junfa Yin, Yehong Han, Yunkai Lv, and Hongyuan Yan. "Selective Extraction and Detection of β-Agonists in Swine Urine for Monitoring Illegal Use in Livestock Breeding." *Food Chemistry* 313, July 2019 (2020b): 126155. 10.1016/j.foodchem.2019.126155

Yuan, Yanan, Ning Sun, Hongyuan Yan, Dandan Han, and Kyung Ho Row. "Determination of Indometacin and Acemetacin in Human Urine via Reduced Graphene Oxide-Based Pipette Tip Solid-Phase Extraction Coupled to HPLC." *Microchimica Acta* 183, no. 2 (2016): 799–804.

Yuan, Yanan, Mingwei Wang, Nan Jia, Chengcheng Zhai, Yehong Han, and Hongyuan Yan. "Graphene/Multi-Walled Carbon Nanotubes as an Adsorbent for Pipette-Tip Solid-Phase Extraction for the Determination of 17 ʏ-Estradiol in Milk Products." *Journal of Chromatography A* 1600 (2019): 73–9. 10.1016/j.chroma.2019.04.055

Zhai, Yujuan, Na Li, Lei Lei, Xiao Yang, and Hanqi Zhang. "Dispersive Micro-Solid-Phase Extraction of Hormones in Liquid Cosmetics with Metal-Organic Framework." *Analytical Methods* 6, no. 23 (2014): 9435–45. 10.1039/c4ay01763c

Zhang, Ruiqi, Siming Wang, Ye Yang, Yulan Deng, Di Li, Ping Su, and Yi Yang. "Modification of Polydopamine-Coated Fe_3O_4 Nanoparticles with Multi-Walled Carbon Nanotubes for Magnetic-μ-Dispersive Solid-Phase Extraction of Antiepileptic Drugs in Biological Matrices." *Analytical and Bioanalytical Chemistry* 410, no. 16 (2018a): 3779–3788.

Zhang, Ying, Wanliang Liao, Yuanyuseidian Dai, Weiping Wang, and Aijun Wang. "Covalent Organic Framework Schiff Base Network-1-Based Pipette Tip Solid Phase Extraction of Sulfonamides from Milk and Honey." *Journal of Chromatography A* 1634 (2020): 461665. 10.1016/j.chroma.2020.461665

Zhang, Yun, Yong G. Zhao, Wei S. Chen, He L. Cheng, Xiu Q. Zeng, and Yan Zhu. "Three-Dimensional Ionic Liquid-Ferrite Functionalized Graphene Oxide Nanocomposite for Pipette-Tip Solid Phase Extraction of 16 Polycyclic Aromatic Hydrocarbons in Human Blood Sample." *Journal of Chromatography A* 1552 (2018b): 1–9. 10.1016/j.chroma.2018.03.039

Zhou, Jingwei, Yaoyao Hu, Peichun Chen, and Hongwu Zhang. "Preparation of Restricted Access Monolithic Tip via Unidirectional Freezing and Atom Transfer Radical Polymerization for Directly Extracting Magnolol and Honokiol from Rat Plasma Followed by Liquid Chromatography Analysis. *Journal of Chromatography A* 1625 (2020): 461238. 10.1016/j.chroma.2020.461238

4

Stir Bar Sorptive Extraction in Analytical Toxicology Studies

Natasa P. Kalogiouri, Elisave-Ioanna Diamantopoulou, and Victoria F. Samanidou
Laboratory of Analytical Chemistry, Department of Chemistry, Aristotle University of Thessaloniki, Thessaloniki, Greece

CONTENTS

4.1 Introduction 61
4.2 SBSE Principles 62
    4.2.1 SBSE Methodology 63
    4.2.2 SBSE Applications in Toxicology Studies 63
    4.2.3 SBSE Optimization Factors 64
    4.2.4 SBSE Advantages, Limitations, and Novel Strategies 64
    4.2.5 Novel Coatings 65
4.3 Conclusions 67
References 68

4.1 Introduction

Bioanalysis refers to the analysis and quantification of compounds (drugs, hormones, metabolites, etc.) in biological samples (blood, blood serum, urine, saliva, hair, tissues, etc.). Methods used for this purpose include: sample preparation, analyte separation, and further detection. Biological samples are complex matrices that are hard to handle due to the presence of a large variety of compounds, such as salts, phospholipids, fats, and proteins. In addition, the determination of analytes in trace levels requires very sensitive and precise analytical methods (Abdel-Rehim et al. 2020; Ocaña-González et al. 2016).

Sample preparation is the most important and time-consuming stage of the analytical process. In fact, choosing the sample treatment technique is often considered more difficult than selecting the detection technique. The importance of this step is inextricably linked to the complexity of the samples analyzed and the detected concentration levels (Camino-Sánchez et al. 2014). Specifically, in real samples, and especially in biological fluids, analytes are often found in trace/ultra-trace amounts. The complexity of the matrix also limits the sensitivity and selectivity of the analysis and is a possible cause of matrix interfering effects. Taking into consideration all of the above, a clean-up process and a pre-concentration method are necessary before analyzing a biological matrix with the existing chromatographic methods (e.g., gas chromatography, GC; high-performance liquid chromatography, HPLC) (Camino-Sánchez et al. 2014; Hasan et al. 2020; Taghvimi and Hamishehkar 2019). Sample preparation methods should ideally be selective, efficient, reliable and robust, and environmentally friendly (Hasan et al. 2020).

Current trends in analytical chemistry center on miniaturization of sample preparation procedures and environmental protection of the following basic principles of green analytical chemistry (GAC) (Kissoudi and Samanidou 2018). 'Solvent-less' or 'solvent-minimized' techniques are preferred over traditional ones (liquid-liquid extraction, LLE, or Soxhlet), due to their many advantages, such as minimum or no emission of pollutants-toxic solvents into the environment, simplicity and miniaturization of the process, enhanced solute selectivity and recovery, and low sample volumes in agreement with the GAC principles. In fact, the reduction of solvent consumption is expected to contribute to environmental sustainability and minimize analytical costs (Hasan et al. 2020; Kissoudi and Samanidou 2018; Nogueira 2012; Ayazi and Matin 2016). Most commonly used techniques in biological matrices for drug extraction, over the years, have been LLE and solid-phase extraction (SPE) (Kassem 2011). However, those techniques use large volumes of organic toxic solvents, require large sample volumes, and are time-consuming (Taghvimi and Hamishehkar 2019). In order to overcome these disadvantages, new solventless sample preparation techniques have been introduced, such as solid-phase microextraction (SPME), in-tube SPME, liquid-phase microextraction (LPME), micro liquid–liquid extraction (MLLE), dispersive liquid–liquid extraction (DLLE), and stir bar sorptive extraction (SBSE) (Camino-Sánchez et al. 2014; Hasan et al. 2020; Kassem 2011).

These techniques reduce both waste and preparation time of the samples, as they combine extraction and concentration of analytes in only one step, allowing the direct extraction of analytes, even from complex matrices, such as biological fluids (blood, urine, hair, etc.), using small volumes of toxic solvents or none at all (Kassem 2011). Moreover, these sorption-based approaches have been demonstrated to be highly sensitive and selective prior to the application of chromatographic techniques. Nowadays, these methods have gained more acceptance throughout the scientific community, especially for trace analysis of volatile and semi-volatile compounds, such as drugs, in biological fluids, due to easy manipulation and cost-effectiveness, with SPME and SBSE being the most effective and commonly used ones (Nogueira 2012).

4.2 SBSE Principles

SBSE has recently become very popular for drug analysis in biological samples at trace levels due to the high sensitivity it has exhibited and other significant advantages. The basic principles of SBSE are similar to SPME. However, it shows simplicity, higher extraction efficiency, sample clean-up, robustness capacity, and rapidity compared to SPME and the classic sample preparation techniques (Marques et al. 2019; Taghvimi 2019).

SPME was developed in 1990, and it was considered a major breakthrough in sample preparation. An externally coated fibre was either immersed in liquid samples (immersion SPME) or exposed to the headspace of a solid or liquid sample (headspace SPME), leading to the extraction of organic compounds. The used external coating was polydimethylsiloxane (PDMS), a non-polar polymer that promotes hydrophobic interactions with target compounds (Nogueira 2012). It was observed that during the extraction of very apolar compounds (logKow > 5), sorption on PDMS was followed by adsorption on the Teflon-coated stir bar used for sample agitation and on the vessel wall. This led to the development of a stir bar coated with PDMS and a sample preparation method known as stir bar sorptive extraction (SBSE) (David et al. 2019), which was first introduced by Baltussen et al. in 1999 (Baltussen et al. 1999). SBSE was developed and commercialized under the trade name Twister by Gerstel GmbH & Co. KG.

In SBSE, a glass-coated magnetic bar, coated with a layer (typically 0.5–1 mm) of sorptive, usually PDMS, as shown in Figure 4.1, is directly added to a vial containing the aqueous sample and is stirred for a certain time, until equilibrium of analytes concentration between the sample matrix and PDMS is reached. The organic compounds to be extracted are absorbed into the stirring bar. The retention mechanism occurs mainly through Van-der-Waals forces. However, the formation of hydrogen bonds with oxygen atoms of PDMS is likely to happen, depending on the molecular structure of the analytes (Nogueira 2012). After the sorption, the bar is rinsed with deionized water, dried, and transferred to a clean vial, where the captured compounds can be desorbed thermally for GC or into a liquid solvent

FIGURE 4.1 Schematic presentation of a stir bar used in SBSE.

(LD) for liquid chromatography (LC) (Hasan et al. 2020; Kissoudi and Samanidou 2018; Talebpour et al. 2012). This technique is based on the partition of the solute between the sample and the absorbent phase, which in this case is the stir bar and not the fibre, like in SPME (Marín-San Román et al. 2020). New applications are constantly being developed, and improvements are still being made (Camino-Sánchez et al. 2014).

4.2.1 SBSE Methodology

The SBSE technique has two extraction modes: immersion SBSE and headspace SBSE (HS-SBSE). Immersion is used in liquid samples. The stir bar is directly introduced into the sample. A certain agitation time is required to maintain the equilibrium between the absorbent phase and the sample. On the other hand, HS-SBSE is used in liquid, solid, and gaseous samples. The stir bar, in this case, is introduced into the vial adapted for the headspace. The sample is agitated and sometimes heated so that equilibrium between the sample and the gas phase can be achieved faster (Marín-San Román et al. 2020).

The methodology requires two steps: extraction and desorption. During the extraction, the Twister is immersed into the sample, using one of the modes mentioned above. Once equilibrium occurs, the stir bar is removed, then inserted into a glass, and transferred to the thermal desorption unit (TDU), or desorbed by a liquid solvent (LD). If the mode used is immersion, the Twister itself works as a stirrer. If the chosen method is HS-SBSE, the sample is stirred with a magnet. After its removal from the sample, the Twister must be cleaned with deionized water so that the remains of proteins, salts, sugars, or other undesirable sample constituents are removed. The parameters that should be optimized at this step are both the kinetic (extraction time, agitation speed, dilution, and volume of the sample) and the thermodynamic parameters (temperature, pH, the addition of salts, and organic modifiers).

Into the TDU, the analytes are thermally desorbed and transferred to the gas chromatography-mass spectrometry (GC-MS). Since absorption is a weaker process than adsorption, heat during thermal desorption (TD) is applied at lower temperatures in order to avoid losses of thermolabile solutes. During TD, desorption temperature, pressure, time, and flow are some variables that should be optimized. On the other hand, during liquid desorption (LD, or back extraction), the immersion of the stir bar into the glass vial must be performed under sonication or mechanical treatment to improve desorption efficiency. Solvent type (e.g., methanol, acetonitrile, mixtures), immersion time, and the number of desorption steps are parameters that should be taken into consideration (Marín-San Román et al. 2020; Nogueira 2012).

Comparing the two desorption approaches, TD is an on-line approach since the direct and quantitative transfer of extracted solutes introduction into the GC system is possible. That leads to higher sensitivity and the possibility of automatization. LD is an off-line, cost-effective approach, with the advantage of combining with GC, HPLC, or capillary electrophoresis (CE) systems (Nogueira 2012). SBSE is mostly combined with TD due to the remarkable thermostability of PDMS. However, novel coatings that have been developed are not as thermal steady as PDMS, so recently, LD has been highly applied in combination with LC. Sensitivity is reduced when using the LD method because only a fraction of the extract is analyzed by LC or GC.

4.2.2 SBSE Applications in Toxicology Studies

SBSE has a broad spectrum of applications to biological matrices (blood, blood serum, urine, hair, etc.) (Camino-Sánchez et al. 2014). It is rapid, simple, cost-effective, easily automated, and in agreement with the principles of GAC, as the use of organic-toxic reagents is the least possible. It requires small sample amounts, which is very useful in the toxicological analysis of biological fluids, where sampling in large amounts is often prohibitive, for obvious reasons. In addition, SBSE is used for trace, or even ultra-trace (parts per trillion, ppt) analysis of semipolar or non-polar species (log $K_{ow} > 3$) with low limits of detection, whereas sampling of hydrophilic or highly polar compounds is still challenging (Camino-Sánchez et al. 2014; Hasan et al. 2020; Marques et al. 2019; Talebpour et al. 2012). Concerning the quantitative extraction, in SBSE, it is acquired at a significantly lower K due to the lower phase ratio β. Moreover, as sampling takes place simultaneously with the stirring, competitive sorption from an additional stirrer, which was the main limitation in SPME and was the fact that led to the development of SBSE, can be avoided (Moein et al. 2014). Calibration in SBSE can still be done, even if the extraction is incomplete with the use of water samples with known concentrations of target analytes, for example (Baltussen et al. 1999). One of the most useful and interesting features of this technique is that each stir bar can be reused several times without leading to any degradation of the PDMS coating. However, before being reused, the stir bars must be cleaned up with suitable solvents (e.g., acetonitrile) or through TD treatment (Nogueira 2012).

Recently, SBSE has been efficiently used for the extraction of carbamazepine from serum samples (Vilarinho et al. 2019; Alvani-Alamdari et al. 2019) and fluoxetine (Marques et al. 2019) in plasma, as well as for the determination of endocrine-disrupting chemicals in biological fluids, including cord blood, placenta, amniotic fluid, maternal urine, and breastmilk, as it has already been reviewed (Jiménez-Díaz et al. 2015).

4.2.3 SBSE Optimization Factors

Several parameters have to be evaluated during sample preparation with SBSE techniques, including the type of coating as well as its thickness, pH, ionic strength, temperature, agitation, extraction time, and analyte desorption. During thermal desorption of the analyte, the flow rate of gas is a parameter that should be taken into consideration. In order to reduce the desorption time, a high flow rate (up to 100 mL/min) is recommended (Lancas et al. 2009).

4.2.4 SBSE Advantages, Limitations, and Novel Strategies

SBSE exhibits several advantages, such as increased sensitivity, simpleness, and rapidity enabling robust extraction and concentration in a single step while minimizing the use of organic solvents and sample volumes.

Although SBSE is considered the most useful and interesting sorption-based technique and has shown significance among other techniques, under certain circumstances it presents some limitations. There are a limited number of commercially available coatings; PDMS, ethylene glycol (EG)-silicone, and polyacrylate (PA) are common commercial coatings for SBSE, with PDMS being the one most used. These coatings restrict the application of SBSE in the analysis of semipolar or non-polar analytes. Regarding the more polar analytes (log K_{ow} < 3), PDMS as a coating, for example, has proved to be inefficient due to the weak hydrophobic interactions between the analytes and PDMS. Ethylene glycol-PDMS copolymer (EG-silicone), as a new trademarked coating, has shown higher recovery for both non-polar and polar analytes, because of the polar nature of EG and the non-polar nature of its silicone base. Although EG is compatible with thermal desorption and it is able to bind to polar compounds, its shelf-life is shorter than PDMS due to its lower stability. EG bars are also less robust than PDMS stir bars and can break more easily (Hashemi and Kaykhaii 2021). However, both of these coatings may not have enough capacity and the expected selectivity, especially for trace amounts of analytes in complex matrices, such as biological fluids (Meng et al. 2021).

The coating is not chemically bonded to the substrate, and this can lead to bleeding at even relatively low temperatures during thermal desorption of the analytes when transferred from or to the GC system. Moreover, the coating is vulnerable to washing away, and proper desorption solvent must be used in order to avoid the coating washing away due to its non-chemical bond with the substrate. Furthermore, during the desorption step, where organic solvents may be used, the memory effect can be present. Recently, in order to overcome this limitation, room-temperature ionic liquids (ILs) have been used, replacing toxic solvents with environmentally friendly solvents. Moreover, there are many parameters that should be taken into consideration during the extraction step in order to obtain the best possible results, and thus, the extraction conditions need to be optimized. It is time-consuming, especially when thick and highly viscous polymeric sorbents are used as coatings; reaching equilibrium requires hours. It is expensive since TD requires the use of an expensive thermal desorption unit. It requires a high volume of back-extraction solvent: This can evidently lead to the dilution of the pre-concentrated analytes. Finally, when compounds of high concentration are extracted, the chromatography column used can get overloaded due to the higher sensitivity of the technique.

TABLE 4.1
Main benefits and drawbacks of SBSE as a sample preparation technique of biological fluids (Baltussen et al. 1999; Marín-San Román et al. 2020)

Technique	Benefits	Drawbacks
SBSE	More sensitive and robust than SPME	A lower number of absorbents available
	Extraction and concentration in a single step	Requires a specific, expensive TDU
	Quick and easy	
	Less handling and less sample volume than SPME	
	Without organic solvents	

In order to overcome these drawbacks, several strategies have been proposed, including the use of new polymeric phases, derivatization procedures, multi-mode assays, and alternative sorption-based approaches (Camino-Sánchez et al. 2014; Marín-San Román et al. 2020; Moein et al.,2014; Nogueira 2012; Talebpour et al. 2012). The main advantages and disadvantages of SBSE are summarized in Table 4.1.

4.2.5 Novel Coatings

Considering the limited range of commercially available SBSE coatings, researchers have turned their attention to developing novel coatings to expand the application of the technique and improve the versatility of the stir bars for analysis of compounds in biological fluids with SBSE. Lab-made stir bar coatings are also being developed to achieve higher extraction efficiency in less time. New SBSE coatings can be prepared using adhesion, molecular imprinting (MIP), sol-gel, monolith coating procedures, and solvent exchange procedures (Hasan et al. 2020). All the novel coatings used in toxicological studies are presented in Table 4.2.

Liu et al. (2004) were the first to use sol-gel technology in stir-bars in order to produce a partially hydroxy-terminated-PDMS coated stir-bar, for extracting organophosphorus and polycyclic aromatic hydrocarbons. The sol-gel process involves the transformation of a colloidal liquid solution (sol) into a solid matrix (gel). Several steps form this method, including hydrolysis-condensation polymerization, with organic ligands, of metal alkoxides that eventually lead to the synthesis of gels. This method produces coatings with thermal, mechanical, and chemical stability; selectivity; and most of all, tuneable porosity (Hasan et al. 2020; Moein et al. 2014). Sol-gel coatings interact well with the surface of the sample due to the presence of functional groups in the procure chemical structure that is added to the sol-gel solution. The most common sol-gel procures are tetraethoxysilane (TEOS) and methyltrimethoxysilane (MTMOS). If carbon-based composites are mixed with sol, carbon-ceramic materials (CCMs) are produced. Graphene oxide (GO), on the other hand, is also a carbon-based material with unique physical and chemical properties. Large interaction between the analyte and the adsorbent can be achieved with this method due to the nano surface area of GO. GO is also a suitable adsorbent for drug adsorption, and thus, it was identified as one of the best possible novel coating materials used in SBSE in toxicology and bioanalysis (Hasan et al. 2020; Nogueira 2012; Taghvimi and Hamishehkar 2019).

Adhesion techniques involve achieving the extraction of target compound materials on SBSE substrates through two approaches: physical adhesion techniques (PAT) or chemical adhesion techniques (CAT). In the first approach, a PDMS sol, or any other polymer acting as a glue, forms the preliminary adhesive film. Then, specific particles, such as octadecyl (C_{18}) silica, are added to the adhesive film by incubation and post-incubation treatments. In fact, two or more sampling, or sorbent, materials with different enrichment capabilities are combined (dual-phase stir bars) in order to improve the recovery of volatile and polar compounds in comparison to the conventional PDMS stir bar. In the second chemical approach, the substrates, such as polyether ether ketone stainless steel wire (SSW), are first chemically modified and then covalently immobilized. PATs are simple and cost-effective techniques in which reproducibility of the preparation of coatings can be easily achieved. However, the lifetime of coatings prepared with this technique is significantly lower than common PDMS stir bars. It can be extended,

TABLE 4.2

Novel coatings by sol-gel technique, used in the toxicological analysis of biological fluids (Hasan et al. 2020)

Coating material	Lifetime (cycles)	Target compounds	Sample	LOD (µg/L, µg/Kg)	Method
AlMBF$_4$ ionic liquid	>50	Ketoprofen Naproxen Fenbufen	Urine	0.23–0.31	SBSE-LD-HPLC-UV
Ni-ZnS-activated carbon	12	Losartan Valsartan	Urine Plasma	0.12–0.15	SBSE-LD-HPLC-UV
Ag (I) imprinted MPTS	15	Ag (I)	Hair Nail	0.04	SBSE-LD-FI-AAS
Nano graphene oxide	NR	Amphetamine Methamphetamine	Urine	10–11	SBSE-LD-HPLC-UV
Layered double hydroxide/graphene	NR	Organochlorine pesticides	Urine	5-8	SBSE-LD-HPLC-UV
PDMS/Ge	NR	4-chloro-1-naphthol	Urine	0.034	SBSE-LD-HPLC-DAD
MWCNTs/polyaniline	50	Propanol	Plasma	0.03	SBSE-LD-HPLC-FLD
Pyrrole	3	Estradiol	Urine	10	SBSE-LD-GC-FID
Zeolitic imidazolate framework-67/ cobalt nanoporous carbon	70	Fluoracil Phenobarbital	Urine Plasma	0.21–1.4	SBSE-LD-HPLC-UV
Zn-Al layered double hydroxides/Zeolitic imidazolate framework-8	NR	Benzylpenicillin	Blood Urine Bovine milk	0.05	SBSE-LD-HPLC-UV
Layered double hydroxide/graphene	NR	Organochlorine pesticides	Urine	0.22–1.38	SBSE-LD-GC-MS

*MPTS: (3-mercaptopropyl)trimethoxysilane
*NR: not reported

though, when the coating is protected by a Polytetrafluoroethylene (PTFE) membrane. Also, reduced mechanical stability has been reported. When the CAT approach is being used, the stir bars produced exhibitchemical and mechanical stability, and their thickness, which may affect the efficiency of the extraction, can be controlled (Bicchi et al. 2005; Hasan et al. 2020; Nogueira, 2012).

Molecularly imprinted polymers (MIPs) have also been used and evaluated over the years. They have proven excellent selectivity and also achieved adsorption equilibrium rather fast. However, they can only be used for very specific matrices and target compounds (Nogueira 2012). Using this technique, stirs bar coatings are prepared in three stages:

1. Covalent or non-covalent chemical reaction between a template molecule and a functional monomer
2. Co-polymerisation (thermal or photo-polymerization) of the produced mixture with a cross-linking agent
3. Removal of the template molecule

The process mentioned leads to the formation of an extremely selective porous polymer, toward the target molecule, used in the first step of the process, in size, shape, and chemical functionality (Hasan et al. 2020).

MIP-produced stir bar coatings have many advantages over common ones, including great selectivity upon the target compound, reproducibility, simple and cost-effective preparation, higher mechanical and chemical stability, and faster adsorption kinetics. However, the polymerization process may affect their efficiency. Also, the removal of the template molecule requires harsh conditions, which can lead to a reduction of the desorption efficiency and bleeding. Last, the templates may be toxic and expensive, or even hard to obtain (Hasan et al. 2020; Wyszomirski and Prus 2012). All novel coatings produced by the MIP technique are presented in Table 4.3.

TABLE 4.3

Novel coatings produced by molecular imprinting technique used in the toxicological analysis of biological fluids (Hasan et al. 2020)

Coating material	Lifetime (cycles)	Target compounds	Sample	LOD (µg/L, µg/Kg)	Method
Propanol imprinted/ graphene oxide	>50	Propanolol	Urine	0.037	SBSE-LD-HPLC-UV
Dopamine imprinted	NR	Dopamine	Urine	0.03	SBSE-LD-HPLC-FLD
Carbamazepine imprinted	8	Carbamazepine	Human blood serum	10	SBSE

*NR: not reported

TABLE 4.4

Novel coatings produced by monolith formation, used in the toxicological analysis of biological fluids (Hasan et al. 2020)

Coating material	Lifetime (cycles)	Target compounds	Sample	LOD (µg/L, µg/Kg)	Method
VPD-EGDMA	15	Losartan Valsartan	Human plasma	7–27	SBSE-LD-HPLC-DAD
Poly (VPD-EGDMA)	>15	Diazepam Nordazepam	Human plasma	10–12	SBSE-LD-HPLC-UV

*DVB: divinyl benzene, EGDMA: ethylene glycol dimethacrylate
*NR: not reported

Another category is polymer monolith coatings. The novel coatings produced by monolith formation are presented in Table 4.4. In general, a polymer monolith refers to a porous polymer containing a network of interconnected pores that is produced through the polymerization of a functional monomer and a crosslinker, with the presence of an initiator. The main advantages are preparation ease of the monomer mixtures, high permeability, favorable mass transfer characteristics, low cost, and suitability from non-polar to polar compounds. These coatings are prepared in three steps:

1. Silylation of the stir bar surface
2. Immersion of the stir bar in a mixture of monomers, crosslinkers, and initiators, which have been ultrasonicated
3. Thermal or photo-polymerization

The advantages of these coatings are high reproducibility, good mechanical and chemical stability, low cost, easy preparation methods, and the ability to produce bimodal porosity (micro-porous and macro-porous) (Hasan et al. 2020; Nogueira 2012).

4.3 Conclusions

Several aspects of SBSE, including the basic theory and methodology, advantages, limitations, and future trends were presented. The majority of the SBSE applications in toxicology studies exhibited high selectivity, good linearity, precision, and high sensitivity. Considering its applications in toxicology studies, SBSE will definitely play a tremendous role in sample preparation, enabling extraction in a single step and reducing solvent extraction, disposal cost, and extraction time. It is crucial to develop novel phases to extend SBSE applications and increase sensitivity. Furthermore, innovative developments in SBSE instrumentation constitute another research area that needs to undergo further exploration in the near future.

REFERENCES

Abdel-Rehim, Mohamed, Stig Pedersen-Bjergaard, Abbi Abdel-Rehim, Rafael Lucena, Mohammad M. Moein, Soledad Cárdenas, and Manuel Miró. "Microextraction Approaches for Bioanalytical Applications: An Overview." *Journal of Chromatography A* 1616 (2020): 460790. 10.1016/j.chroma.2019.460790

Alvani-Alamdari, Sima, Abolghasem Jouyban, Maryam Khoubnasabjafari, Ali Nokhodchi, and Elaheh Rahimpour. "Efficiency Comparison of Nylon-6-Based Solid-Phase and Stir Bar Sorptive Extractors for Carbamazepine Extraction." *Bioanalysis* 11, no. 9 (2019): 899–911. 10.4155/bio-2018-0321

Ayazi, Zahra, and Amir Abbas Matin. "Development of Carbon Nanotube-Polyamide Nanocomposite-Based Stir Bar Sorptive Extraction Coupled to HPLC-UV Applying Response Surface Methodology for the Analysis of Bisphenol A in Aqueous Samples." *Journal of Chromatographic Science* 54, no. 10 (2016): 1841–50. 10.1093/chromsci/bmw135

Baltussen, Erik, Pat Sandra, Frank David, and Carel Cramers. "Stir Bar Sorptive Extraction (SBSE), a Novel Extraction Technique for Aqueous Samples: Theory and Principles." *Journal of Microcolumn Separations* 11, no. 10 (1999): 737–47. 10.1002/(SICI)1520-667X(1999)11:10<737::AIDMCS7>3.0.CO;2-4

Bicchi, Carlo, Chiara Cordero, Erica Liberto, Patrizia Rubiolo, Barbara Sgorbini, Frank David, and Pat Sandra. "Dual-Phase Twisters: A New Approach to Headspace Sorptive Extraction and Stir Bar Sorptive Extraction." *Journal of Chromatography A* 1094, no. 1–2 (2005): 9–16. 10.1016/j.chroma.2005.07.099

Camino-Sánchez, F. J., Rocio Rodríguez-Gómez, A. Zafra-Gómez, Angela Santos-Fandila, and Jose L. Vílchez. "Stir Bar Sorptive Extraction: Recent Applications, Limitations and Future Trends." *Talanta* 130 (2014): 388–99. 10.1016/j.talanta.2014.07.022

David, Frank, Nobuo Ochiai, and Pat Sandra. "Two Decades of Stir Bar Sorptive Extraction: A Retrospective and Future Outlook." *TrAC - Trends in Analytical Chemistry* 112 (2019): 102–11. 10.1016/j.trac.2018.12.006

Hasan, Chowdhury K., Alireza Ghiasvand, Trevor W. Lewis, Pavel N. Nesterenko, & Brett Paull. "Recent Advances in Stir-Bar Sorptive Extraction: Coatings, Technical Improvements, and Applications." *Analytica Chimica Acta* 1139 (2020): 222–40. 10.1016/j.aca.2020.08.021

Hashemi, Sayyed H., and Massoud Kaykhaii. "Nanoparticle Coatings for Stir Bar Sorptive Extraction, Synthesis, Characterization and Application." *Talanta* 221, August 2020 (2021): 121568. 10.1016/j.talanta.2020.121568

Jiménez-Díaz, Immaculada, Fernando Vela-Soria, Rocio Rodríguez-Gómez, Alberto Zafra-Gómez, Oscar Ballesteros, and Alberto Navalón. "Analytical Methods for the Assessment of Endocrine Disrupting Chemical Exposure During Human Fetal and Lactation Stages: A Review." *Analytica Chimica Acta* 892 (2015): 27–48. 10.1016/j.aca.2015.08.008

Kassem, Mohamed G. "Stir Bar Sorptive Extraction for Central Nervous System Drugs from Biological Fluids." *Arabian Journal of Chemistry* 4, no. 1 (2011): 25–35. 10.1016/j.arabjc.2010.06.011

Kissoudi, Maria, and Victoria Samanidou. "Recent Advances in Applications of Ionic Liquids in Miniaturized Microextraction Techniques." *Molecules* 23, no. 6 (2018): 1437. 10.3390/molecules23061437

Lancas, Fernando M., Maria E. C. Queiroz, Paula Grossi, and Igor R. B. Olivares. "Recent Developments and Applications of Stir Bar Sorptive Extraction." *Journal of Separation Science* 32, no. 5–6 (2009): 813–24. 10.1002/jssc.200800669

Liu, Wenmin, Hanwen Wang, and Yafeng Guan. "Preparation of Stir Bars for Sorptive Extraction Using Sol-Gel Technology." *Journal of Chromatography A* 1045, no. 1–2 (2004): 15–22. 10.1016/j.chroma.2004.06.036

Marín-San Román, Sandra, Pilar Rubio-Bretón, Eva P. Pérez-Álvarez, and Teresa Garde-Cerdán. "Advancement in Analytical Techniques for the Extraction of Grape and Wine Volatile Compounds." *Food Research International* 137, April (2020): 109712. 10.1016/j.foodres.2020.109712

Marques, Leticia A., Thais T. Nakahara, Tiago Bervelieri Madeira, Mariana Bortholazzi Almeida, Alessandra Maffei Monteiro, Maria de Almeida Silva, Emanuel Carrilho, Luiz Gustavo Piccoli de Melo, and Suzana L. Nixdorf. "Optimization and Validation of an SBSE–HPLC–FD Method Using Laboratory-Made Stir Bars for Fluoxetine Determination in Human Plasma." *Biomedical Chromatography* 33, no. 1 (2019): e4398. 10.1002/bmc.4398

Meng, Yuan, Weiyi Liu, Xiaohui Liu, Jinlan Zhang, Meng Peng, and Tingting Zhang. "A Review on Analytical Methods for Pharmaceutical and Personal Care Products and Their Transformation Products." *Journal of Environmental Sciences (China)* 101 (2021): 260–81. 10.1016/j.jes.2020.08.025

Moein, Mohammad M., Rana Said, Fatma Bassyouni, and Mohamed Abdel-Rehim. "Solid-Phase Microextraction and Related Techniques for Drugs in Biological Samples." *Journal of Analytical Methods in Chemistry* 2014 (2014): 921350. 10.1155/2014/921350

Nogueira, Jose M. F. "Novel Sorption-Based Methodologies for Static Microextraction Analysis: A Review on SBSE and Related Techniques." *Analytica Chimica Acta* 757 (2012): 1–10. 10.1016/j.aca.2012.10.033

Ocaña-González, Juan A., Rut Fernández-Torres, Miguel Á. Bello-López, and Maria Ramos-Payán. "New Developments in Microextraction Techniques in Bioanalysis. A Review." *Analytica Chimica Acta* 905 (2016): 8–23. 10.1016/j.aca.2015.10.041

Taghvimi, Arezou, and Hamed Hamishehkar. "Developed Nano Carbon-Based Coating for Simultaneous Extraction of Potent Central Nervous System Stimulants from Urine Media by Stir Bar Sorptive Extraction Method Coupled to High-Performance Liquid Chromatography." *Journal of Chromatography B: Analytical Technologies in the Biomedical and Life Sciences* 1125, June (2019): 121701. 10.1016/j.jchromb.2019.06.028

Talebpour, Zahra, Maryam Taraji, and Nuoshin Adib. "Stir Bar Sorptive Extraction and High-Performance Liquid Chromatographic Determination of Carvedilol in Human Serum Using Two Different Polymeric Phases and an Ionic Liquid as Desorption Solvent." *Journal of Chromatography A* 1236 (2012): 1–6. 10.1016/j.chroma.2012.02.063

Vilarhino, F., R. Sendon, and A. van der Kellen. Bisphenol A in Food as a Result of its Migration from Food Packaging, 2019. 10.1016/j.tifs.2019.06.012

Wyszomirski, Miroslaw, and Wojciech Prus. "Molecular Modelling of a Template Substitute and Monomers Used in Molecular Imprinting for Aflatoxin B1 Micro-HPLC Analysis." *Molecular Simulation* 38, no. 11 (2012): 892–5. 10.1080/08927022.2012.667876

5
Microextraction by Packed Sorbent

Tiago Rosado[1,2,3], Eugenia Gallardo[1,2], Duarte Nuno Vieira[4], and Mario Barroso[5]
[1]*Centro de Investigação em Ciências da Saúde, Faculdade de Ciências da Saúde da Universidade da Beira Interior (CICS-UBI), Covilhã, Portugal*
[2]*Laboratório de Fármaco-Toxicologia-UBIMedical, Universidade da Beira Interior, Covilhã, Portugal*
[3]*C4 - Cloud Computing Competence Centre, Universidade da Beira Interior, Covilhã, Portugal*
[4]*Faculdade de Medicina, Universidade de Coimbra, Coimbra, Portugal*
[5]*Serviço de Química e Toxicologia Forenses, Instituto de Medicina Legal e Ciências Forenses - Delegação do Sul, Lisboa, Portugal*

CONTENTS

5.1 Introduction: Fundamental Theory 71
5.2 Configurations and Sorbents 73
5.3 Sample Preparation Process 74
5.4 Applications in Toxicology 75
5.5 New Developments 97
5.6 Perspectives and Future Challenges 106
Acknowledgments 106
References 106

5.1 Introduction: Fundamental Theory

Microextraction by packed sorbent (MEPS) is a sort of miniaturized solid-phase extraction (SPE) technique developed in 2004 by Abdel-Rehim et al. (Abdel-Rehim et al., 2004) and aimed at reducing both sample and solvent volumes, in order to provide an automated procedure by means of its easy coupling to chromatographic systems.

In this sampling approach the sorbent (from 1 to 4 mg) is located in a microsyringe rather than in an isolated extraction cartridge, as occurs in SPE (Figure 5.1).

Another difference relative to the latter, in MEPS the sample flows through the extracting device in a bidirectional fashion (aspirations or strokes), improving the process's efficiency due to the increase in the contact between the sample and the sorbent.

In order to increase the rate of mass transfer from the sample to the sorbent, both the extracting phase and particle size should be small. In addition, as close contact between the sorbent's surface and the sample is relevant, the amount of the sorbent, the loading volume, and the volume of the elution should be carefully optimized in order to avoid exceeding the method's breakthrough point (Abdel-Rehim 2011, 2004).

Activation of the extraction sorbent to facilitate the retention of analytes occurs at a first stage, for which an organic solvent, such as methanol, is used. After this step, the sample is withdrawn using the syringe, and several draw/eject cycles are usually needed in order to concentrate the target compounds in the sorbent. The sorbent is washed by rinsing with water, aiming at eliminating matrix constituents

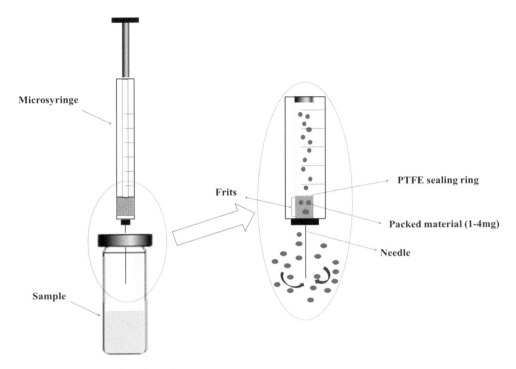

FIGURE 5.1 MEPS manual configuration.

(e.g., proteins). Finally, the analytes are eluted with an organic solvent (e.g., methanol or mobile phase) and directly injected in the analytical instrumentation.

These extraction cycles can be performed in two ways, either by drawing and ejecting several times in the same vial or by discarding the sample to waste after each draw of the syringe (Abdel-Rehim 2004). The whole procedure may be automated using some sort of autosamplers, or it can be connected directly to a gas chromatography (GC) injector using large volume injection approaches. Nevertheless, using liquid chromatography (LC) rather than GC is more prone to adequate automation, as small amounts of water may be introduced in chromatographic instruments due to the difficulty in drying adequately the sorbent prior to elution and to the relatively high polarity of the solvents normally used, which is in general not compatible with GC (Abdel-Rehim 2010, Abdel-Rehim 2011).

This technique is usually aimed at the preparation of liquid samples, so additional steps may be necessary for samples of tissues or hair. In those situations, an organic solvent (e.g., methanol) may be used in order to transfer analytes to the liquid phase prior to MEPS. Nevertheless, complex liquid matrices may also require pre-treatment in order to avoid sorbent clogging and allow extending its use. This is, in addition, important to extract and concentrate analytes present at lower concentrations, providing high sensitivity and selectivity. The influence of matrix interferences may be reduced by sample dilution (to decrease its viscosity, thus facilitating its passage through the sorbent), protein precipitation or filtration using selective filters. It is usually deemed necessary to proceed to pH adjustment to improve the analytes' interaction with the sorbent, and this is particularly important when ion exchange sorbents are involved. Other pre-treatment approaches for MEPS include sample homogenization, by vortex agitation, ultrasounds, or centrifugation (Yang et al. 2017).

Several parameters, namely volume and composition of washing and elution solutions, sorbent amount, and sorbent type, are capable of affecting MEPS performance (Yang et al. 2017). However, selecting the adequate extracting material is the most critical step in optimizing the whole procedure.

When compared to SPE or liquid-liquid extraction (LLE), the MEPS approach is very promising (Altun et al. 2004; Abdel-Rehim 2010), as it reduces sample preparation time and organic solvent consumption, and the cost of analysis is minimal (Abdel-Rehim 2011; Said et al. 2010). Even relative to solid-phase microextraction (SPME), MEPS reduces sample preparation time (<1 min) and sample volume (10–1000 µL) and presents in general a much higher absolute recovery (>50%) (Abdel-Rehim 2011; Barroso et al. 2012; Moein et al. 2015b). Furthermore, the extraction cartridge can be used several times, and more than 50–100 extractions from plasma or 400 extractions from water samples have been described, whereas a conventional SPE column is used once and then discarded (Abdel-Rehim 2011; Barroso et al. 2012; Abdel-Rehim 2010; Altun and Abdel-Rehim 2008).

Although MEPS is a very simple and straightforward extraction technique, it is not free of disadvantages. When its application started increasing, some authors complained about the fact that the available sorbents were scarce, a problem that did not occur with traditional SPE (Páleníková and Hrouzková 2014). Nowadays, and especially in the last five years, a lot of research has been done, and a wide range of options have been developed in the field of solid packing material. These new sorbents have been successfully applied to MEPS syringes, but they seem to be limited to pre-concentrate a small group of analytes. Another disadvantage is the strong dependence of the analytes' recovery on the number of cycles (strokes) that the sample passes through the sorbent (Páleníková and Hrouzková 2014). Commonly, in order to achieve high recovery rates, multiple draw-eject cycles have to be applied, since the analytes' concentration in the sample will decrease after each cycle. Still, this cannot be accepted as a rule, since sorbents can reach a rapid saturation. The increasing number of draw-eject cycles will also increase the mechanical stress on the syringe plunger, resulting in a short life-time of the MEPS syringe (Páleníková and Hrouzková 2014). Another disadvantage, which is usually neglected, is related to solvents that might not be suitable for the procedure. During extraction optimization, it is common practice to mimic SPE procedures, including solvents applied, although reducing their volumes. Yet, it has been described that some solvents, such as dichloromethane and large amounts of isopropanol, can cause sorbent cavitation when passing through the BIN (Rosado et al. 2020a). One cannot forget that the amount of sorbent used in MEPS is around ten times lower than that used in SPE cartridges, and any sorbent loss (even at minimum amounts) can directly affect the extraction efficiency and BIN lifetime. Moreover, these solvents also appear to affect the plunger of the syringe over time.

5.2 Configurations and Sorbents

Several different sorbent materials are available for use in MEPS. These sorbents are essentially silica-based matrices (unmodified silica, C_2, C_8, and C_{18}), strong and weak cation and anion exchange functionalized C_{18} versions (SCX, SAX), and mixed-mode sorbents (80% C_8 and 20% SCX with sulfonic acid-bonded silica) (Table 5.1) (Yang et al. 2017). More recently, new sorbents have been made available, namely porous graphitic carbon and polymeric absorbent polystyrene-divinylbenzene copolymer (PV-DVB), either modified or functionalized, in order to present different retention capabilities for different target analytes (Abuzooda et al. 2015; Karimiyan et al., 2019; Altun and Abdel-Rehim 2008). Table 5.1 summarizes the main types of commercialized sorbents.

A significant number of custom sorbents have been reported for use in MEPS, for instance molecular-imprinted polymers (MIPs), functionalized silica monoliths, based on cyanopropyl hybrid silica, and other restricted access materials (RAM) as well (Daryanavard et al. 2013; Ahmadi et al. 2017; Taghani et al. 2018; Bagheri et al. 2012a,b; Rahimi et al. 2013; Souza et al. 2015). These types of sorbents were developed for specific applications, and as such they are not commercially available. Their use is not yet widespread, but rather still limited to those proof-of-concept applications.

MEPS selectivity obviously depends on the type of sorbent, as different types of interaction (hydrophobic, polar, and ionic) between the analytes and the sorbent may occur (Yang et al. 2017; Pereira et al. 2019).

Particle size obviously influences MEPS performance. The most common particle size in conventional MEPS varies from 30 to 50 µm, but particle sizes of 140 or 3 µm have also been used (Yang et al. 2016; Porto-Figueira et al. 2015). These different sizes can be useful when complex matrices are involved, avoiding sorbent blocking and consequently erratic recoveries. Other formats of sorbents are available, namely graphene aerogel monolith, which does not have particles (Han et al. 2016; Yang et al. 2017).

Different modes are possible when operating MEPS, but the manual syringe is the most widely used format (Table 5.1).

TABLE 5.1
Main types of commercialized sorbents and modes of operation

Type of Sorbent	Characteristics
	Silica-based sorbents
Silica, C_2, C_4, C_8, C_{18}	The retention mechanism is based on normal and reverse phase separation. It is adequate for the extraction of both hydrophobic and hydrophilic analytes from aqueous matrices.
	Ion exchange materials
M1 (80% C_8 and 20% SCX with sulfonic acid-bonded silica), SCX, SAX	The retention mechanism is based on weak cation and anion exchange. It is applicable for easily ionized polar analytes.
Polystyrene copolymer (divinylbenzene, DVB; ENV +)	It is adequate for non-polar compounds.
Modes of Operation	**Characteristics**
Manual syringe	Simplicity, low cost, and ease of operation are the main factors responsible for its increasing popularity. It is a very repetitive process (Abdel-Rehim 2010).
Semi-automatic MEPS devices (e-Vol® syringes, and eXact3 Digital Syringe Driver)	It has sample enrichment and filtering in one single step. It is very easy to use, provides complete customization of extraction procedures, and allows greater precision. These devices could be used with μSPEed cartridges. The μSPEed cartridge design consists of a pressure-driven one-way check valve, allowing ultra-low dead volume connection; the samples and the solvents flow through the sorbent bed in a single direction in every step of the extraction. Therefore, aspiration occurs by pulling back the plunger and bypassing the sorbent when it is discarded. This version uses smaller sorbent particles (3 μm or even smaller, when traditional MEPS uses 50 μm diameter particles) in a small cartridge. These small particles provide a much bigger surface area, enhancing the contact between the sorbent and the analytes and improving a more efficient separation (Porto-Figueira et al. 2015; Pereira et al. 2019)
Automatic approaches	It has sample enrichment and filtering in one single step. It is very easy to use, provides complete customization of extraction procedures, and allows greater precision. These fully automated devices are still considerably expensive. Samples and solvents are loaded and discarded through the same channel, which may be of particular concern for those analytes presenting weak interactions with the sorbent. Indeed, they can be partially eluted and lost during extraction due to sample withdrawal and wash. Whereas it is possible to skip the washing step in a few situations, this strategy will impair selectivity and specificity for most applications, particularly when biological specimens are involved. To overcome this, a two-way valve laterally incorporated into the barrel of the syringe may be used. It is possible to use μSPEed cartridges (Moein et al. 2015b).

5.3 Sample Preparation Process

As already stated, the MEPS procedure usually follows a four-step approach, namely conditioning of the stationary phase, sample aspiration and ejection (strokes), interferences removal (washing), and analytes elution (Figure 5.2).

However, one should not be fooled by this apparent simplicity, as a wide range of optimization steps are deemed necessary in order to maximize extraction efficiency and sensitivity. For instance, selecting adequately the sorbent will be extremely important for a successful sample clean-up and also for analyte recovery.

It is possible to simplify or omit some of the steps depending on the target analytes and the desired degree of cleanliness of the extracts, bearing in mind that the ultimate goal of the procedure is to maximize efficiency.

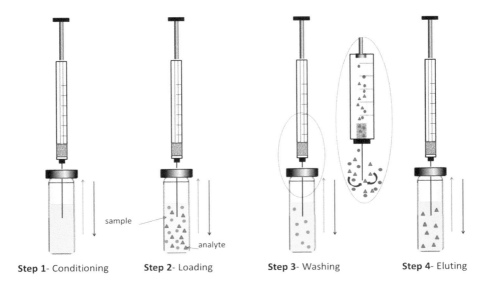

FIGURE 5.2 Operation steps (activation, sample loading, washing, and elution).

For instance, increasing the number of strokes will promote the contact time between the analytes and the sorbent. After the analytes are retained, a washing step is usually performed to remove undesired matrix constituents that are capable of interfering with the analysis. In most published MEPS applications, the wash solvent is the same as was used for sorbent conditioning; the choice of this solvent must be, however, careful and thoroughly optimized in order not to lose analytes in this step. Indeed, incrementing the amount of organic in the wash solvent is useful for efficient removal of matrix interferences, but it also is capable of weakening the analytes' interaction with the sorbent, promoting their early elution.

Analytes are eluted in the last step, which must also be critically optimized to allow their quantitative release from the sorbent in a solvent compatible with the analytical instrumentation that will be used. An organic solvent is generally used, and methanol, isopropanol, or acetonitrile, either by themselves or mixed with acidic or basic solutions (0.1–3%), have been described. In addition, the maximum amount of analyte should be eluted with low solvent volumes whenever possible, in order to increase the enrichment factor and allow direct injection into chromatographic systems if desired. Also, it facilitates the online coupling of extraction and instrumentation, with advantages concerning laboratorial throughput and cost per sample.

Abdel Rehim published in 2011 a tutorial paper on different protocols to use depending on the type of sorbents (Abdel-Rehim 2011). Figure 5.3 summarizes the main steps of MEPS procedures according to the type of sorbent.

Two approaches are usually seen in the optimization of these stages, either using the univariate (one factor is varied at a time) or the multivariate (with the aid of statistical tools allowing multiple factors to be varied simultaneously) ways. Examples of this last approach are the works from Rosado (Rosado et al. 2020b), Prata (Prata et al. 2019), or Oppolzer (Oppolzer et al. 2013), in which they managed to optimize the extraction process in different biological matrices (hair, blood, and urine) with a reduced number of experiments.

5.4 Applications in Toxicology

MEPS has been widely employed, not only in different fields of research, but also in routine analysis in many laboratories. MEPS applicability encompasses clinical, forensic toxicology, food, and environmental analysis applications, with successful implementations to extract a wide range of compounds from different matrices (Pereira et al. 2019).

Regarding clinical toxicology, this field is usually associated with therapeutic drug monitoring (TDM) at designated intervals in order to measure the concentration in the patient's bloodstream. However,

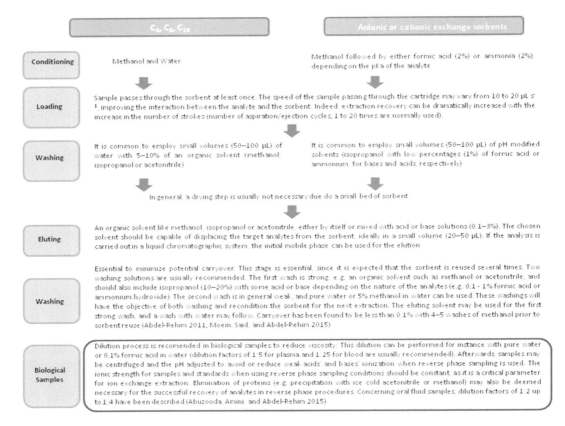

FIGURE 5.3 Main steps of MEPS procedures according to the type of sorbent.

clinical toxicology is a much broader field than just TDM, including catecholamines and metanephrine determination (Xiong and Zhang 2020b; Konieczna et al. 2016; Saracino et al. 2015), measurement of endocrine-disrupting chemical levels that result from human exposure (Silveira et al. 2020; Cristina Jardim et al. 2015), as well as polycyclic aromatic hydrocarbons (PAH) quantification due to their persistence in the environment (Martín Santos et al. 2020) and their effects on humans. Additionally, clinical toxicology has grown to the metabolomic field, and great research has been directed to the diagnostics of several diseases. Examples are the determination of sepsis biomarkers, such as aromatic microbial metabolites (Pautova et al. 2020c; Sobolev et al. 2017), and other biomarkers involved in the pathogenesis and pathophysiology of a wide range of diseases (Biagini et al. 2020; Berenguer et al. 2019). The most used specimen for the determination of this wide range of analytes is urine. Indeed, urine is usually available in sufficient amounts, and metabolites are available at greater concentrations than in other specimens, which makes it a great sample for metabolomics. Regarding sample preparation, proteins and cellular material are not present in urine at high levels, making laboratory analysis a simpler process (Rosado et al. 2017a). Before MEPS extraction, urine has a simple pre-treatment; commonly dilution, filtering, and occasionally a hydrolysis process might be adopted to liberate conjugates of the target analytes. Apart from urine, blood serum is also widely applied in the clinical toxicology field. Besides dilution, this specimen requires a much more thorough pre-treatment, namely centrifugation and/or protein precipitation, to avoid sorbent obstruction during extraction. The same happens with other alternative specimens cleaned up with MEPS for clinical purposes, namely saliva or central cerebrospinal fluid. For all specimens except urine, dilution is almost mandatory, but one should remember that the greater dilution is, the more draw-eject cycles need to be performed to obtain acceptable extraction efficiencies.

Regarding MEPS sorbents adopted in clinical toxicology, C_{18} was by far the most reported, as this sorbent solves problems related to the extraction of non-polar and low polar compounds containing alkyl or aryl groups (Sobolev et al. 2017). Even for phenyl groups present in some aromatic microbial metabolites, this sorbent has proven its suitability (Sobolev et al. 2017). Nevertheless, other sorbents have been also adopted, and Konieczna et al. (Konieczna et al. 2016) stated that individual polar sorbents with surface-displayed amino groups (APS) might be more appropriate for all biogenic amines extraction, resulting in greater recoveries when compared to C_{18} (Konieczna et al. 2016). Most of the biogenic amines analyzed by the authors were very polar compounds; hence, the APS sorbent had the highest affinity (Konieczna et al. 2016). Table 5.2 describes MEPS procedures adopted in the last five years in the field of clinical toxicology.

Furthermore, in the sub-field of TDM, plasma samples are the most common, since therapeutic ranges for the drugs are usually determined in this specimen. Careful plasma pre-treatment will result in easier and faster MEPS procedures, as well as reduced matrix effects. The most adopted pre-treatment involved protein precipitation with trichloroacetic or perchloric acids followed by centrifugation. Once again, C_{18} sorbents were the most adopted for MEPS of fluoroquinolones (e.g., ciprofloxacin and levofloxacin), beta-lactam antibiotics (e.g., meropenem), imidazoles and triazoles, and non-steroidal anti-inflammatory drugs. Fuentes et al. (Fuentes et al. 2019) observed, however, that using a C_8–SCX mixed sorbent allowed better extraction efficiencies for most antidepressant drugs in patients urine samples. This would subsequently result in better sensitivity, allowing detecting lower concentrations.

Although plasma samples are the most used in TDM, other specimens have been cleaned up with MEPS. Locatelli et al. (Locatelli et al. 2015) extracted two fluoroquinolones from human sputum collected from cystic fibrosis patients. This non-conventional sample followed a similar pre-treatment than that of plasma, and even C_{18} has proven to be efficient for the same compounds in previous works. The authors achieved recoveries ranging between 60 and 80% (Locatelli 2015). Oral fluid samples diluted 1:4 were submitted to C_{18} MEPS to pre-concentrate metoprolol en-antiomers, with recoveries ranging between 95 and 98% (Elmongy 2016). Aqueous humour has also been submitted to C_{18} MEPS for the determination of dexamethasone disodium phosphate and dexamethasone in patients with uveitis (Bianchi 2017). Although the authors have achieved great extraction efficiencies for both compounds, the sample load step seems quite laborious with 19 draw-eject cycles (Bianchi et al. 2017). Lastly, dialyzed samples were also used for the extraction of non-steroidal anti-inflammatory drugs, and MEPS has proved to be an excellent alternative to classical SPE (D'Archivio et al. 2016). All MEPS procedures reported in the last five years for TDM purposes are summarized in Table 5.3.

The forensic toxicology field can also be quite challenging. This is a multidisciplinary field that involves the determination and interpretation of the presence of drugs and other potential xenobiotics, usually in biological specimens. Although new MEPS sorbents have not been reported in the last five years in this field, there has been increasing research and application of this miniaturized technique to alternative specimens, especially hair samples. Hair matrix is advantageous due to its longer window for drug detection, hence allowing it to monitor past drug use and users under treatment programs (Rosado et al. 2020a). This represents a challenge in MEPS, since all mentioned biological specimens up to this point were fluid, and hair is solid. It is a very complex, strong, and stable matrix, and for this reason, an appropriate pre-treatment is required to remove the target analytes bound to its inner constituents. This pre-treatment is, actually, considered the extraction step from the sample, after which a further clean-up step can be adopted. Extraction may be carried out with organic solvents, usually methanol, or, depending on the target analytes, by means of weak acid (with hydrochloric acid) or alkaline (with sodium hydroxide) digestions. One should bear in mind, however, that methanol extractions can yield considerable interferences when compared to other procedures, and subsequently provide lower recoveries. After this first sample treatment, MEPS can be applied for the clean-up of the obtained hair extract.

In the last five years, three MEPS procedures have been reported for hair sample clean-up and determination of specific classes of drugs, namely cocaine and metabolites (Rosado et al. 2020b), selected opiates (Rosado et al. 2019), and methadone and EDDP (Rosado et al. 2020a). All three methods used a C_8–SCX mixed sorbent due to the different analytes' properties, and although the

TABLE 5.2

MEPS procedures in clinical toxicology, analytical instrumentation, limits, and recoveries (2015–2020)

Analytes	Sample (Volume/Weight)	Sample Preparation	MEPS Sorbent	MEPS Steps					Analytical Instrumentation	LOD	LOQ	Recoveries	Ref
				Conditioning	Load	Wash	Elution	Sorbent re-use					
Epinephrine (E) Norepinephrine (NE) Dopamine (D) Metanephrine (M) Normetanephrine (N) 3-methoxytyramine (3M)	Urine (0.1 mL)	Dilution with 100 μL of water and 100 μL of DPB buffer solution (pH 9.0)	C$_{18}$ (n.s.); eVol*	Methanol and water (3 × 200 μL)	(7 × 100 μL)	2-amoniethyldip henylboronate buffer (1 × 100 μL) and water/methanol (1/1, v/v) (1 × 50 μL)	Water/ acetonitrile, 95/5, 0.25% formic acid (2 × 25 μL)	Water/ acetonitrile, 95/5, 0.25% formic acid (3 × 100 μL); methanol (1 × 100 μL), DPB buffer (1 × 100 μL), water/ acetonitrile, 95/5, 0.25% formic acid (1 × 100 μL), and water (1 × 100 μL)	LC-MS/MS	0.08 ng/mL (E) 0.30 ng/mL (NE) 0.53 ng/mL (D) 0.176 ng/mL (M) 0.44 ng/mL (N) 0.176 ng/mL (3M)	0.17 ng/mL (E) 0.65 ng/mL (NE) 1.53 ng/mL (D) 1.34 ng/mL (M) 3.43 ng/mL (N) 1.33 ng/mL (3M)	n.s.	(Xiong and Zhang 2020b)
Methylparaben (MeP) Propylparaben (PrP) Bisphenol A (BPA) Bisphenol S (BPS) Benzophenone 1 (BP1) Triclocarban (TCC)	Urine (0.25 mL)	Buffered with 20 μL of 1M ammonium acetate (pH = 5) containing 50 unit of β-glucuronidase; incubated at 37 °C for 12 h; dilution with 250 μL of water	C$_{18}$ (1 mg); Manual	Methanol and water (4 × 100 μL)	(5 × 100 μL)	10% methanol + 0.1% acetic acid (1 × 100 μL)	Methanol (1 × 30 μL)	Methanol and water (10 × 100 μL)	LC-MS/MS	0.10 ng/mL (MeP) 0.01 ng/mL (PrP) 0.02 ng/mL (BPA) 0.01 ng/mL (BPS) 0.005 ng/mL (BP1) 0.01 ng/mL (TCC)	0.33 ng/mL (MeP) 0.03 ng/mL (PrP) 0.07 ng/mL (BPA) 0.03 ng/mL (BPS) 0.02 ng/mL (BP1) 0.03 ng/mL (TCC)	96–104% (MeP) 98–116% (PrP) 104–110% (BPA) 97–105% (BPS) 99–113% (BP1) 101–116% (TCC)	(Silveira et al. 2020)
8-iso prostaglandin F2α (8-isoPGF2α) 8-iso prostaglandin E2 (8-isoPGE2) Prostaglandin E2 (PGE2)	Blood (0.05 mL in DBS)	Whole spot was cut; 350 μL of a methanol:water mixture (70:30 v/v) was added: evaporation down to 50 μL and dilution with water to 100 μL	C$_{18}$ (4 mg); eVol*	Methanol and water (3 × 100 μL)	(5 × 100 μL)	Water:methanol (95:5 v/v) (1 × 100 μL)	Methanol: water (80: 20 v/v) (1 × 100 μL)	n.s.	LC-MS/MS	15 pg/mL (8-isoPGF2α) 20 pg/mL (8-isoPGE2) 15 pg/mL (PGE2)	n.s.	90–100% (8-isoPGF2α) 85–105% (8-isoPGE2) 90–110% (PGE2)	(Biagini et al. 2020)
Benzoic acid (BA) 3-phenylpropionic (PhPA) 3-phenyllactic (PhLA) 4-hydroxybenzoic (p-HBA) 4-hydroxyphenylacetic (p-HPhAA) 4-hydroxyphenylpropionic (p-HPhPA) homovanillic (HVA) 4-hydroxyphenyllactic (p-HPhLA)	Cerebrospinal fluid (0.04 mL)	Dilution with 40 μl of water, pH 7	C$_{18}$ (4 mg); eVol	Methanol, water, 0.3 mM of formic acid (3 × 50 μL)	(20 × 50 μL)	0.3mM of formic acid (2 × 20 μL)	Diethyl ether (10 × 10 μL)	n.s.	GC-MS	0.1 μM (BA) 0.2 μM (PhPA) 0.2 μM (PhLA) 0.3 μM (p-HBA) 0.2 μM (p-HPhAA) 0.2 μM (p-HPhPA) 0.1 μM (HVA) 0.2 μM (p-HPhLA)	0.7 μM (BA) 0.5 μM (PhPA) 0.5 μM (PhLA) 0.6 μM (p-HBA) 0.5 μM (p-HPhAA) 0.5 μM (p-HPhPA) 0.4 μM (HVA) 0.4 μM (p-HPhLA)	60–90% (BA) 80–90% (PhPA) 70–80% (PhLA) 30–40% (p-HBA) 40–60% (p-HPhAA) 60–70% (p-HPhPA) 40–70% (HVA) 35–50% (p-HPhLA)	(Pantova et al. 2020b)

Analytes	Matrix	Pretreatment	Sorbent	Conditioning	Loading	Washing	Elution	Analysis	LOD	LOQ	Recovery	Reference
5-hydroxyindole-3-acetic acid (5HIAA) Indole-3-acetic acid (3IAA) Indole-3-carboxylic (3ICA) Indole-3-lactic acid (3ILA) Indole-3-propionic acid (3IPA)	Cerebrospinal fluid (0.04 mL) Serum (0.04 mL)	Dilution with 40 µl of water	C_{18} (4 mg); eVol*	Methanol, water, 0.3 mM of formic acid (3 × 50 µL)	(20 × 50 µL)	0.3 mM of formic acid (2 × 20 µL)	Diethyl ether (10 × 10 µL)	n.s.	GC-MS	0.3 µM (5HIAA) 0.2 µM (3IAA) 0.3 µM (3ICA) 0.2 µM (3ILA) 0.4 µM (3IPA) 0.4 µM (5HIAA) 0.2 µM (3IAA) 0.4 µM (3ICA) 0.3 µM (3ILA) 0.4 µM (3IPA)	0.4 µM (5HIAA) 0.4 µM (3IAA) 0.5 µM (3ICA) 0.4 µM (3ILA) 0.4 µM (3IPA) 0.4 µM (5HIAA) 0.4 µM (3IAA) 0.5 µM (3ICA) 0.4 µM (3ILA) 0.4 µM (3IPA)	40–80% (n.s) 40–60% (n.s)	(Pautova et al. 2020a)
Naphthalene (1) 2-methylnaphthalene (2) 1-methylnaphthalene (3) Biphenyl (4) Acenaphthylene (5) Acenaphthene (6) 3-phenyltoluene Fluorene (7) Phenanthrene (8) Anthracene (9) Fluoranthene (10) Pyrene (11) Chrysene (12) Benzo(k)fluoranthene (13) Benzo(a)pyrene (14) (15)	Oral fluid (1.5 mL)	Filtration through a 0.45 µm PTFE filter	C_{18} (n.s.); Automated	Ethyl acetate and water (1 × 500 µL)	(3 × 500 µL)	Water (1 × 250 µL)	Ethyl acetate (1 × 50 µL)	GC-MS	17 ng/L (1) 32 ng/L (2) 35 ng/L (3) 79 ng/L (4) 9.4 ng/L (5) 29 ng/L (6) 21 ng/L (7) 11 ng/L (8) 17 ng/L (9) 14 ng/L (10) 4.6 ng/L (11) 16 ng/L (12) 10 ng/L (13) 35 ng/L (14) 10 ng/L (15)	58 ng/L (1) 105 ng/L (2) 117 ng/L (3) 265 ng/L (4) 31 ng/L (5) 96 ng/L (6) 69 ng/L (7) 38 ng/L (8) 55 ng/L (9) 47 ng/L (10) 15 ng/L (11) 52 ng/L (12) 34 ng/L (13) 116 ng/L (14) 34 ng/L (15)	83% (1) 111% (2) 91% (3) 107% (4) 89% (5) 91% (6) 88% (7) 78% (8) 120% (9) 119% (10) 88% (11) 98% (12) 104% (13) 114% (14) 108% (15)	(Martin Santos et al. 2020)
Vanillylmandelic acid (VMA)	Urine (0.010 mL)	Dilution with 90 µL of water	Anion exchange (AX) (n.s.); eVol*	Methanol and water (3 × 100 µL)	(7 × 100 µL)	5% ammonium hydroxide in water and 5% ammonium hydroxide in acetonitrile (1 × 100 µL)	Water/acetonitrile 1:1 (v/v) (2% formic acid) (2 × 50 µL)	Water/ACN 1:1 (v/v) (2% formic acid) (4 × 100 µL) and water (1 × 100 µL)	LC-MS/MS	n.a.	0.5 µg/mL	58–71%	(Xiong and Zhang 2020a)
Leukotriene E (LTE4) Leukotriene B4 (LTB4) 11β-prostaglandin F2α (11β PGF2α)	Urine (n.s.)	pH of 5.1 (n.s.)	Retain anion exchange (R-AX) (n.s); eVol*	Acetonitrile and water at 0.1% formic acid (3 × 250 µL)	(10 × 250 µL)	0.1% formic acid (1 × 100 µL)	Methanol (2 × 50 µL)	Acetonitrile (3 × 250 µL) and water at 0.1% formic acid (1 × 250 µL)	UHPLC-DAD	0.16 ng/mL (LTE4) 0.04 ng/mL (LTB4) 1.12 ng/mL (11βPGF2α)	0.35 ng/mL (LTE4) 0.10 ng/mL (LTB4) 2.11 ng/mL (11βPGF2α)	84–91% (LTE4) 85–92% (LTB4) 96–100% (11βPGF2α)	(Berenguer et al. 2019)
Gamma-aminobutyric acid (GABA) Putrescine (Put) Cadaverine (Cad) L-ornithine (Orn) Spermidine (Spd)	Oral fluid (0.715 mL)	Centrifugation, dilution up to 5.0 mL with water	C_{18} (n.s.); Automated	Ethanol and water (1 × 100 µL)	(5 × 100 µL)	Water:methanol (80:20, v/v) (n.s.)	Ethanol (1 × 10 µL)	n.s.	GC-MS	3.12 ng/mL (GABA) 1.84 ng/mL (Put) 2.63 ng/mL (Cad) 5.15 ng/mL (Orn) 2.60 ng/mL (Spd)	10.40 ng/mL (GABA) 14.30 ng/mL (Put) 8.76 ng/mL (Cad) 17.20 ng/mL (Orn) 8.68 ng/mL (Spd)	89% (GABA) 94% (Put) 103% (Cad) 131% (Orn) 130% (Spd)	(Peña et al. 2019)

(Continued)

TABLE 5.2 (Continued)

MEPS procedures in clinical toxicology, analytical instrumentation, limits, and recoveries (2015–2020)

Analytes	Sample (Volume/Weight)	Sample Preparation	MEPS Sorbent	MEPS Steps					Analytical Instrumentation	LOD	LOQ	Recoveries	Ref
				Conditioning	Load	Wash	Elution	Sorbent re-use					
Benzoic acid (BA) Phenylpropionic acid (PhPA) Phenyllactic acid (PhLA) 4-hydroxybenzoic acid (p-HBA) 4-hydroxyphenylacetic acid (p-HPhAA) 4-hydroxyphenylpropionic acid (p-HPhPA) Homovanillic acid (HVA) 4-hydroxyphenyllactic acid (p-HPhLA)	Serum (0.08 mL)	Concentrated sulphuric acid (2.5 µL) and water (70 µL) were added	C_{18} (4 mg.); n.s.	Methanol, water, 1% formic acid (3 × 50 µL)	(15 × 50 µL)	0.1% formic acid (2 × 20 µL)	Diethylether (10 × 10 µL)	n.s.	GC-MS	n.s.	160 ng/mL (BA) 60–100 ng/mL (PhPA) 60–100 ng/mL (PhLA) 60–100 ng/mL (p-HBA) 60–100 ng/mL (p-HPhAA) 160 ng/mL (p-HPhPA) 160 ng/mL (HVA) 160 ng/mL (p-HPhLA)	50% (BA) 70% (PhPA) 50% (PhLA) 30% (p-HBA) 40% (p-HPhAA) 50% (p-HPhPA) 50% (HVA) 45% (p-HPhLA)	(Pautova et al., 2019)
PS-18 PS-19 PS-20 MOE-18 MOE-19 MOE-20	Serum (0.05 mL)	Deproteinization and dilution with 0.05 mL of ion pair reagent (IPR)	SDVB, C_8, and C_{18} (4 mg); eVol®	Methanol and water (1 × 100 µL); IPR (3 × 50 µL)	(5 × 100 µL)	IPR (1 × 100 µL)	Methanol:IPR (90/10) (1 × 100 µL)	Methanol/IPR 90/10 (3 × 100 µL) Methanol/acetonitrile/NH3 50/50/0.4 (3 × 100 µL) 2-propanol/water/acid formic 90/10/0.1 (3 × 100 µL) Methanol (5 × 100 µL)	UHPLC-UV	n.s.	0.39 µM (PS-18) 0.39 µM (PS-19) 0.39 µM (PS-20) 0.39 µM (MOE-18) 0.39 µM (MOE-19) 0.39 µM (MOE-20)	85% (PS-18) 80% (PS-19) 75% (PS-20) 69% (MOE-18) 84% (MOE-19) 71% (MOE-20)	(Nuckowski et al. 2019)
BIFcis-DCCA trans-DCCA DBCA 3PBA	Urine (0.4 mL)	0.1 mL 1 M sodium acetate buffer (pH = 5.0) with β-glucuronidase; Incubated at 37 °C overnight (at least 12 h) before being acidified with 60 µL of formic acid	C_{18} (4 mg.); eVol®	Methanol (4 × 50 µL); 2% formic acid (3 × 20 µL)	(5 × 100 µL)	30% methanol in water (3 × 50 µL)	1,1,1,3,3,3-hexafluoroisopropanol/diisopropylcarbodiimide/hexane mixture (1/2/97) (2 × 40 µL)	2-propanol (4 × 50 µL) and methanol (4 × 50 µL)	GC-MS	n.s.	0.06 ng/mL (BIF) 0.08 ng/mL (cis-DCCA) 0.08 ng/mL (trans-DCCA) 0.06 ng/mL (DBCA) 0.06 ng/mL (3PBA)	n.s.	(Klimowska and Wielgomas 2018)

Analytes	Matrix	Sorbent (amount; conditioning)	Loading solvent	Sample volume	Washing solvent	Elution solvent	Analysis	LOD	LOQ	Recovery	Reference
Nitrobenzene (NB) 2-Nitrotoluene (2-NT) 3-Nitrotoluene (3-NT) 4-Nitrotoluene (4-NT) 2,6-Dinitrotoluene (2,6-DNT) 1,3-Dinitrobenzene (1,3-DNB) 2,4-Dinitrotoluene (2,4-DNT) 2,4,6-Trinitrotoluene (2,4,6-TNT) 1,3,5-Trinitrobenzene (1,3,5-TNB) 4-Amino-2,6-dinitrotoluene (4-Am-2,6-DNT) 2-Amino-4,6-dinitrotoluene (2-Am-4,6-DNT) 2,4,6-Trinitrophenyl-N-methylnitramine (Tetryl)	Plasma (1 mL)	n.s.	Methanol and water (1 × 100 μL)	(10 × 50 μL)	Water (1 × 50 μL)	Methanol (1 × 30 μL)	Methanol and water (3 × 100 μL)	GC-MS	0.014 to 0.828 ng/mL (n.s.)	0.046 to 2.732 ng/mL (n.s.)	90–95% (NB) 89–94% (2-NT) 90–94% (3-NT) 90–93% (4-NT) 89–93% (2,6-DNT) 91–94% (1,3-DNB) 88–92% (2,4-DNT) 89–94% (2,4,6-TNT) 88–92% (1,3,5-TNB) 88–90% (4-Am-2,6-DNT) 87–90% (2-Am-4,6-DNT) 89–92% (Tetryl)	(Dhingra et al. 2018)
	Urine (1 mL)	C$_{18}$ (4 mg); Manual.							0.014 to 0.828 ng/mL (n.s.)	0.046 to 2.732 ng/mL (n.s.)	92–96% (NB) 91–96% (2-NT) 90–97% (3-NT) 90–96% (4-NT) 90–93% (2,6-DNT) 93–97% (1,3-DNB) 90–93% (2,4-DNT) 90–95% (2,4,6-TNT) 91–95% (1,3,5-TNB) 89–91% (4-Am-2,6-DNT) 89–91% (2-Am-4,6-DNT) 91–92% (Tetryl)	
trans,trans-muconic Acid (ttMA)	Urine (n.s.)	Quaternary ammonium ion exchange resin (SAX column) (4 mg); n.s.	Methanol and water (3 × 100 μL)	(4 × 100 μL)	Water (3 × 100 μL)	10% (v/v) aqueous acetic acid (1 × 100 μL)	10% (v/v) acetic acid and water (4 × 150 μL)	HPLC - UV	0.03 μg/mL	0.1 μg/mL	93–99%	(Soleimani et al. 2017)
Benzoic acid (BA) 3-Phenylpropanoic acid (PhPA) 3-Phenylpropenoic acid (Cinnamic acid) 2-Hydroxy-3-phenylpropanoic acid (PhLA) 4-Hydroxybenzoic acid (HBA) 4-Hydroxyphenylacetic acid (HPhAA) 3-(4-Hydroxyphenyl) propanoic acid (HPhPA) 3-Methoxy-4-hydroxyphenylacetic acid (HVA) 3-(4-Hydroxyphenyl) -2-hydroxypropanoic acid (HPhLA)	Blood (0.08 μL)	C$_{18}$ (1 mg); n.s.	Methanol, water, 1% of formic acid (3 × 50 μL)	(15 × 50 μL)	0.05% Sulphuric acid (2 × 20 μL)	Diethylether (10 × 10 μL)	n.s.	GC-MS	0.5 μM	n.s.	100% (BA) 102% (PhPA) 100% (Cinnamic acid) 58% (PhLA) 34% (HBA) 30% (HPhAA) 64% (HPhPA) 56% (HVA) 21%(HPhLA)	(Sobolev et al. 2017)

(Continued)

TABLE 5.2 (Continued)
MEPS procedures in clinical toxicology, analytical instrumentation, limits, and recoveries (2015–2020)

Analytes	Sample (Volume/Weight)	Sample Preparation	MEPS Sorbent	MEPS Steps					Analytical Instrumentation	LOD	LOQ	Recoveries	Ref
				Conditioning	Load	Wash	Elution	Sorbent re-use					
Gamma-aminobutyric acid (GABA) Putrescine (Put) Cadaverine (Cad) L-ornithine (Orn) Spermidine (Spd)	Urine (0.715 mL)	Dilution 1:6 (n.s.)	C$_{18}$ (4 mg); automated	Ethanol and water (1 × 100 µL)	(5 × 100 µL)	3:1 v/v water:methanol (1 × 100 µL)	Ethanol (1 × 20 µL)	Ethanol (4 × 100 µL)	GC-MS	1.34 ng/mL (GABA) 1.81 ng/mL (Put) 1.10 ng/mL (Cad) 0.18 ng/mL (Orn) 2.70 ng/mL (Spd)	1.34 ng/mL (GABA) 1.81 ng/mL (Put) 1.10 ng/mL (Cad) 0.18 ng/mL (Orn) 2.70 ng/mL (Spd)	90–113% (n.s.)	(Casas Ferreira et al. 2016)
5-hydroxyindole-3-acetic acid (5-HIAA) Homovanilic acid (HVA) 3,4-dihydroxyphenylacetic acid (DOPAC) 3-methoxytyramine (3-MT) 5-hydroxytryptamine (5-HT) Dopamine (DA) Epinephrine (E) Tryptophan (Trp) Norepinephrine (NE) 5-hydroxytryptophane (5-HTrp) Tyrosine (Tyr) 3,4-dihydroxyphenylalanine (L-DOPA)	Plasma (0.1 mL)	0.1% formic acid added to pH 3	APS (4 mg); eVol*	Methanol and water (3 × 100 µL)	(8 × 100 µL)	0.1% formic acid (2 × 50 µL)	Methanol in 0.1% formic acid (3 × 50 µL)	Methanol and water (3 × 100 µL)	LC-MS	2 ng/mL (5-HIAA) 2 ng/mL (HVA) 2 ng/mL (DOPAC) 2 ng/mL (3-MT) 5 ng/mL (5-HT) 5 ng/mL (DA) 5 ng/mL (E) 2 ng/mL (Trp) 5 ng/mL (NE) 5 ng/mL (5-HTrp) 2 ng/mL (Tyr) 5 ng/mL (L-DOPA)	10 ng/mL (5-HIAA) 10 ng/mL (HVA) 10 ng/mL (DOPAC) 10 ng/mL (3-MT) 20 ng/mL (5-HT) 20 ng/mL (DA) 20 ng/mL (E) 10 ng/mL (Trp) 20 ng/mL (NE) 20 ng/mL (5-HTrp) 10 ng/mL (Tyr) 20 ng/mL (L-DOPA)	92% (5-HIAA) 98% (HVA) 94% (DOPAC) 90% (3-MT) 95% (5-HT) 104% (DA) 100% (E) 89% (Trp) 96% (NE) 91% (5-HTrp) 100% (Tyr) 88% (L-DOPA)	(Konieczna et al. 2016)
	Urine (0.05 mL)	0.1% formic acid added to pH 2								2 ng/mL (5-HIAA) 2 ng/mL (HVA) 2 ng/mL (DOPAC) 2 ng/mL (3-MT) 5 ng/mL (5-HT) 5 ng/mL (DA) 5 ng/mL (E) 2 ng/mL (Trp) 5 ng/mL (NE) 5 ng/mL (5-HTrp) 2 ng/mL (Tyr) 5 ng/mL (L-DOPA)	10 ng/mL (5-HIAA) 10 ng/mL (HVA) 10 ng/mL (DOPAC) 10 ng/mL (3-MT) 20 ng/mL (5-HT) 20 ng/mL (DA) 20 ng/mL (E) 10 ng/mL (Trp) 20 ng/mL (NE) 20 ng/mL (5-HTrp) 10 ng/mL (Tyr) 20 ng/mL (L-DOPA)	96% (5-HIAA) 99% (HVA) 97% (DOPAC) 92% (3-MT) 97% (5-HT) 92% (DA) 89% (E) 99% (Trp) 91% (NE) 84% (5-HTrp) 98% (Tyr) 84% (L-DOPA)	

Analyte	Matrix	Sorbent	Sample treatment	Washing	Elution	Analytical technique	LOD	LOQ	Recovery	Reference
Norepinephrine (NE) Epinephrine (E) Dopamine (DA)	Plasma (0.15 mL)	C_{18} (4 mg); n.s.	Pipetted to FTA® cards; cutted; placed into a vial with 100 µL of ultrapure water and 240 µL of buffer solution; vortex agitation (2 min); centrifuged	Methanol and water (3 × 100 µL)	(12 × 100 µL) buffer solution (1 × 100 µL) and water/methanol (50:50; v/v) (1 × 25 µL)	(2.5:97.5, v/v) of methanol, 30.0 mM citric acid in water and 0.5 mM OSA, adjusted to pH 2.92 (1 × 100 µL)	Methanol and water (3 × 100 µL)	HPLC-ECD	0.03 ng/mL (NE) 0.03 ng/mL (E) 0.03 ng/mL (DA)	0.1 ng/mL (NE) 0.1 ng/mL (E) 0.1 ng/mL (DA)	91-93% (NE) 91-95% (E) 85-90% (DA)	(Saracino et al. 2015)
	Urine (0.01 mL)					(2.5:97.5, v/v) of methanol, 30.0 mM citric acid in water and 0.5 mM OSA, adjusted to pH 2.92 (2 × 100 µL)					93-95% (NE) 90-95% (E) 86-89% (DA)	
Methyl paraben (MeP) Ethyl paraben (EtP) Propyl paraben (PrP) Butyl paraben (BuP) Benzyl paraben (BzP)	Urine (0.2 mL)	C_{18} (2 mg); manual	Dilution with 200 µL of phosphate buffer at pH 7.0	Methanol and water (1 × 250 µL)	(4 × 100 µL) 0.1% acetic acid in water (1 × 250 µL) and methanol/water (10:90, v/v) (1 × 100 µL)	Methanol/water (80:20, v/v) (1 × 50 µL)	Methanol and water (5 × 250 µL)	UHPLC-MS/MS	n.s.	0.5 ng/mL (MeP) 0.5 ng/mL (EtP) 0.5 ng/mL (PrP) 0.5 ng/mL (BuP) 0.5 ng/mL (BzP)	n.s.	(Jardim et al. 2015)

DAD (diode array detector); GC (gas chromatography); HPLC (high-performance liquid chromatography); LC (liquid chromatography); LOD (limit of detection); LOQ (limit of quantification); MS (mass spectrometry); MS/MS (tandem mass spectrometry); OSA (ammonium octadecyl sulfate); UHPLC (ultra high-performance liquid chromatography); UV (ultraviolet).

TABLE 5.3

MEPS procedures in drug monitoring, analytical instrumentation, limits, and recoveries (2015–2020)

Analytes	Sample (Volume/Weight)	Sample Preparation	MEPS Sorbent	MEPS Steps					Analytical Instrumentation	LOD	LOQ	Recoveries	Ref
				Conditioning	Load	Wash	Elution	Sorbent re-use					
Lumefantrine (L) desbutyl-lumefantrine (dL)	Plasma (n.s)	1:1 dilution and precipitation of plasma with 0.2% perchloric acid in acetonitrile. Supernatant collection (100 µL)	C$_{18}$ (4 mg); n.s	n.s.	(10 × 70 µL)	Methanol: water (10:90) (n.s.)	Acetonitrile: 0.05% trifluoroacetic acid (90:10) (5 × 50 µL)	n.s.	HPLC-DAD	n.s.	50 ng/mL (L) 50 ng/mL (dL)	92–99% (L) 92–99% (dL)	(Siqueira et al. 2020)
Mirtazapine (MTZ) Venlafaxine (VLX) Escitalopram (ECIT) Fluvoxamine (FVX) Fluoxetine (FLX) Sertraline (SRT)	Urine (0.3 mL)	Dilution with 200 µL of phosphate buffer (0.05 M, pH 7)	Strong cation exchanger (C$_8$ + SCX) (4 mg); eVol®	Methanol and water (4 × 100 µL)	(10 × 100 µL)	Water (1 × 100 µL)	Methanol (1 × 50 µL)	Water, 0.1% formic acid and methanol (4 × 100 µL)	UHPLC-DAD	0.7 ng/mL (MTZ) 4.6 ng/mL (VLX) 1.8 ng/mL (ECIT) 3.8 ng/mL (FVX) 0.5 ng/mL (FLX) 7.0 ng/mL (SRT)	2.1 ng/mL (MTZ) 13.8 ng/mL (VLX) 5.4 ng/mL (ECIT) 11.4 ng/mL (FVX) 1.5 ng/mL (FLX) 21.0 ng/mL (SRT)	92–107% (MTZ) 80–102% (VLX) 92–112% (ECIT) 95–115% (FVX) 98–126% (FLX) 93–122% (SRT)	(Fuentes et al. 2019)
Meropenem (MERO) Levofloxacin (LEVO) Linezolid (LINZ)	Plasma (n.s.)	Dilution with water 1:2 (v/v)	C$_{18}$ (4 mg); semiautomatic	n.s.	(10 × 150 µL)	Water (1 × 100 µL)	Methanol (1 × 50 µL)	Methanol and water (3 × 100 µL)	UHPLC-DAD	4 ng/mL (MERO) 4 ng/mL (LEVO) 7 ng/mL (LINZ)	20 ng/mL (MERO) 10 ng/mL (LEVO) 10 ng/mL (LINZ)	98–99% (MERO) 93–96% (LEVO) 96–97% (LINZ)	(Ferrone et al. 2017)
Ketoconazole (1) Terconazole (2) Voriconazole (3) Bifonazole (4) Clotrimazole (5) Tioconazole (6) Econazole (7) Butoconazole (8) Miconazole (9) Posaconazole (10) Ravuconazole (11) Itraconazole (12)	Plasma (0.17 mL) Urine (0.17 mL)	Dilution with trichloroacetic acid (20 mg/mL) in 1:0.5 (v:v)	C$_{18}$ (n.s.); manual	Methanol and phosphate buffer (40 mM, pH 2.5) (3 × 150 µL)	(8 × 150 µL) (8 × 200 µL)	Phosphate buffer (40 mM, pH 2.5): methanol (90:10, v-v) (1 × 200 µL)	Methanol (8 × 25 µL)	Methanol (5 × 200 µL)	HPLC-DAD	0.017 µg/mL (1) 0.070 µg/mL (2) 0.017 µg/mL (3) 0.017 µg/mL (4) 0.017 µg/mL (5) 0.017 µg/mL (6) 0.017 µg/mL (7) 0.017 µg/mL (8) 0.017 µg/mL (9) 0.017 µg/mL (10) 0.007 µg/mL (11) 0.017 µg/mL (12)	0.05 µg/mL (1) 0.2 µg/mL (2) 0.05 µg/mL (3) 0.05 µg/mL (4) 0.05 µg/mL (5) 0.05 µg/mL (6) 0.05 µg/mL (7) 0.05 µg/mL (8) 0.05 µg/mL (9) 0.05 µg/mL (10) 0.02 µg/mL (11) 0.05 µg/mL (12)	n.s.	(Campestre et al. 2017)
Dexamethasone (DEX) Dexamethasone disodium phosphate (DEX-SP)	Aqueous humour (0.05 mL)	n.s.	C$_{18}$ (n.s.); eVol®	Methanol (4 × 50 µL) and water (2 × 50 µL)	(19 × 50 µL)	n.s.	Methanol (10 × 26 µL)	Methanol (10 × 50 µL)	LC-MS/MS	n.s.	0.5 ng/mL (DEX) 0.7 ng/mL (DEX-SP)	95–105% (DEX) 91–119% (DEX-SP)	(Bianchi et al. 2017)

Analytes	Matrix	Pretreatment	Sorbent; automation	Conditioning	Loading	Washing	Elution	Drying	Detection	LOD	LOQ	Recovery	Reference
Furprofen (FUR), Indoprofen (IND), Ketoprofen (KET), Fenbufen (FEN), Flurbiprofen (FLU), Indomethacin (INM), Ibuprofen (IBU)	Dialyzed samples (0.05 mL)	Centrifugation; dilution to 150 μL of water	C_{18} (n.s.); semi-automated	Methanol and water-methanol (95.5, v/v) (1×250 μL)	(10×50 μL)	10 mM phosphate buffer (pH 2.5) (1×100 μL)	Methanol and 1% sodium hydroxide in water 95.5 (v/v) (1×200 μL)	n.s.	UHPLC-DAD	8 ng/mL (FUR); 9 ng/mL (IND); 9 ng/mL (KET); 9 ng/mL (FEN); 9 ng/mL (FLU); 10 ng/mL (INM); 10 ng/mL (IBU)	25 ng/mL (FUR); 26 ng/mL (IND); 29 ng/mL (KET); 27 ng/mL (FEN); 28 ng/mL (FLU); 29 ng/mL (INM); 33 ng/mL (IBU)	94–99% (FUR); 94–99% (IND); 94–99% (KET); 98–99% (FEN); 94–99% (FLU); 97–99% (INM); 96–99% (IBU)	(D'Archivio et al. 2016)
Levofloxacin (1), Ciprofloxacin (2), Ulifloxacin (3), Moxifloxacin (4), Furprofen (5), Indoprofen (6), Ketoprofen (7), Fenbufen (8), Flurbiprofen (9), Indomethacin (10), Ibuprofen (11)	Plasma and Urine (0.18 mL)	Addition trichloroacetic acid (100 μL); centrifugation	C_{18} (n.s.); n.s.	Methanol and ammonium acetate buffer, 50 mM, pH 2.5 (3×150 μL)	(8×100 μL)	Ammonium acetate buffer: methanol (95.5, v:v) (1×150 μL)	Methanol (8×25 μL)	n.s.	HPLC-DAD	0.03 μg/mL	0.10 μg/mL	n.s.	(D'Angelo et al. 2016)
(R)-Metoprolol(S)-Metoprolol	Plasma (0.1 mL); Oral fluid (0.1 mL)	Dilution with water (1:4)	C_{18} (n.s.); n.s.	Isopropanol and water (1×100 μL)	(4×100 μL)	5% Methanol (2×100 μL)	Isopropanol (2×100 μL)	Isopropanol and water (4×100 μL)	LC-MS/MS	0.5 ng/mL (R); 0.5 ng/mL (S); 0.5 ng/mL (R); 0.5 ng/mL (S)	1.5 ng/mL (R); 1.5 ng/mL (S); 1.5 ng/mL (R); 1.5 ng/mL (S)	94–98% (R); 93–97% (S); 95–98% (R); 96–97% (S)	(Elmongy et al. 2016)
Betaxolol	Urine (0.25 mL)	Dilution 10 times with water and filtered	C_{18} (n.s.); automated	15% Acetonitrile (500 μL)	MEPS-SIC hyphenation (250 μL)	15% Acetonitrile (700 + 500 μL)	Acetonitrile: 0.5% of TEA in water, pH 4.5 (30/70) (2500 μL)	n.s.	HPLC-UV	1.5 ng/mL	5 ng/mL	100–108%	(Šrámková 2015)
Ciprofloxacin (CIP), Levofloxacin (LEV)	Sputum (0.18 mL)	Trichloroacetic acid (20 mg/mL) in 1:0.5 ratio (v:v); centrifugation	C_{18} (n.s.); n.s.	Methanol (3×150 μL) and phosphate buffer (30 mM, pH 2.5) (3×150 μL)	(8×100 μL)	Phosphate buffer (30 mM, pH 2.5) and methanol (95.5, v:v) (1×150 μL)	Methanol (8×25 μL)	n.s.	HPLC-DAD	17 ng/mL	50 ng/mL	60% (CIP); 80% (LEV)	(Locatelli et al. 2015)

DAD (diode array detector); GC (gas chromatography); HPLC (high-performance liquid chromatography); LC (liquid chromatography); LOD (limit of detection); LOQ (limit of quantification); MS (mass spectrometry); MS/MS (tandem mass spectrometry); TEA (triethylamine); UHPLC (ultra high-performance liquid chromatography); UV (ultraviolet).

obtained recoveries were low for some markers of cocaine and opiates consumption, the limits of determination were comparable to those reported for SPE. This is explained not only by the analytical equipment used, but also by the selectivity of MEPS and the clean extracts obtained.

Another alternative specimen that was widely used in this field in the past years was oral fluid. The reported applications of MEPS for oral fluid clean-up and drug pre-concentration are noteworthy, since with a few microlitres and a rapid procedure it was possible to determine up to 30 different analytes (Rocchi et al. 2018). MEPS's potential is proven in this field due to the rapid extraction of a great number of xenobiotics from small amounts of samples. Moreover, its application to blood, plasma, and urine has continued to be reported in the last five years. The most described sorbent in the forensic toxicology field is not C_{18}, like in the previous fields, however. Instead, mixed-mode sorbent appears to be the most suitable to pre-concentrate multi-class drugs on a multi-method. Table 5.4 describes MEPS procedures adopted in the last five years in the field of forensic toxicology.

The field of analysis, for which more developments regarding sorbents were observed, is undoubtedly environmental toxicology. The samples' type do not vary much; all of them are water samples, except one which is soil, and the analytes to be determined are mainly polycyclic aromatic hydrocarbons, phenoxyacetic acid herbicides, other pesticides, endocrine-disrupting chemicals, and trace levels of a few pharmaceutical drugs. Another different aspect in this field is the volume of sample submitted to MEPS. In fact, while in other fields MEPS works with volumes in the order of microlitres, in environmental toxicology volumes of several millilitres are used. The latter is also justified by the fact that the new developed sorbents are packed in larger capacity syringes, commonly insulin syringes (1 mL). Most publications in this field did not specify sample pre-treatment before MEPS, perhaps because their major goal was sorbent development; however, a few mentioned centrifugation or filtration of the samples.

The use of soil as an environmental sample has further proven the great versatility of this miniaturized technique. Serenjeh et al. (Serenjeh et al. 2020) proposed a headspace approach of MEPS for the determination of volatile polycyclic aromatic hydrocarbons in soil. The authors used 2 mg of aminoethyl functionalized SBA-15 (SBA-15-NH$_2$) as sorbent, and after pre-heating the soil sample 15 min at 150 °C, the MEPS syringe sampled the air in the closed vial to concentrate the analytes (Serenjeh et al. 2020). Although the reported extraction efficiencies were not high, this approach appears as an excellent option for other solid samples. The latter procedure and other MEPS applications in the last five years for environmental toxicology are resumed in Table 5.5.

Finally, the food toxicology field has the most heterogeneous types of samples. Specimens used in this field can go from solid (e.g., fruits, flour) to liquid (e.g., milk, wine, juices), and MEPS applications have proven suitable for all of them. Even though many sorbent developments were made in the last five years concerning food toxicology analysis, C_{18} continues to be the most reported sorbent. Indeed, this sorbent has been applied to pre-concentrate fluoroquinolones from bovine milk (Aresta et al. 2019), polybrominated diphenyl ether (Souza et al. 2019), phthalates in cold drinks (Kaur et al., 2016), ochratoxin A and furanic derivatives in wines (Savastano et al. 2016; Perestrelo et al. 2015), and polychlorinated biphenyls in bovine serum (Yang et al. 2016), all of them resulting in recoveries above 70%. Poorer recoveries were reported for this sorbent when applied to pre-concentrate pesticides in sugarcane juice samples (27 to 65%) (Fumes et al. 2016). Noteworthy is the work reported by Di Ottavio et al. (Di Ottavio et al. 2017) that accomplished the extraction of 25 pesticide and fungicide residues in wheat flour. The target analytes are widely used in wheat and present different physico-chemical characteristics; hence, the authors opted for highly cross-linked polystyrene divinylbenzene (HDVB) sorbent.

Depending on the sample type, different pre-treatments should be adopted. For instance, milk and egg samples should undergo a protein precipitation step, whereas fruits and other solid samples should be crushed and solubilized under sonication. For all of them, a further centrifugation step and dilution should be employed to improve sorbent durability. The different procedures are summarized in Table 5.6.

TABLE 5.4

MEPS procedures in forensic toxicology, analytical instrumentation, limits, and recoveries (2015–2020)

Analytes	Sample Preparation	Sample (Volume/Weight)	MEPS Sorbent	MEPS Steps					Analytical Instrumentation	LOD	LOQ	Recoveries	Ref
				Conditioning	Load	Wash	Elution	Sorbent Re-use					
Cocaine (COC), Benzoylecgonine (BEG), Ecgonine methyl ester (EME), Norcocaine (NCOC), Cocaethylene (COET), Anhydroecgonine methyl (AEME)	1 mL of 0.1 M hydrochloric acid incubation overnight at 60°C. Neutralization with 100 µL of 1M sodium hydroxide	Hair (50 mg)	M1 (80% C$_8$ and 20% SCX) (4 mg); Manual	Methanol and water (1 × 250 µL)	(21 × 150 µL)	Water and acetate buffer of pH 4 (1 × 50 µL)	2% ammonium hydroxide in methanol (3 × 100 µL)	1% ammonia in methanol–acetonitrile (50:50, v/v) and 1% formic acid in 2-propanol–water (10:90) (2 × 250 µL)	GC-MS/MS	0.010 ng/mg (COC) 0.025 ng/mg (BEG) 0.025 ng/mg (EME) 0.025 ng/mg (NCOC) 0.010 ng/mg (COET) 0.150 ng/mg (AEME)	0.010 ng/mg (COC) 0.025 ng/mg (BEG) 0.025 ng/mg (EME) 0.025 ng/mg (NCOC) 0.010 ng/mg (COET) 0.150 ng/mg (AEME)	44–65% (COC) 21–28% (BEG) 1–3% (EME) 36–44% (NCOC) 63–73% (COET) 4–6% (AEME)	Rosado et al. (2020b)
Methadone (MET), EDDP	1mL of 1M sodium hydroxide for 45 min at 50°C; neutralization with 100 µL of 20% formic acid	Hair (50 mg)	M1 (80% C$_8$ and 20% SCX) (4 mg); Manual	Methanol and 2% formic acid (3 × 250 µL)	(9 × 150 µL)	3.36% formic acid (3 × 50 µL)	2.36% ammonium hydroxide in methanol (6 × 100 µL)	1% ammonia in methanol–acetonitrile (50:50, v/v) and 1% formic acid in 2-propanol–water (10:90) (4 × 250 µL)	GC-MS/MS	0.01 ng/mg (MET) 0.01 ng/mg (EDDP)	0.01 ng/mg (MET) 0.01 ng/mg (EDDP)	73–109% (MET) 84–111% (EDDP)	Rosado et al. (2020a)
Methylone	10 µL of 1 M carbonate buffer at pH 9.0	Oral fluid (0.5 mL)	C$_{18}$ (4 mg); eVol*	2-propanol and water (1 × 100 µL)	(5 × 100 µL)	0.1 M carbonate buffer at pH 9.0 (1 × 100 µL)	2-propanol (5 × 100 µL)	n.s.	IMS	4 ng/mL	14 ng/mL	78–91%	Sorribes-Soriano et al. (2020)
Fentanyl (F), Sufentanil (SuF), Alfentanil (AlF), Acrylfentanyl (AcryF), Thiofentanyl (ThF), Valerylfentanyl (ValF), Furanylfentanyl (FuF), Acetyl fentanyl (AcetF), Carfentanil (CarF), Norfentanyl (NorF), Acetyl norfentanyl (AcetNorF)	Dilution with 0.6 mL water	Urine (0.2 mL)	C$_{18}$ (n.s.); eVol*	Methanol and water (2 × 50 µL)	(8 × 50 µL)	Water and isopropyl alcohol (95.5 v/v) (2 × 50 µL)	Acetonitrile (1 × 50 µL)	Acetonitrile: methanol mixture (1:1 v/v) and water: methanol mixture (95:5 v/v) (4 × 50 µL)	LC-MS/MS	0.1 ng/mL (F) 0.1 ng/mL (SuF) 0.1 ng/mL (AlF) 0.1 ng/mL (AcryF) 0.1 ng/mL (ThF) 0.1 ng/mL (ValF) 0.1 ng/mL (FuF) 0.1 ng/mL (AcetF) 0.1 ng/mL (CarF) 1 ng/mL (NorF) 1 ng/mL (AcetNorF)	1 ng/mL (F) 1 ng/mL (SuF) 1 ng/mL (AlF) 1 ng/mL (AcryF) 1 ng/mL (ThF) 1 ng/mL (ValF) 1 ng/mL (FuF) 1 ng/mL (AcetF) 1 ng/mL (CarF) 1 ng/mL (NorF) 1 ng/mL (AcetNorF)	28–35% (F) 31–32% (SuF) 30–30% (AlF) 29–34% (AcryF) 27–31% (ThF) 32–35% (ValF) 30–37% (FuF) 24–27% (AcetF) 31–34% (CarF) 13–14% (NorF) 5–6% (AcetNorF)	da Cunha et al. (2020)
Dichloropane	pH adjusted with 10 µL phosphate buffer (1 M, pH 7)	Oral fluid (0.09 mL)	C$_8$ (4mg); eVol*	2-propanol (3 × 100 µL) and water (2 × 100 µL)	(4 × 100 µL)	Water (4 × 100 µL)	2-propanol (10 × 50 µL)	n.s.	IMS	30 ng/mL	90 ng/mL	85–107%	Sorribes-Soriano et al. (2019)
									GC-MS	70 ng/mL	200 ng/mL		

(Continued)

TABLE 5.4 (Continued)

MEPS procedures in forensic toxicology, analytical instrumentation, limits, and recoveries (2015–2020)

Analytes	Sample (Volume/Weight)	Sample Preparation	MEPS Sorbent	MEPS Steps					Analytical Instrumentation	LOD	LOQ	Recoveries	Ref
				Conditioning	Load	Wash	Elution	Sorbent Re-use					
Tramadol (TRM), Codeine (COD), Morphine (MOR), 6-acetylcodeine (6-AC), 6-monoacetylmorphine (6-MAM), Fentanyl (FNT)	Hair (50 mg)	2 mL of methanol incubated overnight at 65°C; evaporation and reconstitution with 500 µL of 2% formic acid	M1 (80% C8 and 20% SCX) (4 mg); manual	Methanol and 2% formic acid (3 × 250 µL)	(15 × 150 µL)	3.36% formic acid (3 × 50 µL)	2.36% ammonium hydroxide in methanol (8 × 100 µL)	1% ammonium hydroxide in acetonitrile:methanol (1:1) and 1% formic acid in isopropanol:water (10:90) (4 × 250 µL)	GC-MS/MS	0.010 ng/mg (TRM) 0.010 ng/mg (COD) 0.025 ng/mg (MOR) 0.010 ng/mg (6-AC) 0.025 ng/mg (6-MAM) 0.025 ng/mg (FNT)	0.010 ng/mg (TRM) 0.010 ng/mg (COD) 0.025 ng/mg (MOR) 0.010 ng/mg (6-AC) 0.025 ng/mg (6-MAM) 0.025 ng/mg (FNT)	74–90% (TRM) 51–59% (COD) 22–35% (MOR) 69–99% (6-AC) 53–61% (6-MAM) 75–86% (FNT)	Gallardo et al. (2019)
Codeine (COD), Morphine (MOR), 6-monoacetylmorphine (6-MAM)	Blood (0.25 mL)	Dilution with 0.4 mL of 0.1 M phosphate buffer (pH 6); protein precipitation with ice-cold acetonitrile; centrifugation, evaporation and addition of 8.5 mL of 2% formic acid	M1 (80% C8 and 20% SCX) (4 mg); manual	Methanol and 2% formic acid (3 × 250 µL)	(20 × 250 µL)	3.36% formic acid (1 × 250 µL)	2.36% ammonium hydroxide in methanol (11 × 250 µL)	Methanol and water (3 × 250 µL)	GC-MS/MS	5 ng/mL (COD) 5 ng/mL (MOR) 5 ng/mL (6-MAM)	5 ng/mL (COD) 5 ng/mL (MOR) 5 ng/mL (6-MAM)	13–20% (COD) 6–8% (MOR) 14–20% (6-MAM)	Prata et al. (2019)
Amphetamine (AMP), Methamphetamine (MAMP), 3,4-methylenedioxyamphetamine (MDA), 3,4-methylenedioxyethyl-methamphetamine (MDMA), 3,4-methylenedioxy-N-methyl-α-ethylfenilethylamine (MBDB), 3,4-methylenedioxy-N-ethylamphetamine (MDE)	Urine (0.2 mL)	Dilution with 0.1 mL of ammonium acetate (pH 6.7)	C18 (4 mg); manual	Methanol and water (1 × 250 µL)	(9 × 100 µL)	Water (1 × 150 µL) and water:methanol (95:5) (1 × 150 µL)	2% ammonium hydroxide in acetonitrile (4 × 100 µL)	Ammonium hydroxide in acetonitrile: methanol (1:1) and 1% formic acid in isopropanol:water (10:90) (4 × 100 µL)	GC-MS	n.s.	35 ng/mL (AMP) 25 ng/mL (MAMP) 50 ng/mL (MDA) 35 ng/mL (MDMA) 25 ng/mL (MBDB) 25 ng/mL (MDE)	32–49% (AMP) 19–38% (MAMP) 30–48% (MDA) 40–52% (MDMA) 34–50% (MBDB) 52–71% (MDE)	Malaca et al. (2019)

Analytes	Matrix	Sample pretreatment	Sorbent and MEPS mode	Washing solution	Elution solution	Drying step	Analytical technique	LOQ	Recovery	References
Azynphos-ethyl (AZP), Diazinon (DZN), Chlorpyrifos (CLP), Chlorfenvinfos (CLF), Parathion-ethyl (PRT), Quinalphos (QLP)	Blood (0.1 mL)	Dilution with 500 μL of ammonium acetate buffer (pH 4.9)	C₁₈ (4 mg); manual	Methanol and water (4 × 250 μL)	2-propanol (1.5%) in 0.1% formic acid in water (1 × 25 μL)	Methanol (4 × 110 μL)	GC-MS/MS	n.s.	2.5 μg/mL (AZP) 0.5 μg/mL (DZN) 0.5 μg/mL (CLP) 0.5 μg/mL (CLF) 0.5 μg/mL (PRT) 0.5 μg/mL (QLP)	61–68% (AZP) 58–78% (DZN) 59–68% (CLP) 64–74% (CLF) 62–76% (PRT) 70–78% (QLP)	Santos et al. (2018)
Piperonyl piperazine (1) Methylone (2) 4-MEOPP (3) Dimethylcathinone (4) Buphedrone (5) Methedrone (6) Buthylone (7) Ethcathinone (8) Mephedrone (9) 4-MEC (10) Methoxetamine (11) alpha-PVP (12) 2C-B (13) 3,4-MDPV (14) AM-1220 (15) JWH-200 (16) AB-005 (17) JWH-018 N-pentanoic acid (18) JWH-018 N-(5-hydroxypentyl) (19) XLR-11 N-(4-hydroxypentyl) (20) MAM-2201 N-pentanoic acid (21) JWH-073 (22) WIN-55 (23) UR-144 N-(5-hydroxypentyl) (24) MAM-2201 (25) JWH-250 (26) XLR-11 (27) JWH-018 (28) JWH-081 (29) JWH-122 (30)	Oral fluid (0.09 mL)	Dilution with 30 μL of methanol and 60 μL of water (total volume 200 μL)	C₁₈ (n.s.); n.s.	Methanol (3 × 250 μL); Water-methanol (75:25, v/v) (3 × 250 μL)	Water-methanol (90:10, v/v) (3 × 200 μL)	10 mM formic acid (5 × 100 μL)	Methanol (3 × 250 μL)	UHPLC-MS/MS	0.850 ng/mL (1) 0.385 ng/mL (2) 0.830 ng/mL (3) 0.045 ng/mL (4) 0.070 ng/mL (5) 0.230 ng/mL (6) 0.126 ng/mL (7) 0.050 ng/mL (8) 0.045 ng/mL (9) 0.012 ng/mL (10) 0.470 ng/mL (11) 0.050 ng/mL (12) 0.053 ng/mL (13) 0.005 ng/mL (14) 0.012 ng/mL (15) 0.020 ng/mL (16) 0.036 ng/mL (17) 0.045 ng/mL (18) 0.035 ng/mL (19) 0.093 ng/mL (20) 0.275 ng/mL (21) 0.005 ng/mL (22) 0.017 ng/mL (23) 0.030 ng/mL (24) 0.020 ng/mL (25) 0.009 ng/mL (26) 0.012 ng/mL (27) 0.060 ng/mL (28) 0.045 ng/mL (29)	2.600 ng/mL (1) 1.150 ng/mL (2) 2.500 ng/mL (3) 0.135 ng/mL (4) 0.210 ng/mL (5) 0.700 ng/mL (6) 0.380 ng/mL (7) 0.150 ng/mL (8) 0.130 ng/mL (9) 0.035 ng/mL (10) 1.400 ng/mL (11) 0.150 ng/mL (12) 0.160 ng/mL (13) 0.015 ng/mL (14) 0.035 ng/mL (15) 0.055 ng/mL (16) 0.110 ng/mL (17) 0.135 ng/mL (18) 0.100 ng/mL (19) 0.280 ng/mL (20) 0.820 ng/mL (21) 0.015 ng/mL (22) 0.050 ng/mL (23) 0.090 ng/mL (24) 0.060 ng/mL (25) 0.025 ng/mL (26) 0.035 ng/mL (27) 0.170 ng/mL (28) 0.135 ng/mL (29)	49–53% (1) 68–77% (2) 85–96% (3) 73–85% (4) 41–46% (5) 40–46% (6) 45–49% (7) 48–55% (8) 31–40% (9) 33–36% (10) 33–38% (11) 65–72% (12) 67–96% (13) 70–84% (14) 63–70% (15) 49–52% (16) 67–68% (17) 78–92% (18) 68–91% (19) 58–68% (20) 75–80% (21) 65–74% (22) 70–89% (23) 72–95% (24) 87–96% (25) 60–84% (26) 78–89% (27) 73–86% (28) 61–69% (29) 76–93% (30)	Rocchi et al. (2018)
Tetrahydrocabahinol (THC), 11-hydroxy-tetrahydrocabahinol (11-OH-THC), 11-Nor-9-carboxy-tetrahydrocabahinol (THC-COOH)	Plasma (0.25 mL)	Protein precipitation; dilution with 5 mL of 0.1 mM potassium phosphate buffer (pH = 6)	M1 (80% C₈ and 20% SCX) (4 mg); manual	Methanol and 0.1% formic acid (4 × 250 μL)	3% acetic acid and 5% methanol (1 × 100 μL)	10% ammonium hydroxide in methanol (6 × 100 μL)	n.s.	GC-MS/MS	0.1 ng/mL (THC) 0.1 ng/mL (11-OH-THC) 0.1 ng/mL (THC-COOH)	0.1 ng/mL (THC) 0.1 ng/mL (11-OH-THC) 0.1 ng/mL (THC-COOH)	53–78% (THC) 57–66% (11-OH-THC) 62–65% (THC-COOH)	Rosado et al. (2017b)

(Continued)

TABLE 5.4 (Continued)

MEPS procedures in forensic toxicology, analytical instrumentation, limits, and recoveries (2015–2020)

Analytes	Sample (Volume/Weight)	Sample Preparation	MEPS Sorbent	MEPS Steps					Analytical Instrumentation	LOD	LOQ	Recoveries	Ref
				Conditioning	Load	Wash	Elution	Sorbent Re-use					
Cocaine (COC), Benzoylecgonine (BEG), Ecgonine methyl ester (EME)	Urine (0.2 mL)	Centrifuged at 4,500 rpm during 15 min; 100 μL of 0.1 mM potassium phosphate buffer	M1 (80% C$_8$ and 20% SCX) (4 mg); manual	Methanol and 0.1% formic acid (1 × 250 μL)	(6 × 150 μL)	0.1% formic acid (4 × 50 μL)	1% ammonium hydroxide in methanol (4 × 100 μL)	1% ammonia in methanol-acetonitrile (50:50, v/v) and 1% formic acid in 2-propanol-water (10:90) (4 × 100 μL)	GC-MS	25 ng/mL (COC) 25 ng/mL (BEG) 25 ng/mL (EME)	25 ng/mL (COC) 25 ng/mL (BEG) 25 ng/mL (EME)	67–83% (COC) 25–44% (BEG) 15–37% (EME)	Rosado et al. (2017a)
Morphine (1), Naloxone (2), Methylone (3), Flephedrone (4), Ethylcathinone (5), Ethylcathinone ephedrine (6), Scopolamine (7) 6-monoacetylmorphine (6-MAM) (8), Ethylone (9), Methylephedrine (10), Butylone (11), Mephedrone (12), Pentedrone (13), Benzoylecgonine (BEG) (14) Cocaine (15), Methylenedioxypyrovalerone (MDPV) (16) Cocaethylene (17) Pyrovalerone (18) EDDP (19) Buprenorphine (20) Methadone (21)	Oral fluid (0.3 mL)	200 μL of methanol; shaken; centrifugation; 300 μL of supernatant dilution with 200 μL of phosphate buffer (50 mM, pH 9)	M1 (80% C$_8$ and 20% SCX) (4 mg); eVol*	Methanol and water (1 × 100 μL)	(6 × 100 μL)	Water/methanol 90:10 (v:v) (1 × 50 μL)	Dichloromethane/2-propanol/ammonium hydroxide (78:20:2, v:v:v) (1 × 90 μL)	Eluent and methanol (1 × 100 μL); water, 0.1% formic acid and methanol (4 × 100 μL)	UHPLC-MS/MS	2.5 ng/mL (1) 0.5 ng/mL (2) 0.25 ng/mL (3) 0.25 ng/mL (4) 0.25 ng/mL (5) 0.25 ng/mL (6) 1 ng/mL (7) 1 ng/mL (8) 0.25 ng/mL (9) 0.25 ng/mL (10) 0.25 ng/mL (11) 0.25 ng/mL (12) 0.25 ng/mL (13) 0.5 ng/mL (14) 0.25 ng/mL (15) 0.25 ng/mL (16) 0.25 ng/mL (17) 0.25 ng/mL (18) 0.25 ng/mL (19) 0.25 ng/mL (20) 0.25 ng/mL (21)	10 ng/mL (1) 1 ng/mL (2) 0.5 ng/mL (3) 0.5 ng/mL (4) 0.5 ng/mL (5) 0.5 ng/mL (6) 2.5 ng/mL (7) 2.5 ng/mL (8) 0.5 ng/mL (9) 0.5 ng/mL (10) 0.5 ng/mL (11) 0.5 ng/mL (12) 0.5 ng/mL (13) 1 ng/mL (14) 0.5 ng/mL (15) 0.5 ng/mL (16) 0.5 ng/mL (17) 0.5 ng/mL (18) 0.5 ng/mL (19) 0.5 ng/mL (20) 0.5 ng/mL (21)	22–87% (1) 94–107% (2) 75–93% (3) 60–100% (4) 100–125% (5) 84–106% (6) 44–86% (7) 86–99% (8) 91–110% (9) 87–101% (10) 84–125% (11) 86–124% (12) 75–99% (13) 85–112% (14) 79–105% (15) 72–92% (16) 87–108% (17) 92–102% (18) 90–104% (19) 92–107% (20) 82–111% (21)	Ares et al. (2017)
Chlordiazepoxide (CLD), Medazepam (MDP), Lorazepam (LZP), Oxazepam (OZP), Diazepam (DZP)	Alcoholic beverage	Dilution with water in 1:5 ratio (v/v)	C$_{18}$ (n.s.); eVol*	Acetonitrile and water (3 × 100 μL)	(6 × 100 μL)	n.s.	Acetonitrile and water 90:10 (v/v), both acidified with 0.1% formic acid (3 × 100 μL)	2-n-propanol (2 × 100 μL)	UHPLC-UV	1 μg/mL (CLD) 0.5 μg/mL (MDP) 0.5 μg/mL (LZP) 0.5 μg/mL (OZP) 0.5 μg/mL (DZP)	2 μg/mL (CLD) 1 μg/mL (MDP) 1 μg/mL (LZP) 1 μg/mL (OZP) 1 μg/mL (DZP)	61–64% (CLD) 61–67% (MDP) 81–87% (LZP) 88–91% (OZP) 70–72% (DZP)	Magrini et al. (2016)

Analytes	Matrix	Sample pretreatment	Sorbent	Washing	Elution	Analysis	LOD	LOQ	Recoveries	Reference
Norketamine (NK), Ketamine (K)	Plasma (0.25 mL)	Dilution with 7 mL of phosphate buffer	M1 (80% C8 and 20% SCX) (4 mg); manual	(26 × 250 μL)	0.1% acetic acid (1 × 100 μL) and 10% methanol (1 × 100 μL)	6% ammonia in methanol (1 × 100 μL)	Methanol (5 × 250 μL) and water (4 × 250 μL)	GC-MS/MS	5 ng/mL (NK) 5 ng/mL (K)	10 ng/mL (NK) 10 ng/mL (K)	63–75%(NK) 73–89% (K)	Moreno et al. (2015)
	Urine (0.2 5mL)	Dilution with 0.25 mL of water		(8 × 250 μL)	5.25% acetic acid (1 × 250 μL) and 5% methanol in water (1 × 100 μL)	3% ammonia in methanol (1 × 100 μL)					73–76%(NK) 89–101% (K)	
Amphetamine (AMP) (1) Benzoylecgonine (BEG) (2) Buprenorphine (3) Cocaine (COC) (4) Codeine (COD) (5) Diacetylmorphine (6) Ecgonine methylester (EME) (7) EDDP (8) Ketamine (9) 3,4-methylenedioxyamphetamine (MDA) (10) Methyldiethanolamine (MDEA) (11) 3,4-methylenedioxyethylmethamphetamine (MDMA) (12) Mescaline (13) Methadone (14) Methamphetamine (MAMP) (15) Morphine (MOR) (16) Norbuprenorphine (17) Norcocaine (NCOC) (18) Phencyclidine (19)	Oral fluid (0.12 mL)	Dilution with 80 μL of water and 40 μL of 25 mM NH3 in methanol; sonication and centrifugation	C18 (n.s.); n.s.	Methanol and water/methanol (80:20, v/v) (2 × 250 μL)	50 mM NH3 in water/methanol (90:10, v/v) (3 × 100 μL)	5 mM formic acid in methanol (5 × 100 μL)	n.s.	LC-MS/MS	1 ng/mL (1) 0.8 ng/mL (2) 2 ng/mL (3) 0.3 ng/mL (4) 2 ng/mL (5) 10 ng/mL (6) 1 ng/mL (7) 0.3 ng/mL (8) 0.5 ng/mL (9) 1 ng/mL (10) 0.5 ng/mL (11) 0.5 ng/mL (12) 2 ng/mL (13) 0.2 ng/mL (14) 1 ng/mL (15) 2 ng/mL (16) 2 ng/mL (17) 0.5 ng/mL (18) 0.2 ng/mL (19)	3 ng/mL (1) 2 ng/mL (2) 5 ng/mL (3) 2 ng/mL (4) 5 ng/mL (5) 30 ng/mL (6) 4 ng/mL (7) 0.8 ng/mL (8) 1.5 ng/mL (9) 3 ng/mL (10) 1 ng/mL (11) 1 ng/mL (12) 5 ng/mL (13) 0.5 ng/mL (14) 3 ng/mL (15) 5 ng/mL (16) 5 ng/mL (17) 1.5 ng/mL (18) 0.6 ng/mL (19)	90% (1) 25% (2) 67% (3) 82% (4) 100% (5) 101% (6) 18% (7) 75% (8) 79% (9) 81% (10) 85% (11) 90% (12) 78% (13) 74% (14) 88% (15) 70% (16) 70% (17) 88% (18) 71% (19)	Montesano et al. (2015)

DAD (diode array detector); GC (gas chromatography); HPLC (high-performance liquid chromatography); IMS (ion mobility spectrometry); LC (liquid chromatography); LOD (limit of detection); LOQ (limit of quantification); MS (mass spectrometry); MS/MS (tandem mass spectrometry); UHPLC (ultra high-performance liquid chromatography); UV (ultraviolet).

TABLE 5.5

MEPS procedures in environmental toxicology, analytical instrumentation, limits, and recoveries (2015–2020)

Analytes	Sample (Volume/Weight)	Sample Preparation	MEPS Sorbent	MEPS Steps - Conditioning	MEPS Steps - Load	MEPS Steps - Wash	MEPS Steps - Elution	MEPS Steps - Sorbent re-use	Analytical Instrumentation	LOD	LOQ	Recoveries	Ref
Naphthalene (N), Fluorene (F), Fluoranthene (FT), Phenanthrene (PT), Pyrene (PY)	Soil (10 g)	15 min pre-heating vial at 150 °C	Aminoethyl functionalized SBA-15 (SBA-15-NH$_2$) (2 mg); Temperature of 0 °C t	Methanol (10 × 100 μL)	Headspace (10 × 100 μL)	n.s.	Methanol (10 × 400 μL)	Methanol (10 × 100 μL)	HPLC–UV	0.083 ng/g (N) 0.025 ng/g (F) 0.014 ng/g (FT) 0.028 ng/g (PT) 0.043 ng/g (PY)	0.250 ng/g (N) 0.075 ng/g (F) 0.042 ng/g (FT) 0.084 ng/g (PT) 0.130 ng/g (PY)	25% (N) 8% (F) 72% (FT) 65% (PT) 78% (PY)	(Serenjeh et al. 2020)
Nitrobenzene (NB), 2-Nitrotoluene (2-NT), 3-Nitrotoluene (3-NT), 4-Nitrotoluene (4-NT), 2,6-Dinitrotoluene (2,6-DNT),1,3-Dinitrobenzene (1,3-DNB), 2,4-Dinitrotoluene (2,4-DNT), 2,4,6-Trinitrotoluene (2,4,6-TNT), 1,3,5-Trinitrobenzene (1,3,5-TNB), 4-Amino-2,6-dinitrotoluene (4-Am-2,6-DNT), 2-Amino-4,6-dinitrotoluene (2-Am-4,6-DNT), 2,4,6-Trinitrophenyl-Nmethylnitramine (Tetryl)	River water (1 mL); Ground water (1 mL)	Filtered n.s.	C$_{18}$ (4 mg); Manual.	Methanol and water (1 × 100 μL)	(10 × 50 μL)	n.s.	Methanol (1 × 30 μL)	Methanol and water (3 × 100 μL)	GC-MS	0.014 to 0.828 ng/mL (n.s.) 0.014 to 0.828 ng/mL (n.s.)	0.046 to 2.732 ng/mL (n.s.) 0.046 to 2.732 ng/mL (n.s.)	93–96% (NB) 92–98% (2-NT) 93–98% (3-NT) 93–97% (4-NT) 93–96% (2,6-DNT) 93–96% (1,3-DNB) 91–96% (2,4-DNT) 92–96% (2,4,6-TNT) 92–94% (1,3,5-TNB) 90–92% (4-Am-2,6-DNT) 91–92% (2-Am-4,6-DNT) 92–95% (Tetryl) 94–98% (NB) 93–98% (2-NT) 93–98% (3-NT) 93–97% (4-NT) 94–98% (2,6-DNT) 94–98% (1,3-DNB) 94–96% (2,4-DNT) 94–96% (2,4,6-TNT) 91–95% (1,3,5-TNB) 90–93% (4-Am-2,6-DNT) 91–92% (2-Am-4,6-DNT) 95–97% (Tetryl)	(Dhingra et al. 2018)

Analytes	Matrix	Sorbent	Conditioning	Loading	Washing	Elution	Analysis	LOD	LOQ	Recoveries	References
tripropyl phosphate (TPP), tributyl phosphate (TBP), tris(2-chloroethyl) phosphate (TCEP), tris(1-chloro-2-propyl) phosphate (TCPP), tris (1,3-dichloro-2-propyl) phosphate (TDCPP), tris(2-butoxyethyl) phosphate (TBEP), triphenyl phosphate (TPhP), (2-ethylhexyl)-diphenyl phosphate (EHDPP), tris(2-ethylhexyl) phosphate (TEHP), tricresylphosphate (TCP)	Tap water (50 mL)	silica-DVB (4 mg); eVol®	Filtrated through 0.45 μm filters	Methanol and water (2 × 250 μL)	(4 × 500 μL)	n.s.	Acetonitrile (3 × 20 μL)	Acetonitrile (7 × 250 μL)	GC-MS/MS	2.7 ng/L (TPP); 11 ng/L (TBP); 12 ng/L (TCEP); 13 ng/L (TCPP); 22 ng/L (TDCPP); 87 ng/L (TBEP); 13 ng/L (TPhP); 23 ng/L (EHDPP); 26 ng/L (TEHP); 99 ng/L (TCP)	10 ng/L (TPP); 25 ng/L (TBP); 25 ng/L (TCEP); 25 ng/L (TCPP); 50 ng/L (TDCPP); 200 ng/L (TBEP); 25 ng/L (TPhP); 50 ng/L (EHDPP); 50 ng/L (TEHP); 200 ng/L (TCP)	88–94% (TPP); 66–79% (TBP); 76–97% (TCEP); 82–97% (TCPP); 82–92% (TDCPP); 71–81% (TBEP); 71–79% (TPhP); 65–72% (EHDPP); 63–67% (TEHP); 73–83% (TCP)	(Naccarato et al. 2017)
	River water (50 mL)									2.9 ng/L (TPP); 10 ng/L (TBP); 12 ng/L (TCEP); 13 ng/L (TCPP); 24 ng/L (TDCPP); 95 ng/L (TBEP); 12 ng/L (TPhP); 24 ng/L (EHDPP); 28 ng/L (TEHP); 97 ng/L (TCP)		81–87% (TPP); 76–88% (TBP); 68–92% (TCEP); 67–79% (TCPP); 72–78% (TDCPP); 67–77% (TBEP); 74–80% (TPhP); 59–68% (EHDPP); 59–64% (TEHP); 71–81% (TCP)	
	Wastewater (50 mL)									3.0 ng/L (TPP); 12 ng/L (TBP); 12 ng/L (TCEP); 13 ng/L (TCPP); 25 ng/L (TDCPP); 101 ng/L (TBEP); 13 ng/L (TPhP); 28 ng/L (EHDPP); 28 ng/L (TEHP); 107 ng/L (TCP)		68–72% (TPP); 62–87% (TBP); 64–78% (TCEP); 61–73% (TCPP); 71–82% (TDCPP); 63–66% (TBEP); 66–68% (TPhP); 62–71% (EHDPP); 58–63% (TEHP); 67–78% (TCP)	

GC (gas chromatography); HPLC (high-performance liquid chromatography); LOD (limit of detection); LOQ (limit of quantification); MS (mass spectrometry); MS/MS (tandem mass spectrometry); UV (ultraviolet).

TABLE 5.6
MEPS procedures in food toxicology, analytical instrumentation, limits, and recoveries (2015–2020)

Analytes	Sample (Volume/Weight)	Sample Preparation	MEPS Sorbent	MEPS Steps					Analytical Instrumentation	LOD	LOQ	Recoveries	Ref
				Conditioning	Load	Wash	Elution	Sorbent Re-use					
Acenaphthylene (1) Acenaphthene (2) Fluorene (3) Phenanthrene (4) Anthracene (5) Fluoranthene (6) Pyrene (7) Benz[a]anthracene (8) Chrysene (9) Benzo[b]fluoranthene (10) Benzo[a]pyrene (11) Indeno[1,2,3-cd]pyrene (12) Dibenz[ah]anthracene (13) Benzo[ghi]perylene (14)	Apple (20 g)	Skin and pulp were crushed in a blender before the extraction procedure; 25 mL ethanol; water bath and sonication at 30 °C and 35 kHz for 15 min; 1.2 mL was used for MEPS	HyperSep Retain polar-enhanced polymer (PEP) consisting of a styrene divinylbenzene copolymer (PS/DVB) modified with urea functional groups (4 mg); manual	n.s.	(6 × 200 µL)	n.s.	Methylene chloride (1 × 50 µL)	n.s.	GC-MS	0.036 µg/Kg (1) 0.048 µg/Kg (2) 0.044 µg/Kg (3) 0.044 µg/Kg (4) 0.045 µg/Kg (5) 0.031 µg/Kg (6) 0.038 µg/Kg (7) 0.059 µg/Kg (8) 0.062 µg/Kg (9) 0.073 µg/Kg (10) 0.087 µg/Kg (11) 0.130 µg/Kg (12) 0.125 µg/Kg (13) 0.124 µg/Kg (14)	0.120 µg/Kg (1) 0.160 µg/Kg (2) 0.147 µg/Kg (3) 0.146 µg/Kg (4) 0.149 µg/Kg (5) 0.135 µg/Kg (6) 0.128 µg/Kg (7) 0.195 µg/Kg (8) 0.206 µg/Kg (9) 0.242 µg/Kg (10) 0.291 µg/Kg (11) 0.433 µg/Kg (12) 0.418 µg/Kg (13) 0.414 µg/Kg (14)	17% (1) 18% (2) 25% (3) 76% (4) 31% (5) 75% (6) 46% (7) 51% (8) 43% (9) 48% (10) 56% (11) 92% (12) 106% (13) 75% (14)	(Paris et al. 2019)
Marbofloxacin (M), Ciprofloxacin (C), Enrofloxacin €	Bovine Serum (0.225 mL)	Dilution 1:2 with a saturated ammonium sulphate solution and centrifugation	C_8 (n.s.); eVol®	Acetonitrile and water (4 × 50 µL)	(5 × 50 µL)	Water (1 × 50 µL)	0.4% formic acid and acetonitrile (50:50, v/v) (1 × 50 µL)	Acetonitrile (10 × 50 µL)	UHPLC-DAD	13 ng/mL (M) 6 ng/mL (C) 10 ng/mL (E)	43 ng/mL (M) 20 ng/mL (C) 32 ng/mL (E)	83–84% (M) 83–86% (C) 80–81% (E)	(Aresta et al. 2019)
	Bovine Milk (0.225 mL)	Dilution 1:2 with a saturated ammonium sulphate solution and centrifugation								10 ng/mL (M) 10 ng/mL (C) 15 ng/mL (E)	33 ng/mL (M) 32 ng/mL (C) 48 ng/mL (E)	79–80% (M) 83–86% (C) 79–80% (E)	
	Bovine Urine (0.225 mL)	Centrifugation								12 ng/mL (M) 2 ng/mL (C) 5 ng/mL (E)	43 ng/mL (M) 7 ng/mL (C) 17 ng/mL (E)	86–86% (M) 88–89% (C) 84–85% (E)	
BDE-28 BDE-47 BDE-99 BDE-100 BDE-153 BDE-154	Egg samples (50 mg of lyophilized)	4 mL hexane, 15 min ultrasound; 2 mL of concentrated sulphuric acid to the supernatant, 15 min in ultrasound; separation of the organic phase, evaporation and reconstitution with 500 µL of acetonitrile	C_{18} (n.s.); eVol®	Acetonitrile (1 × 100 µL)	(4 × 100 µL)	n.s.	Isooctane (n.s)	Isooctane (2 × 100 µL)	GC-MS	0.42 ng/g lw	1.40 ng/g lw	87–110%	(Souza et al. 2019)

Analytes	Matrix	Sorbent	Sample pretreatment	Loading solvent	Washing solvent	Elution solvent	Analytical technique	LOD	LOQ	Recovery	Reference
5-hydroxymethyl-2-furfural (5HMF), 2-furfural (F), 2-furyl methyl ketone (FMK), 5-methyl-2-furfural (5 MF)	Wine (0.2 mL)	C_8 (n.s.); n.s.	n.s.			Methanol (1×200 μL)	Methanol and 0.1% formic acid (1×250 μL)	UHPLC-DAD	n.s.	n.s.	n.s.	(Perestrelo et al. 2017)
Carbendazim (1) Thiabendazole (2) Dimethoate (3) Carbofuran (4) Acetamiprid (5) Tricyclazole (6) Pirimicarb (7) Aldicarb (8) Imazalil (9) Dichlorvos (10) Propoxur (11) Carbofuran (12) Malaoxon (13) Carbaryl (14) Fosthiazate (15) Isoprocarb (16) Methacrifos (17) Methidathion (18) Methiocarb (19) Malathion (20) Primiphos methyl (21) Coumaphos (22) Chlorpyrifos methyl (23) Benfuracarb (24) Chlorpyrifos ethyl (25)	Wheat flour (200 mg)	HDVB (n.s.); n.s.	375 μL of 10 mM acetonitrile-acetate buffer pH = 5. 60:40 v/v; sonication for 5 min; incubation at 40 °C for 5 min under constant stirring; centrifugation	Methanol (3×100 μL) and 90:10 (v/v) water-acetonitrile solution (2×100 μL)	Water (1×100 μL)	Acetonitrile (3×100 μL)	n.s.	UHPLC-MS/MS	3×10^{-3} mg/kg (1) 3×10^{-3} mg/kg (2) 1×10^{-3} mg/kg (3) 3×10^{-3} mg/kg (4) 3×10^{-3} mg/kg (5) 5×10^{-4} mg/kg (6) 1×10^{-3} mg/kg (7) 5×10^{-4} mg/kg (8) 2×10^{-3} mg/kg (9) 3×10^{-4} mg/kg (10) 1×10^{-3} mg/kg (11) 3×10^{-4} mg/kg (12) 1×10^{-3} mg/kg (13) 3×10^{-3} mg/kg (14) 5×10^{-4} mg/kg (15) 5×10^{-3} mg/kg (16) 5×10^{-3} mg/kg (17) 5×10^{-3} mg/kg (18) 5×10^{-4} mg/kg (19) 5×10^{-3} mg/kg (20) 5×10^{-3} mg/kg (21) 3×10^{-4} mg/kg (22) 1×10^{-3} mg/kg (23) 5×10^{-3} mg/kg (24) 1×10^{-3} mg/kg (25)	7.5×10^{-3} mg/kg (1) 7.5×10^{-3} mg/kg (2) 3×10^{-3} mg/kg (3) 7.5×10^{-3} mg/kg (4) 7.5×10^{-3} mg/kg (5) 1.5×10^{-3} mg/kg (6) 3×10^{-3} mg/kg (7) 1.5×10^{-3} mg/kg (8) 5×10^{-3} mg/kg (9) 9×10^{-4} mg/kg (10) 2.5×10^{-3} mg/kg (11) 7.5×10^{-4} mg/kg (12) 3×10^{-3} mg/kg (13) 7.5×10^{-3} mg/kg (14) 1.5×10^{-3} mg/kg (15) 1.5×10^{-3} mg/kg (16) 1.5×10^{-2} mg/kg (17) 1.5×10^{-2} mg/kg (18) 1.5×10^{-3} mg/kg (19) 1.5×10^{-2} mg/kg (20) 1.5×10^{-2} mg/kg (21) 9×10^{-4} mg/kg (22) 2.5×10^{-3} mg/kg (23) 1.5×10^{-2} mg/kg (24) 3×10^{-3} mg/kg (25)	45% (1) 44% (2) 60% (3) 47% (4) 81% (5) 70% (6) 72% (7) 68% (8) 61% (9) 71%(10) 86%(11) 79% (12) 76% (13) 82%(14) 62%(15) 64%(16) 95%(17) 85% (18) 98% (19) 87% (20) 92% (21) 97% (22) 43% (23) 57% (24) 19% (25)	(Di Ottavio et al. 2017)
Diethyl phthalate (DEP), Dipropyl phthalate (DPP), Dibutyl phthalate (DBP), Benzylbutyl phthalate (BBP), Dicyclohexyl phthalate (DCHP)	Cold drinks (1 mL)	C_{18} (4 mg); n.s.	Degasification in ultrasonic bath; dilution 20 times with methanol; filtration	Methanol and water (1×100 μL)	Water (10×50 μL)	Methanol (1×30 μL)	Methanol and water (3×100 μL)	GC-MS	0.012 ng/mL (DEP) 0.008 ng/mL (DPP) 0.003 ng/mL (DBP) 0.015 ng/mL (BBP) 0.011 ng/mL (DCHP)	0.039 ng/mL (DEP) 0.026 ng/mL (DPP) 0.009 ng/mL (DBP) 0.049 ng/mL (BBP) 0.036 ng/mL (DCHP)	94–99% (DEP) 92–96% (DPP) 96–99% (DBP) 94–96% (BBP) 94–97% (DCHP)	(Kaur et al. 2016)
Ochratoxin A (OTA)	Wine (5 mL)	C_{18} (4mg); eVol®	Dilution 1:4 and 1:2 (v/v) with 2% aqueous acetic acid	Acetonitrile, methanol and 2% aqueous acetic acid/ ethanol (88:12, v/v) (1×50 μL)	2% aqueous acetic acid and 2% aqueous acetic acid/methanol (60/40 v/v) (1×20 μL)	Acetonitrile/ 2% aqueous acetic acid (90/10, v/v) (2×25 μL)	Acetonitrile/ 2% aqueous acetic acid (90/10, v/v) (3×50 μL)	HPLC-FLD	0.09 ng/mL	0.28 ng/mL	90%	(Savastano et al. 2016)

(Continued)

TABLE 5.6 (Continued)
MEPS procedures in food toxicology, analytical instrumentation, limits, and recoveries (2015–2020)

Analytes	Sample (Volume/Weight)	Sample Preparation	MEPS Sorbent	MEPS Steps				Analytical Instrumentation	LOD	LOQ	Recoveries	Ref
				Conditioning	Load	Wash	Elution	Sorbent Re-use				
Tebuthiuron (TEB), Carbofuran (CAR), Atrazine (ATR), Metribuzine (MET), Ametryn (AME), Bifenthrin (BIF)	Sugarcane juices (n.s.)	Centrifugation	C$_{18}$ (n.s.); n.s.	Ethyl acetate (4 × 250 μL) and water (3 × 250 μL)	(9 × 250 μL)	Water (6 × 100 μL)	Ethyl acetate (8 × 30 μL)	n.s.	GC-MS	1.5 ng/mL (TEB) 0.3 ng/mL (CAR) 0.3 ng/mL (ATR) 0.8 ng/mL (MET) 0.2 ng/mL (AME) 0.3 ng/mL (BIF)	10 ng/mL (TEB) 2 ng/mL (CAR) 2 ng/mL (ATR) 3 ng/mL (MET) 2 ng/mL (AME) 2 ng/mL (BIF)	44% (TEB) 65% (CAR) 49% (ATR) 60% (MET) 45% (AME) 27% (BIF)	(Fiunes et al. 2016)
PCB 28 PCB 52 PCB 101 PCB 118 PCB 138 PCB 153 PCB 180	Bovine Serum (0.1 mL)	Protein precipitation; centrifugation; dilution 30 times	C$_{18}$ (2 mg); manual	Methanol and water (1 × 100 μL)	(2 × 100 μL)	20% acetonitrile (1 × 100 μL)	Ethyl acetate (1 × 50 μL)	Ethyl acetate (8 × 50 μL)	GC-MS	0.06 ng/mL (PCB 28) 0.11 ng/mL (PCB 52) 0.28 ng/mL (PCB 101) 0.36 ng/mL (PCB 118) 0.53 ng/mL (PCB 138) 0.38 ng/mL (PCB 153) 0.49 ng/mL (PCB 180)	0.20 ng/mL (PCB 28) 0.37 ng/mL (PCB 52) 0.93 ng/mL (PCB 101) 1.20 ng/mL (PCB 118) 1.77 ng/mL (PCB 138) 1.27 ng/mL (PCB 153) 1.63 ng/mL (PCB 180)	87–89% (PCB 28) 79–91% (PCB 52) 66–88% (PCB 101) 60–82% (PCB 118) 69–72% (PCB 138) 61–65% (PCB 153) 66–74% (PCB 180)	(Yang et al. 2016)
5-hydroxymethyl-2-furfural (5HMF), 2-furfural (F), 2-furyl methyl ketone (FMK), 5-methyl-2-furfural (5 MF)	Wine (0.2 mL)	n.s.	C$_{18}$ (4 mg); eVol®	Methanol and 0.1% formic acid (1 × 250 μL)	(3 × 200 μL)	0.1% formic acid (1 × 100 μL)	Methanol (1 × 200 μL)	Methanol and 0.1% formic acid (1 × 250 μL)	UHPLC-DAD	129.3 ng/L (5HMF) 25.8 ng/L (F) 4.5 ng/L (FMK) 21.3 ng/L (5 MF)	431.0 ng/L (5HMF) 86.1 ng/L (F) 14.9 ng/L (FMK) 70.8 ng/L (5 MF)	86–99% (5HMF) 84–98% (F) 74–96% (FMK) 90–99% (5 MF)	(Perestrelo et al. 2015)

DAD (diode array detector); FLD (fluorescence detector); GC (gas chromatography); HPLC (high-performance liquid chromatography); LC (liquid chromatography); LOD (limit of detection); LOQ (limit of quantification); MS (mass spectrometry); MS/MS (tandem mass spectrometry); UHPLC (ultra high-performance liquid chromatography).

5.5 New Developments

The interest in new materials to be used in sample preparation is not new, and this aims at obtaining greater specificity and selective enrichment (da Silva and Lanças 2020). Finding the suitable sorbent to extract compounds that present different polarities (high polarity and non-polar) can be a big challenge (Mehrani et al. 2020). For this reason, different strategies have been developed to try to solve this problem, and several new solid pack materials have been reported in the last five years (da Silva and Lanças 2020; Mehrani et al. 2020).

Carbon nanomaterials are by far those for which greater interest was observed in the development of new sorbents for MEPS. This is justified by their unique physical and chemical properties, namely the large specific surface area, chemical and thermal stabilities, and excellent mechanical strength (Amiri and Ghaemi 2017). Graphene (G) is a two-dimensional carbon nanomaterial widely applied as sorbent, exhibiting a π-electron-rich structure, allowing strong hydrophobic and π-stacking interactions with many molecules (Sun et al. 2019). Nevertheless, the direct use of graphene as a sorbent is not practical since its large surface area may lead to irreversible binding caused by van der Waals interactions (Sun et al., 2019). This will generate a large backpressure during MEPS and may lead to syringe obstruction (Vasconcelos Soares Maciel et al. 2018; Sun et al. 2019). Furthermore, graphene oxide (GO), a precursor of graphene, presents many polar groups in its chemical structure and can be modified with other materials resulting in improved selectivity and better analyte recovery (Karimiyan et al. 2019; da Silva and Lanças 2020).

Ahmadi et al. (Ahmadi et al. 2018) used GO as MEPS sorbent for the extraction of local anesthetics from plasma and saliva. The authors justified the successful application of the sorbent with its high adsorption capacity for aromatic compounds (Ahmadi et al. 2018). On the other hand, Sun et al. (Sun et al. 2019) developed a sorbent consisting of GO coated with ZnO (GO–ZnO) for the extraction of carbamate pesticides from juice samples. This coating not only prevented graphene aggregation, but also provided hydrophilic surfaces for effective adsorption of water-soluble analytes (Ahmadi et al. 2018). Another way of preventing this problem with graphene was adopted by Vasconcelos et al. (Vasconcelos Soares Maciel et al. 2018), who bonded the GO onto a silica surface with its subsequent transformation to reduced graphene (G-Sil). With this sorbent, the authors improved the extraction of tetracyclines residues from milk samples (Vasconcelos Soares Maciel et al. 2018). Similar development was reported by Fumes et al. (Fumes and Lanças 2017), but using supported graphene on aminopropyl silica for the extraction of parabens from water samples. A different strategy was presented by Karimiyan et al. (Karimiyan et al. 2019), who used polyacrylonitrile/graphene oxide (PAN/GO) nanofibers, and successfully applied them for the pre-concentration of several drugs and metabolites from human plasma samples. It was also shown that ionic liquids (ILs) could be used for the extraction of chlorobenzenes (CBs), chlorophenols (CPs), and bromophenols (BPs) from water samples (Darvishnejad and Ebrahimzadeh 2020). These analytes are environmentally disrupting chemicals, and their pre-concentration was accomplished with a graphitic carbon nitride-reinforced polymer IL nanocomposite, a MEPS sorbent developed by Darvishnejad and Ebrahimzadeh (Darvishnejad and Ebrahimzadeh 2020). Recently, new composite graphitic materials have been made commercially available (CarbonX®) and are produced by coating stable substrates with graphene; these materials have been successfully applied to extract β-blockers from human plasma samples (Abuzooda et al. 2015). Further, a new type of graphitic sorbent (Carbon X-COA) was evaluated for the extraction of the local anesthetics lidocaine and ropivacain from plasma samples (Iadaresta et al. 2015).

Also widely explored, although not that novel, are MIPs. MIPs are provided, stereochemically, with specific recognition sites that are either shaped from a template molecule, such as the target analyte, or from dummy template molecules, such as analytes analogues (de Oliveira 2019). These have the advantage of a high recognition ability for the target analytes, to which the extraction becomes very selective (Meng and Wang 2019). Over the last five years many MIPs have been synthetized for MEPS application. Their synthesis commonly occurs by a complex formation between the functional monomer and template molecule (de Oliveira et al. 2019). Oliveira et al. (de Oliveira et al. 2019) employed a new restricted-access MIP for the determination of estrone and estriol in urine samples based on a crosslinking reaction with BSA to obtain surface protein encapsulation of the MIP. In the same year, Meng et al. (Meng and Wang 2019) proposed the use of MIPs for the determination of levofloxacin from plasma samples, using deep eutectic solvents (DESs) as porogen for MIPs preparation to be applied on MEPS syringe. The DESs choice was based on its non-toxic, low cost, and inertness properties. Earlier, Soleimani et al. (Soleimani et al. 2018) reported the use of MIPs as MEPS sorbents for the pre-concentration of mandelic acid from urine samples. The same authors had previously reported MIPs' successful application to extract trans, trans-muconic acid from the same specimen (Soleimani et al. 2017). A different approach of MIPs was, however, developed by Moein et al. (Moein et al. 2015a). These authors used the dummy molecularly imprinted polymer (DMIP) method and obtained good results with its application for sarcosine extraction from both plasma and urine samples (Moein et al. 2015a).

Conducting polymers (π-conjugated polymers), such as polythiophene, polyaniline, and polypyrrole, are also considered promising sorbent materials to be used in MEPS (Florez et al. 2020; Abolghasemi et al. 2018). They present good environmental stability and nontoxicity and are easy to prepare with low cost (Florez et al. 2020). One of the most studied materials is polythiophene (PTh), gathering qualities as hydrophilic stability, redox activity, and an excellent interaction with aromatic groups (Florez et al. 2020). Florez et al. (Florez et al. 2020) reported PTh as an highly efficient sorbent for MEPS, and used it for the pre-concentration of steroids from bovine milk samples. Previously, Abolghasemi et al. (Abolghasemi et al. 2018) reported a nanostructured star-shaped polythiophene dendrimer as an highly efficient sorbent to extract clofentezine from milk and juice samples. The authors claimed that star-shaped and dendritic conductive polymers are great options due to their unique three-dimensional shape and physicochemical properties (Abolghasemi et al. 2018). In addition to the previous, the development of a nanocomposite consisting of polydopamine, silver nanoparticles, and polypyrrole has been described with great application for the microextraction of antidepressant drugs from urine samples (Bagheri et al. 2016).

Nanoclays are promising sorbent materials as well. Although their hydrophilic nature might turn them unsuitable for the extraction of organic compounds, methods such as cation-exchange reactions with alkyl ammonium, phosphonium, and/or imidazolium compounds may change this (Saraji et al. 2018). Montmorillonite (nanoclay) presents an elevated adsorption capacity, surface area, porosity, and swelling behavior (Saraji et al. 2018). Saraji et al. (Saraji et al. 2018) modified nanoclays with cetyltrimethylammonium bromide (CTAB) using a cation exchange reaction, with further modification by alkoxysilanes, and used it as MEPS sorbent to extract diazinon from water samples. More recently, a reinforced montmorillonite into polystyrene (MMT/PS) was prepared and coated onto cellulose filter paper to pre-concentrate fluoxetine from similar environmental samples (Matin et al. 2020).

Other sorbent materials with great potential due to their unique properties are metal-organic frameworks (MOFs) (Jiang et al. 2020). These consist of porous crystal material generated by the self-assembly of metallic ions (or clusters) with a bi- or multipodal organic linker (Jiang et al. 2020). Although MOFs have shown some drawbacks related to SPE applications, producing high resistance because of their sub-micron to micron size, their unique features enable them to be used in small amounts in MEPS (Jiang et al. 2020). Jiang et al. (Jiang et al. 2020) used a MOF to extract parabens from vegetable oils and obtained satisfactory adsorption capacities. Previously, Jiang et al. (Jiang et al. 2018) had already applied a MOF-MIL-101 (Cr) for semi-automated MEPS of six triazine herbicides from corn samples. Among the reported MOFs, MOF-5 is one of the most studied, and this was coated by amino-functionalized Fe_3O_4 and silica mesoporous (SBA-15) and used as MEPS sorbent to determine mandelic acid in urine samples for the first time by Rahimpoor et al. (Rahimpoor et al. 2019). More recently, the same research team successfully applied a MOF of MIL-53-NH_2 (Al) as MEPS sorbent to pre-concentrate urinary methylhippuric acids (Pirmohammadi et al. 2020).

The latest research on sorbent material applied to MEPS has been boosted by the use of natural compounds, hence called green sorbents. Rasolzadeh et al. (Rasolzadeh 2019) described the use of a biosorbent consisting of *Chlorella vulgaris*, a unicellular green microalgae, for the determination of

TABLE 5.7

New sorbents developed for MEPS, procedures applied in analytical toxicology, analytical instrumentation, limits of determination, and recoveries (2015–2020)

Analytes	Sample (Volume/Weight)	Sample Preparation	MEPS Sorbent	MEPS Steps					Analytical Instrumentation	LOD	LOQ	Recoveries	Ref
				Conditioning	Load	Wash	Elution	Sorbent re-use					
Clinical Toxicology													
Benzoic acid (BA), Phenylpropionic acid (PhPA), CinnamicPhenyllactic acid (PhLA), 4-hydroxybenzoic acid (p-HBA), 4-hydroxyphenylaceticacid (p-HPhAA), 4-hydroxyphenylpropionic acid (p-HPhPA), Homovanillic acid (HVA), 4-hydroxyphenyllactic acid (p-HPhLA)	Serum (80 μL)	Dilution with 70 μL of water and 2.5 μL of concentrated sulphuric acid	Hypercrosslinked polystyrene (HCLPS) (2 mg); n.s.	Acetone and water (2 × 80 μL)	(2 × 80 μL)	0.3 mM formic acid solution (1 × 80 μL)	Diethyl ether (3 × 80 μL)	n.s.	GC-MS	n.s.	0.5 μM	45% (BA) 50% (PhPA) 35% Cinnamic 70% (PhLA) 45% (p-HBA) 45% (p-HPhAA) 40% (p-HPhPA) 40% (HVA) 20% (p-HPhLA)	Pautova 2020c
2-methylhippuric acid (2-MHA), 3-methylhippuric acid (3-MHA), 4-methylhippuric acid (4-MHA)	Urine (n.s.)	Adjusted to pH 2.0	Metal-organic frameworks (MOFs); MIL-53-NH2 (Al) (4 mg);	Methanol-water (1:1, v/v); (3 × 100 μL)	(4 × 100 μL)	Methanol-water (1:1, v/v); (1 × 150 μL)	Acetic acid-methanol 1:9 (4 × 80 μL)	n.s.	HPLC-UV	0.115 μg/mL (2-MHA) 0.005 μg/mL (3-MHA) 0.005 μg/mL (4-MHA)	0.380 μg/mL (2-MHA) 0.016 μg/mL (3-MHA) 0.016 μg/mL (4-MHA)	61–98% (2-MHA) 63–99% (3-MHA) 63–99% (4-MHA)	Pirmohammadi et al. (2020)
Estrone (E1), Estriol (E3)	Urine (0.2 mL)	10 mL urine added to 150 μL of hydrochloric acid (1.0 M); 1 h at 65 °C, then the pH was adjusted to 10; centrifugation	Restricted molecularly imprinted polymer (RAM-MIP) (3 mg); n.s.	water (1 × 100 μL)	(1 × 200 μL)	Water (1 × 200 μL)	Methanol: acetic acid (9:1, v/v) (1 × 200 μL)	n.s.	HPLC-UV	n.s.	100 ng/mL (E1) 100 ng/mL (E3)	74–84% (E1) 67–70% (E3)	de Oliveira et al. (2019)
Mandelic acid (MA)	Urine (0.15 mL)	Centrifugation	MOF-5 @ Fe3O4-NH2 (n.s.); n.s. MOF-5 @ SBA-15 (n.s.); n.s.	n.s.	(4 × 100 μL)	Water (1 × 100 μL)	Methanol-nitric acid 8:2 (4 × 80 μL)	n.s.	HPLC-DAD	0.05 μg/mL	0.1 μg/mL	95% 90%	Rahimpoor et al. (2019)

(Continued)

TABLE 5.7 (Continued)

New sorbents developed for MEPS, procedures applied in analytical toxicology, analytical instrumentation, limits of determination, and recoveries (2015–2020)

Analytes	Sample (Volume/Weight)	Sample Preparation	MEPS Sorbent	MEPS Steps - Conditioning	Load	Wash	Elution	Sorbent re-use	Analytical Instrumentation	LOD	LOQ	Recoveries	Ref
Mandelic acid (MA)	Urine (n.s.)	Acidified and adjusted to pH 2.0	MIP (4 mg); manual	Methanol and water (3 × 100 μL)	(8 × 100 μL)	Water (1 × 100 μL)	Methanol–acetic acid (8:2, v/v) (2 × 100 μL)	Methanol–acetic acid (8:2, v/v) and water (4 × 150 μL)	HPLC-UV	n.s.	0.2 μg/mL	92%	Soleimani et al. (2018)
trans,trans-Muconic acid (tt-MA)	Urine (n.s.)	pH 2.0, centrifugation	MIP (4 mg); manual	Ethanol and water (3 × 100 μL)	(5 × 100 μL)	Water (1 × 100 μL)	Ethanol–acetic acid (80:20, v/v) (2 × 100 μL)	Ethanol–acetic acid (8:2, v/v) and water (3 × 150 μL)	HPLC-UV	0.015 μg/mL	0.05 μg/mL	90–92%	Soleimani et al. (2017)
Sarcosine	Plasma and urine (0.1 mL)	n.s.	Dummy molecularly imprinted polymer (DMIP) (2 mg); automated	Water (1 × 100 μL)	(200 μL n.s.)	Water/HCl (0.1 M) (80:20) (1 × 100 μL)	Acetonitrile/Water (80:20) (1 × 100 μL)	Water/HCl (0.1 M) (80:20) (200 μL n.s.)	LC-MS/MS	1 ng/mL	3 ng/mL	87–89%	Moein et al. (2015a)
							Drug monitoring						
Chlorpromazine (CLOR), Clozapine (CLOZ), Olanzapine (OLA), Quetiapine (QUET)	Plasma (0.1 mL)	Dilution with 400 μL of borate buffer solution (10 mM, pH 9)	Restricted access carbon nanotube (RACNT) (n.s.); manual	Acetonitrile and water (2 × 100 μL)	(3 × 100 μL)	Water (1 × 150 μL)	Acetonitrile (2 × 100 μL)	Acetonitrile and water (2 × 100 μL)	UHPLC-MS/MS	n.s.	10 ng/mL (CLOR) 10 ng/mL (CLOZ) 10 ng/mL (OLA) 10 ng/mL (QUET)	34% (CLOR) 69% (CLOZ) 58% (OLA) 28% (QUET)	Cruz et al. (2020)
o-Toluidine (TOL), Prilocaine (PRL), 2,6-Xylidine (XYL), Lidocaine (LID)	Plasma (0.2 mL)	Dilution (1:5) (n.s.)	Polyacrylonitrile/Graphene Oxide (PAN/GO)	1% formic acid in methanol (5 × 100 μL) and water (3 × 100 μL)	(5 × 100 μL)	5% methanol in water (2 × 100 μL)	1% formic acid in methanol (3 × 100 μL)	1% formic acid in methanol (5 × 100 μL) and 1% formic acid in water (3 × 100 μL)	LC-MS/MS	1.25 nmol/L (TOL) 0.50 nmol/L (PRL) 2.50 nmol/L (XYL) 0.25 nmol/L (LID)	10 nmol/L (TOL) 2 nmol/L (PRL) 10 nmol/L (XYL) 2 nmol/L (LID)	69% (TOL) 96% (PRL) n.s. (XYL) 93% (LID)	Karimiyan et al. (2019)
Levofloxacin	Plasma (0.1 mL)	Addition of 300 μL of methanol; vortexed; centrifugation; supernatant was filtered	Deep eutectic solvents – molecularly imprinted polymers (DESs-MIPs) (4 mg); n.s.	Methanol and water (1 × 1 mL)	(20 × 400 μL)	Water-methanol (50:50, v/v) (1 × 200 μL)	Acetonitrile/ammonia (95/5, v/v) (1 × 400 μL)	n.s.	UHPLC-DAD	0.012 μg/mL	0.04 μg/mL	95–100%	Meng and Wang (2019)
Nitrofurantoin (NFT)	Urine (0.3 mL)	Filtration; pH 8	Dried *C. vulgaris* biomass (4 mg); n.s.	Water (2 × 100 μL)	(14 × 300 μL)	Water (1 × 100 μL)	30% acetone: water (6 × 150 μL)	30% acetone: water (3 × 100 μL)	Spectrophotometry	0.039 μg/mL	0.5 μg/mL	98%	Rasolzadeh et al. (2019)

Analytes	Matrix	Sample pretreatment	Sorbent	Conditioning	Loading	Washing	Elution	Reconstitution	Technique	LOD	LOQ	Recovery	Reference
Lidocaine (LID), Prilocaine (PRL), Ropivacaine (ROP)	Saliva (0.2 mL); Plasma (0.2 mL)	Dilution with 200 µL water; protein precipitation; centrifugation	Reduced graphene oxide (RGO) (2 mg); n.s.	Methanol and water (1×200 µL)	(6×200 µL)	Water: methanol (95:5, %v/v) (1×200 µL)	Methanol: formic acid (90:10, %v/v) (2×100 µL)	n.s.	LC-MS/MS	n.s.	4 nM (LID), 4 nM (PRL), 2 nM (ROP)	97-106% (LID), 95-106% (PRL), 99-106% (ROP); 99-105% (LID), 97-107% (PRL), 98-100% (ROP)	Ahmadi et al. (2018)
Dexamethasone (DEX), Carbamazepine (CBZ), Naproxen (NPX)	Urine (1 mL)	1 mL urine diluted to 5 mL with water; adjusted at pH 7.5	Imprinted interpenetrating polymer network (IPN) (2 mg); n.s.	Methanol and water (3 mL)	(20×1 mL)	n.s.	Methanol (230 µL)	Methanol (4×250 µL) and water (6×250 µL)	HPLC-UV	1.3 ng/mL (DEX), 1.5 ng/mL (CBZ), 1.4 ng/mL (NPX)	4.2 ng/mL (DEX), 5.0 ng/mL (CBZ), 4.7 ng/mL (NPX)	83% (DEX), 91% (CBZ), 89% (NPX)	Asgari et al. (2017)
Amitriptyline (AMP), Imipramine (IMP), Citalopram (CIT)	Urine (1 mL)	Dilution with 4 mL water; pH 5	Nanocomposite consisting of polydopamine, silver nanoparticles and polypyrrole (PDA-Ag-PPy) (2 mg); n.s.	Methanol, acetonitrile and water (1 mL)	(25×1 mL)	Water (1×1 mL)	Acetonitrile (3×100 µL)	Acetonitrile and water (5×250 µL)	GC-MS	0.03 ng/mL (AMP), 0.05 ng/mL (IMP), 0.05 ng/mL (CIT)	0.10 ng/mL (AMP), 0.20 ng/mL (IMP), 0.15 ng/mL (CIT)	91-104% (AMP), 88-96% (IMP), 88-94% (CIT)	Bagheri et al. (2016)
Haloperidol (HAL), Olanzapine (OLZ), Clonazepam (CLN), Mirtazapine (MTZ), Paroxetine (PXT), Citalopram (CIT), Sertraline (SRT), Chlorpromazine (CHP), Imipramine (IMP), Clomipramine (CLO), Quetiapine (QTP), Diazepam (DZP), Fluoxetine (FLX), Clozapine (CLZ), Carbamazepine (CBZ), Lamotrigine (LMT)	Plasma (0.2 mL)	Vortexed, centrifugation, the supernatant was diluted with 300 µL of ammonium acetate (5 mM)	Hybrid silica monolith (n.s.); n.s.	Methanol/acetonitrile mixture (50:50 v/v) and water (4×200 µL)	(4×100 µL)	Water (1×150 µL)	Methanol/acetonitrile (50:50 v/v) (1×100 µL)	Methanol/acetonitrile mixture (50:50 v/v) and water (4×200 µL)	LC-MS/MS	n.s.	0.05 ng/mL (HAL), 0.05 ng/mL (OLZ), 0.10 ng/mL (CLN), 0.05 ng/mL (MTZ), 0.05 ng/mL (PXT), 1 ng/mL (CIT), 0.05 ng/mL (SRT), 0.10 ng/mL (CHP), 0.05 ng/mL (IMP), 0.10 ng/mL (CLO), 0.05 ng/mL (QTP), 0.05 ng/mL (DZP), 0.05 ng/mL (FLX), 0.05 ng/mL (CLZ), 0.10 ng/mL (CBZ), 1 ng/mL (LMT)	n.s.	de Souza et al. (2015)
Metoprolol (MET), Acebutolol (ACE)	Plasma (0.5 mL)	Centrifugation; dilution by water 4 times	New graphitic material (Carbon-XCOS) (2 mg); n.s.	n.s.	n.s.	Water (1×250 µL)	0.1% formic acid in methanol (n.s.)	n.s.	LC-MS/MS	n.s.	10 nM	80-90% (all)	Ahuzooda et al. (2015)
Lidocaine (LID), Ropivacaine (ROP)	Plasma (0.2 mL)	Dilution with 0.1% formic acid in water (0.8 mL); centrifugation	Graphitized carbon (CarbonX®COA) (2 mg); n.s.	0.1% formic acid in acetonitrile and water (5×250 µL)	(4×250 µL)	Water (2×200 µL)	0.1% formic acid in acetonitrile (1×200 µL)	0.1% formic acid in acetonitrile and water (5×250 µL)	LC-MS/MS	1 nM	5 nM	79-82% (all)	Iadaresta et al. (2015)

(Continued)

TABLE 5.7 (Continued)

New sorbents developed for MEPS, procedures applied in analytical toxicology, analytical instrumentation, limits of determination, and recoveries (2015–2020)

Analytes	Sample (Volume/Weight)	Sample Preparation	MEPS Sorbent	MEPS Steps					Analytical Instrumentation	LOD	LOQ	Recoveries	Ref
				Conditioning	Load	Wash	Elution	Sorbent re-use					
Environmental toxicology													
Naphthalene (Naph), Anthracene (A), Acenaphthene (Ace), Phenanthrene (Phe), 2,4-dichlorophenoxyacetic acid (2,4-D), 2-methyl-4-chlorophenoxyacetic acid (MCPA)	Environmental, farm, and industrial water samples (n.s.)	n.s.	Rosin/PAN and aloin/PAN electrospun nanofibers (n.s.); n.s.	2-propanol:methanol (50:50) and water (5 mL)	MEPS Hyphenation 12 adsorption cycles (12 min for 20 mL of sample solution)	Water (2 mL)	3 min for 600 µL of 2-propanol: methanol (50:50)	n.s	GC-FID	0.1–0.3 ng/mL (Naph), 0.1–0.3 ng/mL (A), 0.1–0.3 ng/mL (Ace), 0.1–0.3 ng/mL (Phe) 0.3–0.5 ng/mL (2,4-D) 0.3–0.5 ng/mL (MCPA)	0.33–0.99 ng/mL (Naph), 0.33–0.99 ng/mL (A), 0.33–0.99 ng/mL (Ace), 0.33–0.99 ng/mL (Phe) 0.99–1.65 ng/mL (2,4-D) 0.99–1.65 ng/mL (MCPA)	85–96% (Naph) 85–96% (A), 85–96% (Ace), 85–96% (Phe) 86–98% (2,4-D) 86–98% (MCPA)	Mehrani et al. (2020)
4-Nitrophenol (1) 2-Bromophenol (2) 2-Chlorophenol (3) 1,4-Dichlorobenzen (4) Pentachlorophenol (5) 1,2,3-Trichlorobenzen (6)	Tap water, river water, well water and a sample of cave water (10 mL)	Filtration	Graphitic carbon nitride/polymer ionic liquid connected to halloysite nanotubes (g-C3N4-IL@HNT) (10 mg); n.s.	n.s.	(10 × 1 mL)	Water (1 × 1 mL)	Acetonitrile (n.s.)	Methanol (2 × 2 mL) and water (2 × 5 mL)	HPLC-UV	1 ng/mL (1) 1 ng/mL (2) 0.5 ng/mL (3) 0.5 ng/mL (4) 0.5 ng/mL (5) 0.5 ng/mL (6)	4 ng/mL (1) 4 ng/mL (2) 2 ng/mL (3) 2 ng/mL (4) 2 ng/mL (5) 2 ng/mL (6)	82–97% (1) 82–99% (2) 77–104% (3) 81–94% (4) 95–101% (5) 76–98% (6)	Darvishnejad and Ebrahimzadeh (2020)
La3+Tb3+	Sea water, Power plant water, Industrial wastewater and Radiological wastewater (50 mL)	Filtration	Gelatin/ sodium triphosphate hydrogel nanofiber mat (GT/STP HNFM) (12 mg)	1M nitric acid, methanol and water (1 × 2 mL)	50 mL (n.s.)	n.s.	1M nitric acid (1 × 1 mL)	n.s.	ICP-OES	0.1 ng/mL (La3+) 0.2 ng/mL (Tb3+)	0.3 ng/mL (La3+) 0.6 ng/mL (Tb3+)	85–102% (La3+) 85–100% (Tb3+)	Moradi et al. (2020)
Fluoxetine	Wastewater, river and dam water samples (10 mL)	n.s.	montmorillonite-polystyrene nanocomposite coated on cellulosic paper (MMT/PS/Cell)	Methanol and water (n.s.)	(5 × n.s.)	n.s.	Methanol (25 × 300 µL)	Methanol and water (20 × 300 µL)	Fluorescence spectroscopy (FL)	2 ng/mL	7 ng/mL	76–10%	Matin et al. (2020)
2-Methyl-4-chlorophenoxyacetic acid (MCPA), 2,4-dichlorophenoxyacetic acid (2,4-D)	Farm water samples (2 mL)	n.s.	Bifunctional periodic mesoporous organosilica with imidazolium framework (BFPMO-IL) (4 mg)	Methanol and water pH2 (1 × 300 µL)	(5 × 2 mL)	Water (3 × 50 µL)	Methanol (1 × 150 µL)	Methanol and water (3 × 300 µL)	GC-FID	0.1 ng/mL (MCPA) 0.5 ng/mL (2,4-D)	0.5 ng/mL (MCPA) 1.2 ng/mL (2,4-D)	62% (MCPA) 67% (2,4-D)	Mousavi et al. (2019)

Analytes	Matrix	Sample pretreatment	Sorbent (amount); number of extraction cycles	Loading (volume)	Washing (volume)	Elution (volume)	Reconstitution (volume)	Analytical technique	LOD	LOQ	Recovery	References
Diazinon (DZN)	Agricultural wastewater, river water and well water (2 mL)	n.s.	Montmorillonite (nanoclay) modified by cetyltrimethylammonium bromide (CTAB) (2 mg); n.s.	Methanol and water (1 × 200 μL)	Water (1 × 200 μL)	Methanol (7 × 50 μL)	Water and metanol (7 × 200 μL)	CD-IMS	n.s.	0.2 ng/mL	95–106%	Saraji et al. (2018)
Chlorpyrifos (CLP), Fenthion (FEN), Fenitrothion (FET), Ethion (ETN), Edifenphos (EDI), Phosalone (PSL)	River water, dam water and tap water samples (1 mL)	n.s.	Graphene oxide reinforced polyamide nanocomposite (GO/PA NC) (6 sorbent layers, n.s.); manual	n.s.	n.s.	Methanol (25 × 150 μL)	n.s.	GC-FID	0.3 ng/mL (CLP) 0.3 ng/mL (FEN) 1 ng/mL (FET) 0.2 ng/mL (ETN) 0.3 ng/mL (EDI) 0.3 ng/mL (PSL)	1 ng/mL (CLP) 1 ng/mL (FEN) 3 ng/mL (FET) 1 ng/mL (ETN) 1 ng/mL (EDI) 1 ng/mL (PSL)	96–113% (CLP) 88–111% (FEN) 78–99% (FET) 90–113% (ETN) 85–98% (EDI) 102–105% (PSL)	Ayazi et al. (2018)
Diazinon (DZN), Malathion (MLT), Ethion (ETN)	Water samples (10 mL)	n.s.	Natural nanopertite (n.s.); n.s.	Methanol/acetone (1:1) and water (2 or 3 × 2 mL)	Water (1 × 1 mL)	Dichloromethane (1 × 100 μL)	n.s.	GC-MS	0.07 ng/mL (DZN) 0.38 ng/mL (MLT) 0.13 ng/mL (ETN)	0.2 ng/mL (DZN) 0.2 ng/mL (MLT) 0.4 ng/mL (ETN)	85–96% (DZN) 81–92% (MLT) 94–103% (ETN)	Taghani et al. (2018)
Methyl paraben (MeP), Ethyl paraben (EtP), Propylparaben (PrP), Butyl paraben (BuP), Benzyl paraben (BeP)	Lake water, domestic wastewater, a swimming pool and tapwater	Centrifugation	Graphene supported on aminopropyl silica (Si-G) (7 mg); (n.s.)	n.s.	n.s.	Acetonitrile (10 × 100 μL)	Acetonitrile (4 × 500 μL) and water (4 × 1 mL)	LC-MS/MS	0.06 ng/mL (MeP) 0.06 ng/mL (EtP) 0.06 ng/mL (PrP) 0.06 ng/mL (BuP) 0.09 ng/mL (BeP)	0.2 ng/mL (MeP) 0.2 ng/mL (EtP) 0.2 ng/mL (PrP) 0.2 ng/mL (BuP) 0.3 ng/mL (BeP)	n.s.	Fumes and Lanças (2017)
Dimethyl phthalate (DMP), Diethyl phthalate (DEP), Di-isobutyl phthalate (DIBP), Di-n-butyl phthalate (DnBP), Di-2-ethylhexyl phthalate (DEHP)	Tap water, river water and mineral water (8 mL)	n.s.	Hydroxyapatite [HAP] (2 mg); manual	Methanol and water (1 × 0.5 mL)	Water (1 mL)	Dichloromethane (1 × 60 μL)	Methanol (1 × 200 μL) and water (1 mL)	GC-FID	0.02 ng/mL (DMP) 0.05 ng/mL (DEP) 0.04 ng/mL (DIBP) 0.05 ng/mL (DnBP) 0.1 ng/mL (DEHP)	0.07 ng/mL (DMP) 0.15 ng/mL (DEP) 0.14 ng/mL (DIBP) 0.1 ng/mL (DnBP) 0.25 ng/mL (DEHP)	88–99% (DMP) 88–98% (DEP) 87–99% (DIBP) 88–99% (DnBP) 86–97% (DEHP)	Amiri et al. (2017)
Dimethyl phthalate (DMP), Diethyl phthalate (DEP), Di-isobutyl phthalate (DIBP), Di-n-butyl phthalate (DnBP), Di-2-ethylhexyl phthalate (DEHP)	Tap water, river water and mineral water (10 mL)	n.s.	3D carbon nanotube/carbon nanofiber–graphene nanostructures (CNT/CNF–G) (2 mg); manual	Methanol and water (1 × 1 mL)	Water (1 × 1 mL)	Methanol (0.75 mL)	Methanol (3× 100 μL) and water (1 mL)	GC-FID	0.006 ng/mL (DMP) 0.01 ng/mL (DEP) 0.007 ng/mL (DIBP) 0.001 ng/mL (DnBP) 0.005 ng/mL (DEHP)	0.02 ng/mL (DMP) 0.03 ng/mL (DEP) 0.02 ng/mL (DIBP) 0.003 ng/mL (DnBP) 0.015 ng/mL (DEHP)	93–97% (DMP) 90–99% (DEP) 92–97% (DIBP) 92–98% (DnBP) 94–99% (DEHP)	Amiri and Ghaemi (2017)
Food toxicology												
Progesterone (PGN), Prednisolone (PRE), Estradiol (ESD)	Bovine milk (0.25 mL)	5 mL of bovine milk added to 10 mL of acetonitrile; centrifugation; dilution with 25 mL water	Polythiophene (PTh) (4 mg); n.s.	Water (1 × 250 μL)	(2 × 250 μL)	Methanol: formic acid (7: 3, v/v) (1 × 200 μL)	n.s.	HPLC-DAD	n.s.	16 ng/mL (PGN) 16 ng/mL (PRE) 16 ng/mL (ESD)	99% (PGN) 88% (PRE) 97% (ESD)	Florez et al. (2020)

(Continued)

TABLE 5.7 (Continued)

New sorbents developed for MEPS, procedures applied in analytical toxicology, analytical instrumentation, limits of determination, and recoveries (2015–2020)

Analytes	Sample (Volume/ Weight)	Sample Preparation	MEPS Sorbent	MEPS Steps					Analytical Instrumentation	LOD	LOQ	Recoveries	Ref
				Conditioning	Load	Wash	Elution	Sorbent re-use					
Methyl 4-hydroxybenzoate (MP), Ethyl 4-hydroxybenzoate (EP), Propyl 4-hydroxybenzoate (PP)	Vegetable oil samples (1 mL)	Dilution with 5 mL of n-hexane	Metal-organic frameworks (MOFs); HKUST-1 (20 mg); semi-automated	n.s.	Sonication in the syringe for 11 min at room temperature	Hexane (1 mL)	Methanol (1.1 mL)	n.s.	LC-MS/MS	n.s.	1.97 ng/mL (MP) 3.03 ng/mL (EP) 4.57 ng/mL (PP)	82–112% (MP) 74–102% (EP) 87–120% (PP)	Jiang et al. (2020)
Metolcarb (MCB), Carbaryl (CBL), Isoprocarb (ICB), Diethofencarb (DCB)	Juice samples (5 mL)	Centrifugation; filtration; NaCl (0.75 g)	Reduced graphene oxide coated with ZnO (RGO–ZnO) (5 mg); manual	Acetonitrile (2 mL) and water (4 mL)	(5 × 1 mL)	Acetonitrile/water (5:95, v/v); (1 × 1 mL)	Acetonitrile (6 × 200 μL)	n.s.	HPLC-UV	0.45 ng/mL (MCB) 0.23 ng/mL (CBL) 1.21 ng/mL (ICB) 1.14 ng/mL (DCB)	1.45 ng/mL (MCB) 0.75 ng/mL (CBL) 3.52 ng/mL (ICB) 3.46 ng/mL (DCB)	97% (MCB) 95–99% (CBL) 93–102% (ICB) 98–102% (DCB)	Sun et al. (2019)
Prometryn (P), Atrazine (A), Terbumeton (T), Secbumeton (S)	Maize powder (1 g)	5 mL of n-hexane and sonication; centrifugation	Au/LDH nanohybrids (60 mg); n.s.	n.s.	4 mL of supernatant, addition 60 mg of Au/LDH nanohybrids; ultrasound and the mixture was drawn into a 5 mL syringe	n-hexane (800 μL)	Ethyl acetate (400 μL)	n.s.	HPLC-DAD	0.035 ng/g (P) 0.056 ng/g (A) 0.085 ng/g (T) 0.108 ng/g (S)	0.12 ng/g (P) 0.19 ng/g (A) 0.28 ng/g (T) 0.36 ng/g (S)	92–99% (P) 92–100% (A) 94–100% (T) 94–101% (S)	Li et al. (2018)
Clofentezine	Milk and juice samples (0.4 mL)	Adjusted pH 4	Nanostructured star-shaped polythiophene dendrimer (S-PTh dendrimer) (3 mg); manual	Methanol, acetone and acetonitrile (n.s.) and 5 mL water	n.s.	Water (2 × 100 μL)	Methanol (4 × 100 μL)	Acetonitrile (6 × 100 μL)	HPLC-DAD-UV	2 ng/mL	n.s.	n.s.	Abolghasemi et al. (2018)
Chlortetracycline (CTC), Tetracycline (TTC), Oxytetracycline (OXT), Doxycycline (DOX)	Milk samples (5 mL)	5 mL of milk sample and 2 mL of trifluoroacetic acid 20% (v/v); 20 mL of a McIlvaine buffer solution/EDTA and centrifugation	Graphene particles supported on silica (G-Sil) (7 mg); n.s.	Methanol (8 × 0.5 mL) and McIlvaine/EDTA buffer (2 × 1 mL)	(6 × 1 mL)	McIlvaine/EDTA buffer (2 × 1 mL)	Methanol (9 × 100 μL)	n.s.	LC-MS/MS	0.03 ng/mL (CTC) 0.13 ng/mL (TTC) 0.11 ng/mL (OXT) 0.21 ng/mL (DOX)	0.05 ng/mL (CTC) 0.42 ng/mL (TTC) 0.95 ng/mL (OXT) 0.69 ng/mL (DOX)	n.s.	Vasconcelos Soares Maciel et al. (2018)

Analyte	Sample	Sample preparation	Sorbent	Conditioning	Loading	Washing	Elution	Other	Detection	LOD	LOQ	Recovery	Reference
Triazine herbicides (n.s.)	Corn samples (1 g)	5 mL of Acetonitrile; sonicated; centrifuged; 4 mL supernatant was used	Metal-organic frameworks (MOFs); MIL-101(Cr) (9 mg); semi-automated	n.s.	Supernatant was transferred to tube containing sorbent under sonication at room temperature for 9 min	n-hexane (1 mL)	Acetonitrile (2 mL)	n.s.	LC-MS/MS	0.01–0.12 ng/g (n.s.)	0.05–0.35 ng/g (n.s.)	73–107% (n.s.)	Jiang et al. (2018)
Norfloxacin (NOR), Ofloxacin (OFL), Fleroxacin (FLE), Ciprofloxacin (CIP), Danofloxacin (DAN), Lomefloxacin (LOM), Enrofloxacin (ENR), Orbifloxacin (ORB)	Milk samples (0.01 g)	Protein precipitation; centrifugation; interlayer was used	Immunoaffinity; glass beads bound with QN monoclonal antibodies (0.2 g)	n.s.	Interlayer was passed through	Water (3 × 1 mL) and	Methanol-PBS solution (9:1, v/v) (1 × 600 μL)	n.s.	HPLC-FLD	0.1 ng/g (NOR); 0.1 ng/g (OFL); 0.1 ng/g (FLE); 0.1 ng/g (CIP); 0.05 ng/g (DAN); 0.1 ng/g (LOM); 0.1 ng/g (ENR); 0.1 ng/g (ORB)	0.3 ng/g (NOR); 0.3 ng/g (OFL); 0.3 ng/g (FLE); 0.3 ng/g (CIP); 0.15 ng/g (DAN); 0.3 ng/g (LOM); 0.3 ng/g (ENR); 0.3 ng/g (ORB)	63% (NOR); 73% (OFL); 91% (FLE); 58% (CIP); 86% (DAN); 63% (LOM); 67% (ENR); 54% (ORB)	Zhang et al. (2017)

CD-IMS (Corona discharge ion mobility spectrometry); DAD (diode array detector); FID (flame ionization detector); FLD (fluorescence detector); GC (gas chromatography); HPLC (high-performance liquid chromatography); ICP-OES (inductively plasma optical emission spectrometry); LC (liquid chromatography); LOD (limit of detection); LOQ (limit of quantification); MS (mass spectrometry); MS/MS (tandem mass spectrometry); UHPLC (ultra high-performance liquid chromatography); UV (ultraviolet).

nitrofurantoin in urine samples. Not fully green, but still pertinent, was the work published by Mehrani et al. (Mehrani et al. 2020) in which natural compounds extracted from aloe vera plants and gum of pine trees were used to synthesize the sorbents. These compounds were aloin (polar compound) and rosin (non-polar compound). After their coupling with polyacrylonitrile (PAN), aloin and rosin formed aloin/PAN and rosin/PAN nanofibers used as sorbents to pre-concentrate polycyclic aromatic hydrocarbons and phenoxyacetic acid herbicides from water samples (Mehrani et al. 2020).

Over the last five years, MEPS applicability has been greatly explored in all fields of analytical toxicology (Table 5.7). These new sorbent developments have represented the majority of the published articles regarding MEPS, justifying the importance of the solid material packed in the syringe to improve method selectivity.

5.6 Perspectives and Future Challenges

MEPS emerged in accordance with green chemistry principles and aimed to improve the sustainable development for chemists in both the research and routine analysis fields. Although MEPS is still limited to research, the last five years have been very productive, with a large number of new sorbents developed and new approaches tested, but their application for routine analysis at an industrial scale remains scarce. Therefore, it is urgent to implement techniques such as MEPS that provide great enrichment factors, are rapid and automated, minimize sample volumes required, and reduce toxic wastes.

The commercially available sorbents do not seem to cover all necessities, hence the constant look for new solid materials. Nevertheless, new solid materials developed and reported are restricted to few classes of target analytes and are not suitable for a multi method approach. Interesting enough is all the new research dedicated to green sorbents, namely microalgae and vegetable materials. More studies should be performed in this field, including sorbent stability and broader application. Ion liquids continue being explored in this matter and appear as a great option for future sorbent developments, revealing low toxicity and wide applicability.

Finally, MEPS coupling with more recent MS technology should be considered. Over the last five years no linear ion trap, orbitrap, and quadrupole time-of-flight mass analyzers were described with MEPS. The coupling with the mentioned mass analyzers would offer the possibility to surpass the limitations of multi-target screening.

Acknowledgments

The authors acknowledge Fundação para a Ciência e a Tecnologia (FCT) and Community Funds (UIDB/00709/2020). T. Rosado acknowledges the Centro de Competências em Cloud Computing in the form of a fellowship (C4_WP2.6_M1 – Bioinformatics; Operação UBIMEDICAL – CENTRO-01-0145-FEDER-000019 – C4 – Centro de Competências em Cloud Computing), supported by Fundo Europeu de Desenvolvimento Regional (FEDER) through the Programa Operacional Regional Centro (Centro 2020).

REFERENCES

Abdel-Rehim, Mohamed. "New Trend in Sample Preparation: On-Line Microextraction in Packed Syringe for Liquid and Gas Chromatography Applications: I. Determination of Local Anaesthetics in Human Plasma Samples Using Gas Chromatography–Mass Spectrometry." *Journal of Chromatography B* 801, no. 2 (2004): 317–21. 10.1016/j.jchromb.2003.11.042.

Abdel-Rehim, Mohamed. "Recent Advances in Microextraction by Packed Sorbent for Bioanalysis." *Journal of Chromatography A* 1217, no. 16 (2010): 2569–80. 10.1016/j.chroma.2009.09.053.

Abdel-Rehim, Mohamed. "Microextraction by Packed Sorbent (MEPS): A Tutorial." *Analytica Chimica Acta* 701, no. 2 (2011): 119–28. 10.1016/j.aca.2011.05.037

Abdel-Rehim, Mohamed, Zeki Altun, and Lars Blomberg. "Microextraction in Packed Syringe (MEPS) for Liquid and Gas Chromatographic Applications. Part II – Determination of Ropivacaine and Its Metabolites in Human Plasma Samples Using MEPS with Liquid Chromatography/Tandem Mass Spectrometry." *Journal of Mass Spectrometry* 39, no. 12 (2004): 1488–93. 10.1002/jms.731

Abolghasemi, Mir Mahdi, Hoda Taheri, Mehdi Jaymand, and Marzieh Piryaei. " Nanostructured Star-Shaped Polythiophene Dendrimer as a Highly Efficient Sorbent for Microextraction in Packed Syringe for HPLC Analysis of the Clofentezine in Milk and Juice Samples." *Separation Science Plus* 1, no. 3 (2018): 202–8. 10.1002/sscp.201700036

Abuzooda, Thana, Ahmad Amini, and Mohamed Abdel-Rehim. "Graphite-Based Microextraction by Packed Sorbent for Online Extraction of β-Blockers from Human Plasma Samples." *Journal of Chromatography. B, Analytical Technologies in the Biomedical and Life Sciences* 992, June (2015): 86–90. 10.1016/j.jchromb.2015.04.027

Ahmadi, Mazaher, Hatem Elmongy, Tayyebeh Madrakian, and Mohamed Abdel-Rehim. "Nanomaterials as Sorbents for Sample Preparation in Bioanalysis: A Review." *Analytica Chimica Acta* 958, March (2017): 1–21. 10.1016/j.aca.2016.11.062

Ahmadi, Mazaher, Mohammad Mahdi Moein, Tayyebeh Madrakian, Abbas Afkhami, Soleiman Bahar, and Mohamed Abdel-Rehim. "Reduced Graphene Oxide as an Efficient Sorbent in Microextraction by Packed Sorbent: Determination of Local Anesthetics in Human Plasma and Saliva Samples Utilizing Liquid Chromatography-Tandem Mass Spectrometry." *Journal of Chromatography. B, Analytical Technologies in the Biomedical and Life Sciences* 1095, September (2018): 177–82. 10.1016/j.jchromb.2018.07.036

Altun, Zeki, and Mohamed Abdel-Rehim. "Study of the Factors Affecting the Performance of Microextraction by Packed Sorbent (MEPS) Using Liquid Scintillation Counter and Liquid Chromatography-Tandem Mass Spectrometry." *Analytica Chimica Acta* 630, no. 2 (2008): 116–23. 10.1016/j.aca.2008.09.067

Altun, Zeki, Mohamed Abdel-Rehim, and Lars G. Blomberg. "New Trends in Sample Preparation: On-Line Microextraction in Packed Syringe (MEPS) for LC and GC Applications Part III: Determination and Validation of Local Anaesthetics in Human Plasma Samples Using a Cation-Exchange Sorbent, and MEPS-LC-MS-MS." *Journal of Chromatography B, Analytical Technologies in the Biomedical and Life Sciences* 813, no. 1–2 (2004): 129–35. 10.1016/j.jchromb.2004.09.020

Amiri, Amirhassan, Mohammad Chahkandi, and Azadeh Targhoo. "Synthesis of Nano-Hydroxyapatite Sorbent for Microextraction in Packed Syringe of Phthalate Esters in Water Samples." *Analytica Chimica Acta* 950 (2017): 64–70. 10.1016/j.aca.2016.11.027

Amiri, Amirhassan, and Ferial Ghaemi. "Microextraction in Packed Syringe by Using a Three-Dimensional Carbon Nanotube/Carbon Nanofiber–Graphene Nanostructure Coupled to Dispersive Liquid-Liquid Microextraction for the Determination of Phthalate Esters in Water Samples." *Microchimica Acta* 184, no. 10 (2017): 3851–8. 10.1007/s00604-017-2416-8

Ares, Ana María, Purificación Fernández, María Regenjo, Ana María Fernández, Antonia María Carro, and Rosa Antonia Lorenzo. "A Fast Bioanalytical Method Based on Microextraction by Packed Sorbent and UPLC–MS/MS for Determining New Psychoactive Substances in Oral Fluid." *Talanta* 174 (2017): 454–61.

Aresta, Antonella, Pietro Cotugno, and Carlo Zambonin. "Determination of Ciprofloxacin, Enrofloxacin, and Marbofloxacin in Bovine Urine, Serum, and Milk by Microextraction by a Packed Sorbent Coupled to Ultra-High Performance Liquid Chromatography." *Analytical Letters* 52, no. 5 (2019): 790–802. 10.1080/00032719.2018.1496093

Asgari, Sara, Habib Bagheri, Ali Es-haghi, and Roya AminiTabrizi. "An Imprinted Interpenetrating Polymer Network for Microextraction in Packed Syringe of Carbamazepine." *Journal of Chromatography A* 1491 (2017): 1–8. 10.1016/j.chroma.2017.02.033

Ayazi, Zahra, Fatemeh Shekari Esfahlan, and Parastou Matin. "Graphene Oxide Reinforced Polyamide Nanocomposite Coated on Paper as a Novel Layered Sorbent for Microextraction by Packed Sorbent."

International Journal of Environmental Analytical Chemistry 98, no. 12 (2018): 1118–34. 10.1080/03067319.2018.1535061

Bagheri, Habib, Noshin Alipour, and Zahra Ayazi. "Multiresidue Determination of Pesticides from Aquatic Media Using Polyaniline Nanowires Network as Highly Efficient Sorbent for Microextraction in Packed Syringe." *Analytica Chimica Acta* 740 (2012a): 43–9. 10.1016/j.aca.2012.06.026

Bagheri, Habib, Zahra Ayazi, Ali Aghakhani, and Noshin Alipour. "Polypyrrole/Polyamide Electrospun-Based Sorbent for Microextraction in Packed Syringe of Organophosphorous Pesticides from Aquatic Samples." *Journal of Separation Science* 35, no. 1 (2012b): 114–20. 10.1002/jssc.201100509

Bagheri, Habib, Solmaz Banihashemi, and Faezeh Karimi Zandian. "Microextraction of Antidepressant Drugs into Syringes Packed with a Nanocomposite Consisting of Polydopamine, Silver Nanoparticles and Polypyrrole." *Microchimica Acta* 183, no. 1 (2016): 195–202. 10.1007/s00604-015-1606-5

Barroso, Mário, Ivo Moreno, Beatriz Da Fonseca, João António Queiroz, and Eugenia Gallardo. "Role of Microextraction Sampling Procedures in Forensic Toxicology." *Bioanalysis* 4, no. 14 (2012): 1805–26. 10.4155/bio.12.139

Berenguer, Pedro H., Irene C. Camacho, Rita Câmara, Susana Oliveira, and José S. Câmara. "Determination of Potential Childhood Asthma Biomarkers Using a Powerful Methodology Based on Microextraction by Packed Sorbent Combined with Ultra-High Pressure Liquid Chromatography. Eicosanoids as Case Study." *Journal of Chromatography A* 1584 (2019): 42–56. 10.1016/j.chroma.2018.11.041

Biagini, Denise, Shaula Antoni, Tommaso Lomonaco, Silvia Ghimenti, Pietro Salvo, Francesca Giuseppa Bellagambi, Rosa Teresa Scaramuzzo, Massimiliano Ciantelli, Armando Cuttano, Roger Fuoco, and Fabio Di Francesco. "Micro-Extraction by Packed Sorbent Combined with UHPLC-ESI-MS/MS for the Determination of Prostanoids and Isoprostanoids in Dried Blood Spots." *Talanta* 206, May 2019 (2020): 120236. 10.1016/j.talanta.2019.120236

Bianchi, Federica, Monica Mattarozzi, Nicolò Riboni, Paolo Mora, Stefano A. Gandolfi, and Maria Careri. "A Rapid Microextraction by Packed Sorbent – Liquid Chromatography Tandem Mass Spectrometry Method for the Determination of Dexamethasone Disodium Phosphate and Dexamethasone in Aqueous Humor of Patients with Uveitis." *Journal of Pharmaceutical and Biomedical Analysis* 142 (2017): 343–7. 10.1016/j.jpba.2017.05.025

Campestre, Cristina, Marcello Locatelli, Paolo Guglielmi, Elisa De Luca, Giuseppe Bellagamba, Sergio Menta, Gokhan Zengin, Christian Celia, Luisa Di Marzio, and Simone Carradori. "Analysis of Imidazoles and Triazoles in Biological Samples after MicroExtraction by Packed Sorbent." *Journal of Enzyme Inhibition and Medicinal Chemistry* 32, no. 1 (2017): 1–11. 10.1080/14756366.2017.1354858

Casas Ferreira, Ana María, Bernardo Moreno Cordero, Ángel Pedro Crisolino Pozas, and José Luis Pérez Pavón. "Use of Microextraction by Packed Sorbents and Gas Chromatography-Mass Spectrometry for the Determination of Polyamines and Related Compounds in Urine." *Journal of Chromatography A* 1444 (2016): 32–41. 10.1016/j.chroma.2016.03.054

Cristina Jardim, Valeria, Lidervan de Paula Melo, Diego Soares Domingues, and Maria Eugênia Queiroz. "Determination of Parabens in Urine Samples by Microextraction Using Packed Sorbent and Ultra-Performance Liquid Chromatography Coupled to Tandem Mass Spectrometry." *Journal of Chromatography B: Analytical Technologies in the Biomedical and Life Sciences* 974 (2015): 35–41. 10.1016/j.jchromb.2014.10.026

Cruz, Jonas Carneiro, Henrique Dipe de Faria, Eduardo Costa Figueiredo, and Maria Eugênia Costa Queiroz. "Restricted Access Carbon Nanotube for Microextraction by Packed Sorbent to Determine Antipsychotics in Plasma Samples by High-Performance Liquid Chromatography-Tandem Mass Spectrometry." *Analytical and Bioanalytical Chemistry* 412, no. 11 (2020): 2465–75. 10.1007/s00216-020-02464-4

Cunha, Kelly Francisco da, Leonardo Costalonga Rodrigues, Marilyn A. Huestis, and Jose Luiz Costa. "Miniaturized Extraction Method for Analysis of Synthetic Opioids in Urine by Microextraction with Packed Sorbent and Liquid Chromatography—Tandem Mass Spectrometry." *Journal of Chromatography A* 1624 (2020): 461241. 10.1016/j.chroma.2020.461241

D'Angelo, Veronica, Francesco Tessari, Giuseppe Bellagamba, Elisa De Luca, Roberta Cifelli, Christian Celia, Rosita Primavera, Martina Di Francesco, Donatella Paolino, Luisa Di Marzio, Marcello Locatelli. "Microextraction by Packed Sorbent and HPLC–PDA Quantification of Multiple Anti-Inflammatory Drugs and Fluoroquinolones in Human Plasma and Urine." *Journal of Enzyme Inhibition and Medicinal Chemistry* 31 (2016): 110–6. 10.1080/14756366.2016.1209496

D'Archivio, Angelo Antonio, Maria Anna Maggi, Fabrizio Ruggieri, Maura Carlucci, Vincenzo Ferrone, and Giuseppe Carlucci. "Optimisation by Response Surface Methodology of Microextraction by Packed Sorbent of Non Steroidal Anti-Inflammatory Drugs and Ultra-High Performance Liquid Chromatography Analysis of Dialyzed Samples." *Journal of Pharmaceutical and Biomedical Analysis* 125 (2016): 114–21. 10.1016/j.jpba.2016.03.045

Darvishnejad, Mohammad, and Homeira Ebrahimzadeh. "Graphitic Carbon Nitride-Reinforced Polymer Ionic Liquid Nanocomposite: A Novel Mixed-Mode Sorbent for Microextraction in Packed Syringe." *International Journal of Environmental Analytical Chemistry* (2020): 1–14. 10.1080/03067319.2020.1770243

Daryanavard, Seyed Mosayeb, Amin Jeppsson-Dadoun, Lars I. Andersson, Mahdi Hashemi, Anders Colmsjö, and Mohamed Abdel-Rehim. "Molecularly Imprinted Polymer in Microextraction by Packed Sorbent for the Simultaneous Determination of Local Anesthetics: Lidocaine, Ropivacaine, Mepivacaine and Bupivacaine in Plasma and Urine Samples." *Biomedical Chromatography: BMC* 27, no. 11 (2013): 1481–8. 10.1002/bmc.2946

Dhingra, Gaurav, Pooja Bansal, Nidhi Dhingra, Susheela Rani, and Ashok Kumar Malik. "Development of a Microextraction by Packed Sorbent with Gas Chromatography-Mass Spectrometry Method for Quantification of Nitroexplosives in Aqueous and Fluidic Biological Samples." *Journal of Separation Science* 41, no. 3 (2018): 639–47. 10.1002/jssc.201700470

Elmongy, Hatem, Hytham Ahmed, Abdel Aziz Wahbi, Ahmad Amini, Anders Colmsjö, and Mohamed Abdel-Rehim. "Determination of Metoprolol Enantiomers in Human Plasma and Saliva Samples Utilizing Microextraction by Packed Sorbent and Liquid Chromatography–Tandem Mass Spectrometry." *Biomedical Chromatography* 30, no. 8 (2016): 1309–17. 10.1002/bmc.3685

Ferrone, Vincenzo, Roberto Cotellese, Lorenzo Di Marco, Simona Bacchi, Maura Carlucci, Annadomenica Cichella, Paolo Raimondi, and Giuseppe Carlucci. "Meropenem, Levofloxacin and Linezolid in Human Plasma of Critical Care Patients: A Fast Semi-Automated Micro-Extraction by Packed Sorbent UHPLC-PDA Method for Their Simultaneous Determination." *Journal of Pharmaceutical and Biomedical Analysis* 140 (2017): 266–73. 10.1016/j.jpba.2017.03.035

Florez, Diego Hernando Ângulo, Hanna Leijoto de Oliveira, and Keyller Bastos Borges. "Polythiophene as Highly Efficient Sorbent for Microextraction in Packed Sorbent for Determination of Steroids from Bovine Milk Samples." *Microchemical Journal* 153, December (2020): 104521. 10.1016/j.microc.2019.104521

Fuentes, Ana María Ares, Purificación Fernández, Ana Maria Fernández, Antonia M. Carro, and Rosa Antonia Lorenzo. "Microextraction by Packed Sorbent Followed by Ultra High Performance Liquid Chromatography for the Fast Extraction and Determination of Six Antidepressants in Urine." *Journal of Separation Science* 42, no. 11 (2019): 2053–61. 10.1002/jssc.201900060

Fumes, Bruno Henrique, Felipe Nascimento Andrade, Álvaro José dos Santos Neto, and Fernando Mauro Lanças. "Determination of Pesticides in Sugarcane Juice Employing Microextraction by Packed Sorbent Followed by Gas Chromatography and Mass Spectrometry." *Journal of Separation Science* 39, no. 14 (2016): 2823–30. 10.1002/jssc.201600077

Fumes, Bruno Henrique, and Fernando Mauro Lanças. "Use of Graphene Supported on Aminopropyl Silica for Microextraction of Parabens from Water Samples." *Journal of Chromatography A* 1487 (2017): 64–71. 10.1016/j.chroma.2017.01.063

Tiago Rosado, Mário Barroso, Duarte Nuno Vieira, and Eugenia Gallardo. "Determination of Selected Opiates in Hair Samples Using Microextraction by Packed Sorbent: A New Approach for Sample Clean-Up." *Journal of Analytical Toxicology* 43 (2019): 465–76. 10.1093/jat/bkz029

Han, Qiang, Qionglin Liang, Xiaoqiong Zhang, Liu Yang, and Mingyu Ding. "Graphene Aerogel Based Monolith for Effective Solid-Phase Extraction of Trace Environmental Pollutants from Water Samples." *Journal of Chromatography A* 1447 (2016): 39–46. 10.1016/j.chroma.2016.04.032

Iadaresta, Francesco, Carlo Crescenzi, Ahmad Amini, Anders Colmsjö, Hirsh Koyi, and Mohamed Abdel-Rehim. "Application of Graphitic Sorbent for Online Microextraction of Drugs in Human Plasma Samples." *Journal of Chromatography. A* 1422, November (2015): 34–42. 10.1016/j.chroma.2015.10.025

Jiang, Yanxiao, Pinyi Ma, Xinpei Li, Huilan Piao, Dan Li, Ying Sun, Xinghua Wang, and Daqian Song. "Application of Metal-Organic Framework MIL-101(Cr) to Microextraction in Packed Syringe for Determination of Triazine Herbicides in Corn Samples by Liquid Chromatography-Tandem Mass Spectrometry." *Journal of Chromatography A* 1574 (2018): 36–41. 10.1016/j.chroma.2018.09.008

Jiang, Yanxiao, Zucheng Qin, Xiaona Song, Huilan Piao, Jingkang Li, Xinghua Wang, Daqian Song, Pinyi Ma, and Ying Sun. "Facile Preparation of Metal Organic Framework-Based Laboratory Semi-Automatic Micro-Extraction Syringe Packed Column for Analysis of Parabens in Vegetable Oil Samples." *Microchemical Journal* 158, June (2020): 105200. 10.1016/j.microc.2020.105200

Karimiyan, Hanieh, Abdusalam Uheida, Mohammadreza Hadjmohammadi, Mohammad Mahdi Moein, and Mohamed Abdel-Rehim. "Polyacrylonitrile/Graphene Oxide Nanofibers for Packed Sorbent Microextraction of Drugs and Their Metabolites from Human Plasma Samples." *Talanta* 201, August (2019): 474–9. 10.1016/j.talanta.2019.04.027

Kaur, Ramandeep, Heena, Ripneel Kaur, Susheela Rani, and Ashok Kumar Malik. "Simple and Rapid Determination of Phthalates Using Microextraction by Packed Sorbent and Gas Chromatography with Mass Spectrometry Quantification in Cold Drink and Cosmetic Samples." *Journal of Separation Science* 39, no. 5 (2016): 923–31. 10.1002/jssc.201500642

Klimowska, Anna, and Bartosz Wielgomas. "Off-Line Microextraction by Packed Sorbent Combined with on Solid Support Derivatization and GC-MS: Application for the Analysis of Five Pyrethroid Metabolites in Urine Samples." *Talanta* 176, January (2018): 165–71. 10.1016/j.talanta.2017.08.011

Konieczna, Lucyna, Anna Roszkowska, Anna Synakiewicz, Teresa Stachowicz-Stencel, Elzbieta Adamkiewicz-Drozyńska, and Tomasz Baczek. "Analytical Approach to Determining Human Biogenic Amines and Their Metabolites Using EVol Microextraction in Packed Syringe Coupled to Liquid Chromatography Mass Spectrometry Method with Hydrophilic Interaction Chromatography Column." *Talanta* 150 (2016): 331–9. 10.1016/j.talanta.2015.12.056

Li, Xinpei, Ying Sun, Liming Yuan, Li Liang, Yanxiao Jiang, Huilan Piao, Daqian Song, Aimin Yu, and Xinghua Wang. "Packed Hybrids of Gold Nanoparticles and Layered Double Hydroxide Nanosheets for Microextraction of Triazine Herbicides from Maize." *Microchimica Acta* 185, no. 7 (2018): 336. 10.1007/s00604-018-2862-y

Locatelli, Marcello, Maria Teresa Ciavarella, Donatella Paolino, Christian Celia, Ersilia Fiscarelli, Gabriella Ricciotti, Arianna Pompilio, Giovanni Di Bonaventura, Rossella Grande, Gokhan Zengin, and Luisa Di Marzio. "Determination of Ciprofloxacin and Levofloxacin in Human Sputum Collected from Cystic Fibrosis Patients Using Microextraction by Packed Sorbent-High Performance Liquid Chromatography Photodiode Array Detector." *Journal of Chromatography A* 1419 (2015): 58–66. 10.1016/j.chroma.2015.09.075

Magrini, Laura, Achille Cappiello, Giorgio Famiglini, and Pierangela Palma. "Microextraction by Packed Sorbent (MEPS)-UHPLC-UV: A Simple and Efficient Method for the Determination of Five Benzodiazepines in an Alcoholic Beverage *Journal of Pharmaceutical and Biomedical Analysis* 125 (2016): 48–53. 10.1016/j.jpba.2016.03.028

Malaca, Sara, Tiago Rosado, José Restolho, Jesus M Rodilla, Pedro M M Rocha, Lúcia Silva, Cláudia Margalho, Mário Barroso, and Eugenia Gallardo. "Determination of Amphetamine-Type Stimulants in Urine Samples Using Microextraction by Packed Sorbent and Gas Chromatography-Mass Spectrometry." *Journal of Chromatography. B, Analytical Technologies in the Biomedical and Life Sciences* 1120 (2019): 41–50. 10.1016/j.jchromb.2019.04.052

Martín Santos, Patricia, Camilo Jiménez Carracedo, Miguel del Nogal Sánchez, José Luis Pérez Pavón, and Bernardo Moreno Cordero. "A Sensitive and Automatic Method Based on Microextraction by Packed Sorbents for the Determination of Polycyclic Aromatic Hydrocarbons in Saliva Samples" *Microchemical Journal* 152, October 2019 (2020): 104274. 10.1016/j.microc.2019.104274

Matin, Parastou, Zahra Ayazi, and Kazem Jamshidi-Ghaleh. "Montmorillonite Reinforced Polystyrene Nanocomposite Supported on Cellulose as a Novel Layered Sorbent for Microextraction by Packed Sorbent for Determination of Fluoxetine Followed by Spectrofluorimetry Based on Multivariate

Optimisation." *International Journal of Environmental Analytical Chemistry* (2020): 1–16. 10.1080/03067319.2020.1791333

Mehrani, Zahra, Homeira Ebrahimzadeh, and Ebrahim Moradi. "Use of Aloin-Based and Rosin-Based Electrospun Nanofibers as Natural Nanosorbents for the Extraction of Polycyclic Aromatic Hydrocarbons and Phenoxyacetic Acid Herbicides by Microextraction in Packed Syringe Method Prior to GC-FID Detection." *Microchimica Acta* 187, no. 7 (2020): 401–12. 10.1007/s00604-020-04374-9

Meng, Jia, and Xu Wang. "Microextraction by Packed Molecularly Imprinted Polymer Combined Ultra-High-Performance Liquid Chromatography for the Determination of Levofloxacin in Human Plasma." *Journal of Chemistry* 2019 (2019): 1–9. 10.1155/2019/4783432

Moein, Mohammad Mahdi, Abbi Abdel-Rehim, and Mohamed Abdel-Rehim. "On-Line Determination of Sarcosine in Biological Fluids Utilizing Dummy Molecularly Imprinted Polymers in Microextraction by Packed Sorbent." *Journal of Separation Science* 38, no. 5 (2015a): 788–95. 10.1002/jssc.201401116

Moein, Mohammad Mahdi, Rana Said, and Mohamed Abdel-Rehim. "Microextraction by Packed Sorbent." *Bioanalysis* 7, no. 17 (2015b): 2155–61. 10.4155/bio.15.154.

Montesano, Camilla, Maria Chiara Simeoni, Roberta Curini, Manuel Sergi, Claudio Lo Sterzo, and Dario Compagnone. "Determination of Illicit Drugs and Metabolites in Oral Fluid by Microextraction on Packed Sorbent Coupled with LC-MS/MS." *Analytical and Bioanalytical Chemistry* 407, no. 13 (2015): 3647–58. 10.1007/s00216-015-8583-8

Moradi, Ebrahim, Zahra Mehrani, and Homeira Ebrahimzadeh. "Gelatin/Sodium Triphosphate Hydrogel Electrospun Nanofiber Mat as a Novel Nanosorbent for Microextraction in Packed Syringe of La3+ and Tb3+ Ions Prior to Their Determination by ICP-OES." *Reactive and Functional Polymers* 153, December 2019 (2020): 104627. 10.1016/j.reactfunctpolym.2020.104627

Moreno, Ivo, Mário Barroso, Ana Martinho, Angelines Cruz, and Eugenia Gallardo. "Determination of Ketamine and Its Major Metabolite, Norketamine, in Urine and Plasma Samples Using Microextraction by Packed Sorbent and Gas Chromatography-Tandem Mass Spectrometry." *Journal of Chromatography B: Analytical Technologies in the Biomedical and Life Sciences* 1004 (2015): 67–78. 10.1016/j.jchromb.2015.09.032

Mousavi, Kobra Zavar, Yadollah Yamini, Babak Karimi, Shahram Seidi, Mojtaba Khorasani, Mostafa Ghaemmaghami, and Hojatollah Vali. "Imidazolium-Based Mesoporous Organosilicas with Bridging Organic Groups for Microextraction by Packed Sorbent of Phenoxy Acid Herbicides, Polycyclic Aromatic Hydrocarbons and Chlorophenols." *Mikrochimica Acta* 186, no. 4 (2019): 239. 10.1007/s00604-019-3355-3

Naccarato, Attilio, Rosangela Elliani, Giovanni Sindona, and Antonio Tagarelli. "Multivariate Optimization of a Microextraction by Packed Sorbent-Programmed Temperature Vaporization-Gas Chromatography–Tandem Mass Spectrometry Method for Organophosphate Flame Retardant Analysis in Environmental Aqueous Matrices." *Analytical and Bioanalytical Chemistry* 409, no. 30 (2017): 7105–20. 10.1007/s00216-017-0669-z

Nuckowski, Łukasz, Anna Kaczmarkiewicz, Sylwia Studzińska, and Bogusław Buszewski. "A New Approach to Preparation of Antisense Oligonucleotide Samples with Microextraction by Packed Sorbent." *Analyst* 144, no. 15 (2019): 4622–32. 10.1039/c9an00740g

Oliveira, Hanna Leijoto de, Leila Suleimara Teixeira, Laíse Aparecida Fonseca Dinali, Bruna Carneiro Pires, Nathália Soares Simões, and Keyller Bastos Borges. "Microextraction by Packed Sorbent Using a New Restricted Molecularly Imprinted Polymer for the Determination of Estrogens from Human Urine Samples." *Microchemical Journal* 150, July (2019): 104162. 10.1016/j.microc.2019.104162

Oppolzer, David, Ivo Moreno, Beatriz da Fonseca, Luis Passarinha, Mario Barroso, Suzel Costa, Joao A. Queiroz, and Eugenia Gallardo. "Analytical Approach to Determine Biogenic Amines in Urine Using Microextraction in Packed Syringe and Liquid Chromatography Coupled to Electrochemical Detection." *Biomedical Chromatography* 27, no. 5 (2013): 608–14. 10.1002/bmc.2835

Ottavio, Francesca Di, Flavio Della Pelle, Camilla Montesano, Rossana Scarpone, Alberto Escarpa, Dario Compagnone, and Manuel Sergi. "Determination of Pesticides in Wheat Flour Using Microextraction on Packed Sorbent Coupled to Ultra-High Performance Liquid Chromatography and Tandem Mass Spectrometry." *Food Analytical Methods* 10, no. 6 (2017): 1699–708. 10.1007/s12161-016-0720-2

Páleníková, Agneša, and Svetlana Hrouzková. "Microextraction in Packed Syringe: Solvent-Minimized Sample Preparation Technique." *Monatshefte Fur Chemie* 145, no. 4 (2014): 537–49. 10.1007/s00706-013-1119-z

Paris, Alice, Jean Luc Gaillard, and Jérôme Ledauphin. "Rapid Extraction of Polycyclic Aromatic Hydrocarbons in Apple: Ultrasound-Assisted Solvent Extraction Followed by Microextraction by Packed Sorbent." *Food Analytical Methods* 12, no. 10 (2019): 2194–204. 10.1007/s12161-019-01568-7

Pautova, Alisa, Zoya Khesina, Maria Getsina, Pavel Sobolev, Alexander Revelsky, and Natalia Beloborodova. "Determination of Tryptophan Metabolites in Serum and Cerebrospinal Fluid Samples Using Microextraction by Packed Sorbent, Silylation and GC–MS Detection." *Molecules* 25, no. 14 (2020a): 3258. Multidisciplinary Digital Publishing Institute.

Pautova, Alisa K., Zoya B. Khesina, Tatiana N. Litvinova, Alexander I. Revelsky, and Natalia V. Beloborodova. "Metabolic Profiling of Aromatic Compounds in Cerebrospinal Fluid of Neurosurgical Patients Using Microextraction by Packed Sorbent and Liquid–Liquid Extraction with Gas Chromatography–Mass Spectrometry Analysis." *Biomedical Chromatography* 35, no. 2 (2020b): e4969. 10.1002/bmc.4969

Pautova, Alisa K., Pavel D. Sobolev, and Alexander I. Revelsky. "Analysis of Phenylcarboxylic Acid-Type Microbial Metabolites by Microextraction by Packed Sorbent from Blood Serum Followed by GC–MS Detection." *Clinical Mass Spectrometry* 14 (2019): 46–53. The Association for Mass Spectrometry: Applications to the Clinical Lab (MSACL). 10.1016/j.clinms.2019.05.005

Pautova, Alisa K., Pavel D. Sobolev, and Alexander I. Revelsky. "Microextraction of Aromatic Microbial Metabolites by Packed Hypercrosslinked Polystyrene from Blood Serum." *Journal of Pharmaceutical and Biomedical Analysis* 177 (2020c): 112883. 10.1016/j.jpba.2019.112883

Peña, Javier, Ana María Casas-Ferreira, Marcos Morales-Tenorio, Bernardo Moreno-Cordero, and José Luis Pérez-Pavón. "Determination of Polyamines and Related Compounds in Saliva via in Situ Derivatization and Microextraction by Packed Sorbents Coupled to GC-MS." *Journal of Chromatography B: Analytical Technologies in the Biomedical and Life Sciences* 1129, October (2019): 121821. 10.1016/j.jchromb.2019.121821

Pereira, Jorge A. M., João Gonçalves, Priscilla Porto-Figueira, José A. Figueira, Vera Alves, Rosa Perestrelo, Sonia Medina, and José S. Câmara. "Current Trends on Microextraction by Packed Sorbent - Fundamentals, Application Fields, Innovative Improvements and Future Applications." *The Analyst* 144, no. 17 (2019): 5048–74. 10.1039/c8an02464b

Perestrelo, Rosa, Enderson Rodriguez, and José S. Câmara. "Impact of Storage Time and Temperature on Furanic Derivatives Formation in Wines Using Microextraction by Packed Sorbent Tandem with Ultrahigh Pressure Liquid Chromatography." *LWT – Food Science and Technology* 76 (2017): 40–7. 10.1016/j.lwt.2016.10.041.

Perestrelo, Rosa, Catarina L. Silva, and José S. Câmara. "Quantification of Furanic Derivatives in Fortified Wines by a Highly Sensitive and Ultrafast Analytical Strategy Based on Digitally Controlled Microextraction by Packed Sorbent Combined with Ultrahigh Pressure Liquid Chromatography." *Journal of Chromatography A* 1381 (2015): 54–63. 10.1016/j.chroma.2015.01.020

Pirmohammadi, Zahra, Abdulrahman Bahrami, Davood Nematollahi, Saber Alizadeh, Farshid Ghorbani Shahna, and Razzagh Rahimpoor. "Determination of Urinary Methylhippuric Acids Using MIL-53-NH2 (Al) Metal–Organic Framework in Microextraction by Packed Sorbent Followed by HPLC–UV Analysis." *Biomedical Chromatography* 34, no. 1 (2020): 1–10. 10.1002/bmc.4725

Porto-Figueira, Priscilla, José A. Figueira, Jorge A. M. Pereira, and José S. Câmara. "A Fast and Innovative Microextraction Technique, MSPEed, Followed by Ultrahigh Performance Liquid Chromatography for the Analysis of Phenolic Compounds in Teas." *Journal of Chromatography. A* 1424, December (2015): 1–9. 10.1016/j.chroma.2015.10.063

Prata, Margarida, Andreia Ribeiro, David Figueirinha, Tiago Rosado, David Oppolzer, José Restolho, André R. T. S. Araújo, Suzel Costa, Mário Barroso, and Eugenia Gallardo. "Determination of Opiates in Whole Blood Using Microextraction by Packed Sorbent and Gas Chromatography-Tandem Mass Spectrometry." *Journal of Chromatography. A* 1602, September (2019): 1–10. 10.1016/j.chroma.2019.05.021

Rahimi, Akram, Payman Hashemi, Alireza Badiei, Mehdi Safdarian, and Marzieh Rashidipour. "Microextraction of Rosmarinic Acid Using CMK-3 Nanoporous Carbon in a Packed Syringe." *Chromatographia* 76, no. 13 (2013): 857–60. 10.1007/s10337-013-2471-1

Rahimpoor, Razzagh, Abdulrahman Bahrami, Davood Nematollahi, Farshid Ghorbani Shahna, and Maryam Farhadian. "Facile and Sensitive Determination of Urinary Mandelic Acid by Combination of Metal Organic Frameworks with Microextraction by Packed Sorbents." *Journal of Chromatography B:*

Analytical Technologies in the Biomedical and Life Sciences 1114–1115, January (2019): 45–54. 10.1016/j.jchromb.2019.03.023

Rasolzadeh, Fahimeh, Payman Hashemi, Maryam Madadkar Haghjou, and Mehdi Safdarian. "Chlorella Vulgaris Microalgae as a Green Packing for the Microextraction by Packed Sorbent of Nitrofurantoin in Urine." *Analytical and Bioanalytical Chemistry Research* 6, no. 2 (2019): 419–29.

Rocchi, Rachele, Maria Chiara Simeoni, Camilla Montesano, Gabriele Vannutelli, Roberta Curini, Manuel Sergi, and Dario Compagnone. "Analysis of New Psychoactive Substances in Oral Fluids by Means of Microextraction by Packed Sorbent Followed by Ultra-High-Performance Liquid Chromatography–Tandem Mass Spectrometry." *Drug Testing and Analysis* 10, no. 5 (2018): 865–73. 10.1002/dta.2330

Rosado, Tiago, Eugenia Gallardo, Duarte Nuno Vieira, and Mário Barroso. "Microextraction by Packed Sorbent as a Novel Strategy for Sample Clean-Up in the Determination of Methadone and EDDP in Hair." *Journal of Analytical Toxicology* (2020a). 10.1093/jat/bkaa040

Rosado, Tiago, Eugenia Gallardo, Duarte Nuno Vieira, and Mário Barroso. "New Miniaturized Clean-up Procedure for Hair Samples by Means of Microextraction by Packed Sorbent: Determination of Cocaine and Metabolites." *Analytical and Bioanalytical Chemistry* 412, no. 28 (2020b): 7963–76. 10.1007/s00216-020-02929-6

Rosado, Tiago, Alexandra Gonçalves, Cláudia Margalho, Mário Barroso, and Eugenia Gallardo. "Rapid Analysis of Cocaine and Metabolites in Urine Using Microextraction in Packed Sorbent and GC/MS." *Analytical and Bioanalytical Chemistry* 409, no. 8 (2017a): 2051–63. 10.1007/s00216-016-0152-2

Rosado, T., L. Fernandes, M. Barroso, and E. Gallardo. "Sensitive Determination of THC and Main Metabolites in Human Plasma by Means of Microextraction in Packed Sorbent and Gas Chromatography–Tandem Mass Spectrometry." *Journal of Chromatography B* 1043 (2017b): 63–73.

Said, Rana, Mohamed Kamel, Aziza El-Beqqali, and Mohamed Abdel-Rehim. "Microextraction by Packed Sorbent for LC-MS/MS Determination of Drugs in Whole Blood Samples." *Bioanalysis* 2, no. 2 (2010): 197–205. 10.4155/bio.09.187

Santos, Catarina, David Oppolzer, Alexandra Gonçalves, Mário Barroso, and Eugenia Gallardo. "Determination of Organophosphorous Pesticides in Blood Using Microextraction in Packed Sorbent and Gas Chromatography–Tandem Mass Spectrometry." *Journal of Analytical Toxicology* 42, no. 5 (2018): 321–29. 10.1093/jat/bky004

Saracino, Maria Addolorata, Laura Santarcangelo, Maria Augusta Raggi, and Laura Mercolini. "Microextraction by Packed Sorbent (MEPS) to Analyze Catecholamines in Innovative Biological Samples." *Journal of Pharmaceutical and Biomedical Analysis* 104 (2015): 122–9. 10.1016/j.jpba.2014.11.003

Saraji, Mohammad, Mohammad Taghi Jafari, and Mohammad Mehdi Amooshahi. "Sol–Gel/Nanoclay Composite as a Sorbent for Microextraction in Packed Syringe Combined with Corona Discharge Ionization Ion Mobility Spectrometry for the Determination of Diazinon in Water Samples." *Journal of Separation Science* 41, no. 2 (2018): 493–500. 10.1002/jssc.201700967

Savastano, Maria Luisa, Ilario Losito, and Sandra Pati. "Rapid and Automatable Determination of Ochratoxin A in Wine Based on Microextraction by Packed Sorbent Followed by HPLC-FLD." *Food Control* 68, no. 1881 (2016): 391–8. 10.1016/j.foodcont.2016.04.016

Serenjeh, Fariba Nazari, Payman Hashemi, Ali Reza Ghiasvand, Fahimeh Rasolzadeh, Nahid Heydari, and Alireza Badiei. "Cooling Assisted Headspace Microextraction by Packed Sorbent Coupled to HPLC for the Determination of Volatile Polycyclic Aromatic Hydrocarbons in Soil." *Analytica Chimica Acta* 1125 (2020): 128–34. 10.1016/j.aca.2020.05.067

Silva, Luis Felipe da, and Fernando Mauro Lanças. "β-Cyclodextrin Coupled to Graphene Oxide Supported on Aminopropyl Silica as a Sorbent Material for Determination of Isoflavones." *Journal of Separation Science* 43, no. 3 (2020): 4347–55. 10.1002/jssc.202000598

Silveira, Romena Sanglard, Bruno Alves Rocha, Jairo Lisboa Rodrigues, and Fernando Barbosa. "Rapid, Sensitive and Simultaneous Determination of 16 Endocrine-Disrupting Chemicals (Parabens, Benzophenones, Bisphenols, and Triclocarban) in Human Urine Based on Microextraction by Packed Sorbent Combined with Liquid Chromatography Tandem Mass Spectrom." *Chemosphere* 240 (2020): 124951. 10.1016/j.chemosphere.2019.124951

Siqueira, Sarah Aparecida, Christian Fernandes, and Isabela Costa César. "Microextraction by Packed Sorbent and High Performance Liquid Chromatography for Simultaneous Determination of Lumefantrine and

Desbutyl-Lumefantrine in Plasma Samples." *Journal of Pharmaceutical and Biomedical Analysis* 190 (2020): 113486. 10.1016/j.jpba.2020.113486

Sobolev, Pavel D., Alisa K. Pautova, and Alexander I. Revelsky. "Microextraction of Aromatic Microbial Metabolites by Packed Sorbent (MEPS) from Model Solutions Followed by Gas Chromatography/Mass Spectrometry Analysis of Their Silyl Derivatives." *Journal of Analytical Chemistry* 72, no. 14 (2017): 1426–33. 10.1134/S1061934817140131

Soleimani, Esmaeel, Abdulrahman Bahrami, Abbas Afkhami, and Farshid Ghorbani Shahna. "Determination of Urinary Trans,Trans-Muconic Acid Using Molecularly Imprinted Polymer in Microextraction by Packed Sorbent Followed by Liquid Chromatography with Ultraviolet Detection." *Journal of Chromatography. B, Analytical Technologies in the Biomedical and Life Sciences* 1061–1062, September (2017): 65–71. 10.1016/j.jchromb.2017.07.008

Soleimani, Esmaeel, Abdulrahman Bahrami, Abbas Afkhami, and Farshid Ghorbani Shahna. "Selective Determination of Mandelic Acid in Urine Using Molecularly Imprinted Polymer in Microextraction by Packed Sorbent." *Archives of Toxicology* 92, no. 1 (2018): 213–22. 10.1007/s00204-017-2057-z

Sorribes-Soriano, Aitor, Silvia Sánchez-Martínez, Rocío Arráez-González, Francesc Albert Esteve-Turrillas, and Sergio Armenta. "Methylone Determination in Oral Fluid Using Microextraction by Packed Sorbent Coupled to Ion Mobility Spectrometry." *Microchemical Journal* 153, October 2019 (2020): 104504. 10.1016/j.microc.2019.104504

Sorribes-Soriano, Aitor, Amparo Monedero, Francesc A. Esteve-Turrillas, and Sergio Armenta. "Determination of the New Psychoactive Substance Dichloropane in Saliva by Microextraction by Packed Sorbent – Ion Mobility Spectrometry." *Journal of Chromatography A* 1603 (2019): 61–6. 10.1016/j.chroma.2019.06.054

Souza, Israel D. de, Diego S. Domingues, and Maria E. C. Queiroz. "Hybrid Silica Monolith for Microextraction by Packed Sorbent to Determine Drugs from Plasma Samples by Liquid Chromatography-Tandem Mass Spectrometry." *Talanta* 140, August (2015): 166–75. 10.1016/j.talanta.2015.03.032

Souza, Marilia Cristina Oliveira, Bruno Alves Rocha, Juliana Maria Oliveira Souza, Andresa Aparecida Berretta, and Fernando Barbosa. "A Fast and Simple Procedure for Polybrominated Diphenyl Ether Determination in Egg Samples by Using Microextraction by Packed Sorbent and Gas Chromatography–Mass Spectrometry." *Food Analytical Methods* 12, no. 7 (2019): 1528–35. 10.1007/s12161-019-01484-w

Šrámková, Ivana, Petr Chocholouš, Hana Sklenářová, and Dalibor Šatínský. "On-Line Coupling of Micro-Extraction by Packed Sorbent with Sequential Injection Chromatography System for Direct Extraction and Determination of Betaxolol in Human Urine." *Talanta* 143 (2015): 132–7. 10.1016/j.talanta.2015.05.048

Sun, Ting, Yuwan Fan, Peizheng Fan, Fengyun Geng, Peiyu Chen, and Feng Zhao. "Use of Graphene Coated with ZnO Nanocomposites for Microextraction in Packed Syringe of Carbamate Pesticides from Juice Samples." *Journal of Separation Science* 42, no.12 (2019): 2131–9. 10.1002/jssc.201900257

Taghani, Abdollah, Nasser Goudarzi, Ghadam Ali Bagherian, Mansour Arab Chamjangali, and Amir Hossein Amin. "Application of Nanoperlite as a New Natural Sorbent in the Preconcentration of Three Organophosphorus Pesticides by Microextraction in Packed Syringe Coupled with Gas Chromatography and Mass Spectrometry." *Journal of Separation Science* 41, no. 10 (2018): 2245–52. 10.1002/jssc.201701276

Vasconcelos Soares Maciel, Edvaldo, Bruno Henrique Fumes, Ana Lúcia de Toffoli, and Fernando Mauro Lanças. "Graphene Particles Supported on Silica as Sorbent for Residue Analysis of Tetracyclines in Milk Employing Microextraction by Packed Sorbent." *Electrophoresis* 39 (2018): 2047–55. 10.1002/elps.201800051

Xiong, Xin, and Yuanyuan Zhang. "A New Approach for Urinary Vanillylmandelic Acid Determination Using EVol Microextraction by Packed Sorbent Coupled to Liquid Chromatography-Tandem Mass Spectrometry." *Journal of Analytical Science and Technology* 11, no. 28 (2020a). 10.1186/s40543-020-00226-6

Xiong, Xin, and Yuanyuan Zhang. "Simple, Rapid, and Cost-Effective Microextraction by the Packed Sorbent Method for Quantifying of Urinary Free Catecholamines and Metanephrines Using Liquid Chromatography-Tandem Mass Spectrometry and Its Application in Clinical Analysis." *Analytical and Bioanalytical Chemistry* 412, no. 12 (2020b): 2763–75. 10.1007/s00216-020-02436-8

Yang, Liu, Qiang Han, Shuya Cao, Junchao Yang, Jiang Zhao, Molin Qin, and Mingyu Ding. "Self-Made Microextraction by Packed Sorbent Device for the Cleanup of Polychlorinated Biphenyls from Bovine Serum." *Journal of Separation Science* 39, no. 8 (2016): 1518–23. 10.1002/jssc.201501009

Yang, Liu, Rana Said, and Mohamed Abdel-Rehim. "Sorbent, Device, Matrix and Application in Microextraction by Packed Sorbent (MEPS): A Review." *Journal of Chromatography. B, Analytical Technologies in the Biomedical and Life Sciences* 1043, February (2017): 33–43. 10.1016/j.jchromb.2016.10.044

Zhang, Xinda, Cuicui Wang, Linyan Yang, Wei Zhang, Jing Lin, and Cun Li. "Determination of Eight Quinolones in Milk Using Immunoaffinity Microextraction in a Packed Syringe and Liquid Chromatography with Fluorescence Detection." *Journal of Chromatography B: Analytical Technologies in the Biomedical and Life Sciences* 1064, September (2017): 68–74. 10.1016/j.jchromb.2017.09.004

6

Thin-Film Solid-Phase Microextraction: Applications in Analytical Toxicology

Lucas Morés[1], Josias Merib[2], and E. Carasek[1]
[1]*Departamento de Química, Universidade Federal de Santa Catarina, Florianópolis, Brazil*
[2]*Departamento de Farmacociências, Universidade Federal de Ciências da Saúde de Porto Alegre, Porto Alegre, Brazil*

CONTENTS

6.1 Theory of TF-SPME 117
    6.1.1 Introduction and Fundamentals 117
    6.1.2 Main Optimizations for TF-SPME 119
6.2 Coating Preparation Methods for TF-SPME 119
    6.2.1 Dip Coating 119
    6.2.2 Spin Coating 120
    6.2.3 Spray Coating 120
    6.2.4 Electrospinning Coating 120
6.3 Application of TF-SPME in Analytical Toxicology 120
    6.3.1 Urine Samples 121
    6.3.2 Blood (Plasma/Serum) 123
    6.3.3 Other Matrices Evaluated by TF-SPME 125
6.4 Conclusion 126
References 126

6.1 Theory of TF-SPME

6.1.1 Introduction and Fundamentals

The first format of thin-films employed for extraction procedures was introduced in 2001 by J. B. Wilcockson and F. A. P. Gobas. This approach involved an alternative configuration of solid-phase microextraction (SPME) to measure fugacities of organic chemicals in biological samples (Wilcockson and Gobas 2001). The main differences between thin-film SPME (TF-SPME) compared to the traditional SPME approaches consist of the rectangular-shaped film of the extraction phase, which implies a higher surface-to-volume ratio, improving the sensitivity and efficiency of the procedure. Another significant improvement is the stainless-steel supports used in TF-SPME, which can overcome some limitations regarding the fragility of the traditional SPME fibres (Jiang and Pawliszyn 2012; Grandy et al. 2016; Cudjoe et al. 2009; RiaziKermani 2012).

    Some theoretical aspects related to the extraction capacity of TF-SPME in equilibrium conditions can be described by Equation 6.1 (Pawliszyn 2012).

$$n_{ep}^{eq} = C_S^O \frac{K_{es} V_s V_e}{K_{es} V_e + V_s} \quad (6.1)$$

This equation considers the intrinsic correlation between the amount of analyte extracted in equilibrium conditions (n_{ep}^{eq}) and the initial concentration of the analyte in the sample (C_S^O). K_{es} is the distribution constant of the analyte between the extraction phase and the sample matrix; V_s and V_e are the volume of the sample and extraction phase, respectively (Pawliszyn 2012).

Another theoretical approach is related to the kinetics of the extraction process in equilibrium conditions. In this case, the time required to extract 95% of the analyte ($t_{95\%}$) from the sample can be obtained through Equation 6.2 (Bruheim et al. 2003).

$$t_{95\%} = 3 \frac{\delta K_{es}(b-a)}{D_s} \quad (6.2)$$

According to this equation, time depends on the thickness of the boundary layer (δ), the distribution coefficient of the analytes (K_{es}), the thickness of the coating (b – a), and the diffusion coefficient of the analytes (D_s).

Moreover, a correlation between time and film thickness can also be described, and lesser thickness requires a shorter time to reach equilibrium. In addition, the rate of analyte extracted as a function of time $\left(\frac{dn}{dt}\right)$ is proportional to the surface area of the coating ($D_s A$) and the concentration of the analyte in the sample (C_s). Consequently, larger surface areas allow for a higher amount of analytes extracted, as can be shown in Equation 6.3 (Bruheim et al. 2003).

$$\frac{dn}{dt} = \left(\frac{D_s A}{\delta}\right) C_s \quad (6.3)$$

Basically, TF-SPME can be used in direct immersion and headspace modes. Bruheim et al. employed a thin-film of polydimethylsiloxane (PDMS) immobilized on a tip of a stainless-steel support as an extraction phase, according to Figure 6.1. Both extraction modes were studied and compared with a typical SPME fibre for the determination of semi-volatile compounds. The thin-film was inserted in the sample flask and removed after the extraction process; then, the PDMS sheet was rolled up and inserted in another vial for the liquid desorption step, or directly into an injector for thermal desorption. The authors compared the extraction efficiency of SPME and TF-SPME techniques, and higher amounts of analytes were extracted using the thin-film approach (Bruheim et al. 2003; Mirnaghi et al. 2013). Different

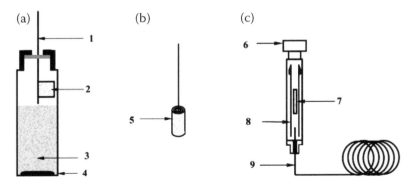

FIGURE 6.1 Scheme of the headspace membrane SPME system. 1. Deactivated stainless-steel rod. 2. Flat sheet membrane. 3. Sample solution. 4. Teflon-coated stirring bar. 5. Rolled membrane. 6. Injector nut. 7. Rolled membrane. 8. Glass liner. 9. Capillary column. (Source: Bruheim et al. 2003).

experimental setups involving TF-SPME have been applied in several fields, particularly in analytical toxicology (Olcer et al. 2019).

6.1.2 Main Optimizations for TF-SPME

In order to achieve satisfactory extraction efficiency and reliable results, TF-SPME exhibits some specific parameters that need to be carefully studied and optimized. Variables such as extraction time, concentration of salt in the sample (salting-out effect), sample pH, temperature, and agitation rate can directly affect the extraction performance.

Regarding the extraction step, one of the main parameters to be evaluated is the extraction time. This factor depends on the nature of the analytes (chemical structure and physicochemical properties), the type of sorbent phase, and the complexity of the matrix. Another variable frequently studied is the salting-out effect. In this case, NaCl is frequently added to the aqueous sample in order to reduce the solubility of the analytes and facilitate the mass transfer from the sample matrix to the extraction phase.

Sample pH can also influence the extraction efficiency since neutral analytes are preferably extracted. Therefore, considering acidic analytes, pH should be maintained 1 to 2 units below the pKa of the compounds studied. On the other hand, considering basic analytes, sample pH should be kept 1 or 2 units higher than pKa of the compounds. Moreover, the extraction phase should be chemically stable at the pH used in the extraction step (Carasek and Merib 2015; Morés et al. 2018).

Temperature can also be an important variable, mainly for volatile analytes since it directly affects the distribution constant of the analytes between the matrix and the extraction phase. Therefore, this variable can be optimized according to the properties of the analytes and the sample matrix, in order to promote efficient distribution of the analytes from the sample matrix to the extraction phase (Moradi et al. 2019). Sample stirring can also be evaluated, since high stirring rates can reduce the boundary layer around the sorbent phase. This fact can influence the kinetics of the extraction procedure, particularly in the direct immersion mode (Qin et al. 2008).

Related to the desorption step, thermal and liquid desorption can be adopted depending on analytes and the analytical instrument used. Regarding thermal desorption, temperature and time of exposure of the TF-SPME into the desorption port can be optimized (Olcer et al. 2019). In liquid desorption, time and type of solvent are often evaluated. In both cases, attention needs to be paid in order to allow satisfactory desorption without damaging the TF-SPME coating (Olcer et al. 2019; Carasek and Merib 2015; Corazza et al. 2017).

6.2 Coating Preparation Methods for TF-SPME

Different strategies can be adopted for the preparation of thin-films used as extraction phases. In this section, some of these strategies are briefly discussed, including dip coating, spin coating, electrospinning coating, and spray coating (Olcer et al. 2019).

6.2.1 Dip Coating

Dip coating is the methodology most used for synthesizing thin-films in TF-SPME (Gómez-Ríos et al. 2017). In this case, physical and chemical procedures can be employed. Regarding the physical procedure, the support is immersed directly into a solution, usually named slurry, that consists of a mixture of organic solvent, binder, dispersant, and the extraction phase; therefore, the immobilization occurs by physical accumulation on the surface of the support, without the need for chemical reactions (Mohammadzadeh et al. 2019). Related to the chemical process, the support is usually activated or pre-functionalized in order to form chemical bonds in the first layer. Subsequently, the other layers are accumulated physically over the others (Olcer et al. 2019; Zargar et al. 2017). Afterward, the substrate can be evaporated, and the thin-film layers can be controlled. Factors such as immersion time, number of dipping cycles, composition, concentration of the coating, etc., can influence the quality of the coating. This strategy can provide thin-films of several to hundreds of micrometres (Mohammadzadeh et al. 2019; Lončarević and Čupić 2019).

6.2.2 Spin Coating

Spin coating is a relatively simple strategy, which consists of adding the material (extraction phase) to a substrate. This mixture (extraction phase and solvent) is placed on a plate and subjected to rotation; therefore, due to the centrifugal force, the extraction phase can be dispersed and uniformly deposited on the surface on the plate, and the solvent can be evaporated (Boudriouna et al. 2017). Afterward, the thin-film is formed, and some variables are adjusted to ensure the best uniformity of the layer, such as rotation speed, solvent evaporation temperature, solution viscosity, and surface tension (Mishra et al. 2019; Sahoo et al. 2018). This method provides a fast and relatively low-cost process, in addition to allowing the deposition of several mixed or layered materials, providing films from micrometres to nanometres (Olcer et al. 2019; Carasek and Merib 2015; Morés et al. 2018; Moradi et al. 2019; Qin et al. 2008; Corazza et al. 2017; Gómez-Ríos et al. 2017; Mohammadzadeh et al. 2019; Zargar et al. 2017; Lončarević and Čupić 2019; Boudriouna et al. 2017; Mishra et al. 2019).

6.2.3 Spray Coating

This strategy employs a mixture containing the extraction phase dissolved in an appropriate solvent. The mixture passes through a nozzle and, with the aid of an inert gas, forms fine droplets of the aerosol (Olcer et al. 2019; Azis and Ismail 2015). This aerosol is sprayed on the surface of the support, and the film is formed with the evaporation of the solvent. Some parameters can be evaluated to ensure a uniform formation of the film, such as the composition of the extraction phase, mixture viscosity, drying rate, and surface tension (Olcer et al. 2019; Lončarević and Čupić 2019).

6.2.4 Electrospinning Coating

Electrospinning is one of the most recent methods of preparing TF-SPME films. The technique employs a syringe loaded with a solution containing the extraction phase; the system is then connected to a power supply to produce the thin-films (Olcer et al. 2019; Greiner and Wendorff 2007; Ficai et al. 2016). Some factors are crucial for the formation of the films, such as the solution viscosity, voltage, surface tension, flow, and temperature, among others. It is worth mentioning that the electrospinning time is important for the film thickness (Ficai et al. 2016; Reddy et al. 2016). This method provides thin-films with high surface areas and high stability without the need for glue. This technique can provide thin-films that present thicknesses from micro to nanometres (Olcer et al. 2019; Reddy et al. 2016).

An illustration of some strategies adopted to fabricate thin-films is shown in Figure 6.2. In addition, some features regarding cost, homogeneity, stability, etc., are mentioned.

6.3 Application of TF-SPME in Analytical Toxicology

TF-SPME is a promising technique for sample preparation due to the high extraction efficiency and versatility of producing thin-films of different materials with varied thicknesses. This versatility can significantly increase the range of compounds that can be extracted, as well as the applicability in different matrices. The development of methodologies based on TF-SPME has been growing in recent years, especially with the use in analytical toxicology. In the next sections, some applications in different matrices are highlighted, and discussions regarding the main features of those analytical methodologies are also mentioned.

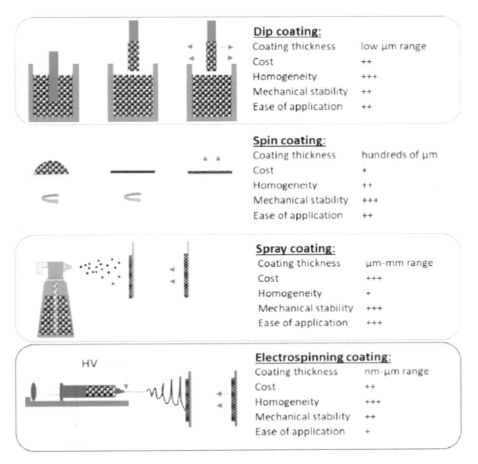

FIGURE 6.2 Techniques used for the preparation of thin-films in TFME. (Cost: +++ economic, + expensive; Homogeneity: +++ homogeneous, + poorly homogeneous; Mechanical stability: +++ stable, + unstable; Ease of application: +++ easy, + difficult, ++ moderate in all cases). (Source: Olcer et al. 2019).

6.3.1 Urine Samples

Human urine is an important biological matrix used for the analysis of a number of compounds of interest to analytical toxicology. This matrix exhibits some advantages, such as non-invasive collection and the possibility of obtaining large volumes, and it also contains a number of important metabolites. Further, the detection window of this matrix is longer than other biological matrices, such as blood or plasma. On the other hand, its composition is heterogeneous and comprised of different amounts of components, such as salts and proteins (Alemayehu et al. 2020; Montesano et al. 2014). In order to overcome some issues regarding the complexity of urine samples, TF-SPME has been successfully applied in this type of matrix.

In a recent study, an analytical methodology was proposed to determine some hormones using TF-SPME coupled with a 96-well plate system using a biosorbent as an extraction phase (do Carmo et al. 2019). In this particular case, the thin-films were produced from bracts obtained from *Aracucaria angustifolia* trees. This natural product was crushed in a knife mill, and particles with 200 mesh were selected and fixed with adhesive tape over the stainless-steel pins of a 96-well plate system. A scheme of the strategy for producing these thin-films is shown in Figure 6.3. In this study, the hormones estrone, 17-β estradiol, estriol, and 17-α ethinylestradiol were extracted and determined in human urine samples by high-performance liquid chromatograph with a fluorescence detector. In this case, limits of detection (LODs) from 0.3 to 3.03 µg L^{-1} and recoveries from 71 to 107% were obtained in different urine samples. This method consisted of a high-throughput alternative, since up to 96 samples can be processed simultaneously.

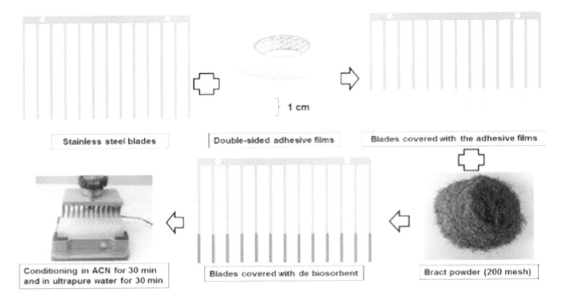

FIGURE 6.3 Diagram of the preparation of the blades with bract for the 96-well plate system for use in the extraction of @@oestrogens in urine. (Source: do Carmo et al. 2019).

In another application, an analytical method was proposed to determine 6 aldehyde biomarkers in patients with lung cancer (Liu and Xu 2017). Thin-films comprised of a metal-organic framework (PS/MOF-199) were produced to obtain nanofibers using the electrospinning strategy. Some micrographs of the films produced are shown in Figure 6.4. The films were fixed in plastic cartridges with two sieves, and the urine samples containing the previously derivatized analytes were analyzed using this experimental setup. The aldehydes were extracted by the metal-organice framework (MOF) nanofiber with subsequent desorption in a mixture of solvents and analysis by high-performance chromatography with UV/Vis detector. The developed method exhibited LOD from 4.2 to 17.3 nmol L^{-1} with recoveries from 82 to 112% for the aldehydes under study, having been applied to eight urine samples. The methodology with MOF/TF-SPME nanofibers was seen to be an interesting alternative for the early diagnosis of this disease.

Quetiapine and clozapine are two compounds that have been used in the treatment of schizophrenia. However, misuse can cause some problems, including depression and adverse effects on the cardiovascular system. A TF-SPME-based analytical methodology was developed for the determination of the drugs quetiapine and clozapine by high-performance liquid chromatography with UV/Vis detector (Li et al. 2016). In this study, magnetic octadecylsilane(ODS)-polyacrylonitrile(PAN)thin-films were produced using SiO_2@$Fe3O4$ nanoparticles through a spraying strategy. The synthetic procedure and the steps of the analytical method are presented in Figure 6.5. The method allowed for recoveries of 99% for

FIGURE 6.4 Scanning electron microscope images obtained from the nanofiber thin-film PS/MOF-199 in 3000x magnification and PS in 5000x magnification, respectively. (Source: Liu and Xu 2017).

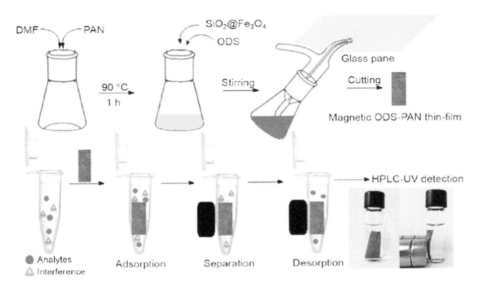

FIGURE 6.5 Schematic diagram of the preparation of the thin-films of the ODS-PAN magnetic particles and the extraction/desorption process of the analytes. (Source: Li et al. 2016).

clozapine and 104–110% for quetiapine, with LOD of 0.003 μg mL^{-1} for both compounds. Moreover, other applications and the analytical features of TF-SPME applied in urine samples are shown in Table 6.1.

6.3.2 Blood (Plasma/Serum)

Blood (plasma/serum) is a complex matrix comprising carbohydrates, lipids, amino acids, salts, metabolites, and molecules that can be monitored as biomarkers. This high complexity is a formidable challenge for the development of analytical methodologies. In addition, collection and storage are more difficult compared to urine, and other components often need to be added to the samples to prevent matrix degradation (Anderson and Anderson 2002; Byrne et al. 2020). In this context, sample preparation techniques are highly necessary, and TF-SPME is an interesting alternative for use with this matrix. Blood is made up of cells suspended in blood plasma. Plasma constitutes about 55% of the blood, being mostly water (92%), in addition to proteins, minerals, hormones, and others. To obtain the plasma, centrifugation is performed to separate it from the blood cells, and an anticoagulant is added, such as ethylenediaminetetraacetic acid (EDTA), heparin, or sodium fluoride, etc. The serum, on the other hand, has a composition like plasma, but it is obtained from coagulated blood before the centrifugation step (Luque-Garcia and Neubert 2007; Niu et al. 2018).

An analytical method that combined TF-SPME with desorption corona beam ionization-mass spectrometry (DCBI) to determine citalopram, sertraline, and fluoxetine in plasma samples was also proposed (Chen et al. 2016). The film used in this approach consisted of sub-micron-sized highly ordered mesoporous silica-carbon composite fibres (OMSCFs) produced by the electrospinning method followed by carbonization. The films achieved pore sizes between 10 and 100 μm, and the methodology is shown in Figure 6.6. LODs were 1 ng mL^{-1} for citalopram, 0.2 ng mL^{-1} for sertraline, and 0.3 ng mL^{-1} for fluoxetine, and recoveries ranging from 83.6 to 116.9% for the three antidepressants were achieved. The use of TF-SPME proved to be advantageous, providing satisfactory enrichment factors, eliminating matrix interference, and allowing for extractions within 5 min.

Recently, an analytical method was developed using a thin-film combining zeolitic imidazole framework (ZIF-8) with polypyrene nanocomposite, layered double hydroxide (LDH), and cotton yarn to obtain the thin-film of a cotton yarn-polypyrrole-layered double hydroxide-zeolitic imidazole framework-8 composite (CY-PPy-LDH-ZIF-8). This extraction phase was used for the extraction of

TABLE 6.1

TF-SPME application in analytical toxicology in human samples.

Extraction Phase	Matrix	Analytes	Coating Preparation Method	LOD	Instrumental Technique	Reference
Bract	Urine	Oestrogen Hormones	Powder added on double-sided tape	0.03–3.03 µg L^{-1}	HPLC-FLD	(do Carmo et al. 2019)
PS/MOF-199	Urine	Aldehydes (C$_4$-C$_9$)	Electrospinning	4.2–17.3 nmol L^{-1}	HPLC-UV	(Liu and Xu 2017)
ODS-PAN	Urine	Quetiapine Clozapine	Spraying	0.003 µg mL^{-1}	HPLC-UV	(Li et al. 2016)
OMSCF	Plasma	Citalopram Sertraline Fluoxetine	Electrospinning	0.3–1 ng mL^{-1}	DCBI-MS	(Chen et al. 2016)
CY-PPy-LDH-ZIF-8	Plasma and Food	Quercetin	Chemical Modification on Cotton Yarn	0.21 µg L^{-1}	HPLC-UV	(Jafari and Hadjmohammadi 2020)
NF/NiONWs/Co$_3$O$_4$	Urine and Plasma	Diclofenac	Electrodeposition/Cyclic Voltammetry	0.21–0.42 µg L^{-1}	HPLC-UV	(Ghani et al. 2019)
Tenax TA	Epithelial Cervical Carcinoma	18 VOCs	Dipping	Not Reported.	GC-MS	(Nozoe et al. 2015)
C18	Liver Microsomes	RepaglinideDA-RPGDC-RPG	Spraying	2.0 ng mL^{-1} (1)	LC-MS/MS	(Simões et al. 2015)
PS/G	Exhaled Breath Condensed	Aldehydes (C$_4$-C$_9$)	Electrospinning	4.2–19.4 nmol L^{-1}	HPLC-VWD	(Huang et al. 2015)
HBL	Saliva	49 Prohibited Substances6 Endogenous Steroids	Spraying	(49) 0.004–0.5 ng mL^{-1} (6) 16–185 pg mL^{-1} (1)	LC-MS/MS	(Bossonneau et al. 2014)
C18-PAN	Plasma	Benzodiazepines	Spraying	0.1–0.3 ng mL^{-1}	LC-MS/MS	(Mirnaghi et al. 2011)
C18-PAN	Blood	Diazepam	Dipping	0.3 µg mL^{-1}	LC-MS/MSDRAT-MS/MS	(Mirnaghi and Pawliszyn 2012)
HLB-PAN	Plasma	25 Prohibited Substances	Spraying	0.25–10 ng mL^{-1} (1)	LC-MS/MS	(Reyes-Garcés et al. 2014)
HLB-PAN	Plasma	NicotineN,N-diethyl-meta-toluamideDiclofenac	Dipping	5.0–10 ng mL^{-1} (1)	LC-MS/MS	(Boyacı et al. 2018)
PA-EG	Plasma	Carvedilol	Chemical reaction with subsequent film formation after polymerization	4.5 ng mL^{-1}	Spectrofluorimeter	(Karimi et al. 2016)
C18-silica glass	Plasma	Benzodiazepines	Dipping	0.4–0.7 ng mL^{-1}	LC-MS/MS	(Mirnaghi et al. 2012)
Cellulose-phenyl isocyanate	Urine	Oestrogenic Hormones	Chemical Modification	0.05–0.23 µg L^{-1}	HPLC-UV	(Saraji and Farajmand 2013)
PS/OCNTs	Urine	Benzo(a)pyrene1-Hydroxypyrene	Electrospinning	Not Reported.	MALDI-TOF-MS	(He et al. 2014)
Polystyrene	Urine	Methadone	Dipping	10 µg L^{-1} (1)	GC-MS	(Ríos-Gómez et al. 2017)

(1) These studies did not show LODs. However, the LOQs are mentioned in this Table.

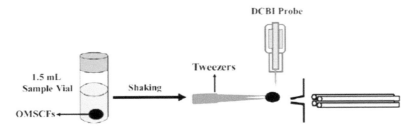

FIGURE 6.6 Diagram of experimental procedure of TF-SPME extraction and DCBI desorption, followed by injection in mass spectrometry system.

quercetin in human plasma and food samples (Jafari and Hadjmohammadi 2020). The LOD obtained was 0.21 µg L^{-1}, and reproducibility was considered satisfactory, with intraday precision from 3.8 to 5.9% and interday from 2.9 to 5.1%. The method proved to be reliable for the determination of quercetin in plasma, with short extraction time (20 min) and simple production of the thin-films of CY-PPy-LDH-ZIF8.

Another approach for the analysis of human plasma was based on a thin nickel film (NF) obtained with electrochemical deposition of nickel oxide nanoworms (NiONWs) (Ghani et al. 2019). This material was treated with Co$_3$O$_4$ by cyclic voltammetry, forming an NF/NiONWs/Co$_3$O$_4$ film. This extraction phase was evaluated in TF-SPME to determine diclofenac by high-performance liquid chromatography with UV detector. LOD of 0.42 µg L^{-1} and recoveries between 90 and 95% were achieved in plasma samples. Other applications of TF-SPME in blood (plasma/serum) are presented in Table 6.1.

6.3.3 Other Matrices Evaluated by TF-SPME

The application of TF-SPME in other complex matrices of interest in analytical toxicology has also been reported. A study involving the determination of volatile metabolites from cancer cells obtained from human epithelial cervical carcinoma was developed using TF-SPME and gas chromatography-mass spectrometry (GC-MS) (Nozoe et al. 2015). In this case, a Tenax TA film (poly (2,6-diphenyl-p-phenylene oxide)) was used. This extraction phase consisted of a porous material with a good adsorption capacity and prepared by dip-coating strategy. Some micrographs obtained by SEM (scanning electron microscopy) and AFM (atomic force microscopy) are shown in Figure 6.7. This phase was an interesting alternative since it exhibited high extraction capacity for the volatile compounds under study. In addition, *in vitro* evaluation of the Tenax TA film proved to be a reliable analytical method for identifying possible cancer biomarkers.

In another study, a thin-film composed of C18 was combined with a 96-well plate system and used for the determination of repaglinide (RPG) and its metabolites (DA-RPG and DC-RPG) in human liver microsomes, with analysis performed by liquid chromatography mass spectrometry (LC-MS)/MS

FIGURE 6.7 Images obtained by SEM in (a) surface of Tenax TA thin-film and (b) cross-section with film deposited in silicon wafer. Image obtained by AFM in (c) Tenax TA thin-film surface. (Source: Nozoe et al. 2015).

(Simões et al. 2015). RPG is especially important for treatment regulating type 2 diabetes; therefore, the monitoring of these compounds in patients with this disease is of great importance. The developed method allowed for high-throughput analysis, obtaining lower limits of quantification (LLOQs) of 0.2 ng mL^{-1} for RPG and its metabolites, with recoveries varying from 83 to 99% for the three analytes.

Moreover, a polystyrene/graphene (PS/G) nanofiber film obtained by electrospinning method was applied in TF-SPME for the determination of 6 aldehydes (C_4-C_9) in exhaled breath condensate (EBC) as a possible early diagnosis of patients with lung cancer, using high-performance liquid chromatography with variable wavelength detector for quantification (Huang et al. 2015). Using this analytical methodology, LODs ranging from 4.2 nmol L^{-1} for pentanal to 19.4 nmol L^{-1} for nonanal were obtained. Moreover, recoveries from 79.8% for heptanal to 105.6% for hexanal were determined. The application was successfully assessed in 15 individuals, 7 patients diagnosed with lung cancer and 8 healthy patients. The concentrations of pentanal and hexanal in sick patients were higher than those obtained from healthy people.

In another application, a metabolomic profile of saliva was obtained using TF-SPME and LC-MS/MS. The method was applied to *ex vivo* and *in vivo* analysis in human saliva samples using a hydrophilic lipophilic balanced (HLB) particle film (Bossonneau et al. 2014). This methodology allowed for the quantification of up to 49 substances, such as cannabinoids, steroids, narcotics, β blockers, stimulants, $β_2$ agonists, glucocorticosteroids, hormones, and other anabolics with LOQs ranging from 0.004 ng mL^{-1} for methadone to 0.98 ngmL^{-1} for budesonide, with recoveries ranging from 4 to 101%. Using this approach, it was also possible to determine endogenous steroids, including cortisol, testosterone, progesterone, estrone, estradiol, and estriol with LOQs ranging from 16 pg mL^{-1} for estriol to 178 pg mL^{-1} for cortisol. The applicability in saliva samples was tested, and the method exhibited satisfactory analytical performance in detecting all the compounds in the samples. The method consisted of an important alternative for analytical toxicology to determine these analytes in a challenging matrix. Other features of TF-SPME based methodologies applied in analytical toxicology are shown in Table 6.1.

6.4 Conclusion

This chapter explored some of the theoretical aspects and applications of TF-SPME in analytical toxicology. Since its proposal, different configurations, formats, and coatings have been developed and applied in order to determine a wide range compounds in biological matrices. Moreover, some of the main parameters of optimization were discussed, and recent applications were also highlighted. The toxicological interests of using TF-SPME were mainly focused on matrices, such as urine and blood (plasma/serum). However, other matrices were also explored, such as saliva, cells from different tissues, and exhaled breath. Due to its versatility and unique features, TF-SPME exhibits great potential to be explored and improved, particularly related to the development of alternative sorbent materials. Moreover, this technique allows for the possibility of full automation, which is highly desirable in analytical toxicology.

REFERENCES

Alemayehu, Yitayal Addis, Seyoum Leta Asfaw, and Tadesse Alemu Terfie. "Nutrient Recovery Options From Human Urine: A Choice For Large Scale Application." *Sustainable Production and Consumption* 24, (2020): 219–31.

Anderson, N. Leigh, and Norman G. Anderson. "The Human Plasma Proteome: History, Character, and Diagnostic Prospects." *Molecular and Cell Proteomics* 1, (2002): 845–67.

Azis, Farhana, and Ahmad Fauzi Ismail. "Spray Coating Methods for Polymer Solar Cells Fabrication: A Review." *Materials Science in Semiconductor Processing* 39, (2015): 416–25.

Bossonneau, Vincent, Ezel Boyaci, Malgorzata Maciazek-Jurczyk, and Janusz Pawliszyn. "In Vivo Solid Phase Microextraction Sampling of Human Saliva for Non-invasive and On-site Monitoring." *Analytica Chimica Acta* 26, (2014): 35–45.

Boudriouna, Azzedine, Mahmoud Chakaroun, Alexis Fischer. 2017. "Organic Light-emitting Diodes." In *An Introduction to Organic Lasers*. France: ISTE Press – Elsevier.

Boyaci, Ezel, Barbara Bojko, Nathaly Reyes-Garcés, Justen J. Poole, Germán Augusto, Gómez-Ríos, Alexandre Teixeira, Beate Nicol, and Janusz Pawliszyn. "High-Throughput Analysis Using Non-depletive Spme: Challenges and Applications to the Determination of Free and Total Concentrations in Small Sample Volumes." *Scientific Reports* 8, (2018): 1167.

Bruheim, Inge, Xiaochuan Liu, and Janusz Pawliszyn. "Thin-Film Microextraction." *Analytical Chemistry* 75, (2003): 1002–10.

Byrne, Hugh J., Franck Bonnier, Jennifer McIntyre, and Drishya Rajan Parachalil. "Quantitative Analysis of Human Blood Serum Using Vibrational Spectroscopy." *Clinical Spectroscopy* 2, (2020): 100004.

Carasek, Eduardo, and Josias Merib. "Membrane-based Microextraction Techniques in Analytical Chemistry: A Review." *Analytica Chimica Acta* 880, (2015): 8–25.

Chen, Di, Yu-Ning Hu, Dilshad Hussain, Gang-Tian Zhu, Yun-Qing Huang, and Yu-Qi Feng. "Electrospun Fibrous Thin Film Microextraction Coupled with Desorption Corona Beam Ionization-mass Spectrometry for Rapid Analysis of Antidepressants In Human Plasma." *Talanta* 152, (2016): 188–95.

Corazza, Gabriela, Josias Merib, Hérica A. Magosso, Otávio R. Bittencourt, and Eduardo Carasek. "A Hybrid Material as a Sorbent Phase for the Disposable Pipette Extraction Technique Enhances Efficiency for the Determination of Phenolic Endocrine-disrupting Compounds." *Journal of Chromatography A*. 1513, (2017): 42–50.

Cudjoe, Erasmus, Dietmar Hein, Vuckovic Dajana, and Janusz Pawliszyn. "Investigation of the Effect of the Extraction Phase Geometry on the Performance of Automated Solid-Phase Microextraction." *Analytical Chemistry* 81, (2009): 4226–32.

do Carmo, Sângela Nascimento, Josias Merib, and Eduardo Carasek. "Bract as Novel Extraction Phase in Thin-film Spme Combined with 96-well Plate System for the High-throughput Determination of Estrogens in Human Urine by Liquid Chromatography Coupled to Fluorescence Detection." *Journal of Chromatography B* 1118–1119, (2019): 17–24.

Ficai, Denisa, Madalina Georgiana Albu, Maria Sonmez, Anton Ficai, and Ecaterina Andronescu. 2016. "Advances in the FIeld of Soft Tissue Engineering: From Pure Regenerative to Integrative Solutions." In *Nanobiomaterials in Soft Tissue Engineering*, edited by Alexandru Mihai Grumezescu. Romania: William Andrew.

Ghani, Milad, Sayed Mehdi Ghoreishi, Shima Salehinia, Narjessadat Mousavi, and Hanieh Ansarinejad. "Electrochemically Decorated Network-like Cobalt Oxide Nanosheets on Nickel Oxide Nanoworms Substrate as a Sorbent for the Thin Film Microextraction of Diclofenac." *Microchemical Journal* 146, (2019): 149–56.

Gómez-Ríos, Germán Augusto, Marcos Tascon, Nathaly Reyes-Garcés, Ezel Boyaci, Justen Poole, and Janusz Pawliszyn. "Quantitative Analysis of Biofluid Spots by Coated Blade Spray Mass Spectrometry, A New Approach to Rapid Screening." *Science Reports* 7, (2017): 16104.

Grandy, Jonathan J., Ezel Boyac, and Janusz Pawliszyn. "Development of a Carbon Mesh Supported Thin Film Microextraction Membrane as a Means to Lower the Detection Limits of Benchtop and Portable GC/MS Instrumentation." *Analytical Chemistry* 88, (2016): 1760–7.

Greiner, Andreas, and Joachim H. Wendorff. "Electrospinning: A Fascinating Method for the Preparation of Ultrathin Fibers." *Angewandte Chemie International Edition* 46, (2007): 5670–703.

He, Xiao-Mei, Gang-Tian Zhu, Jia Yin, Qin Zhao, Bi-Feng Yuan, Yu-Qi Feng. "Electrospun Polystyrene/Oxidized Carbon Nanotubes Film as Both Sorbent for Thin Film Microextraction and Matrix-assisted Laser Desorption/Ionization Time-of-Flight Mass Spectrometry." *Journal of Chromatography A* 1351, (2014): 29–36.

Huang, Jing, Hongtao Deng, Dandan Song, and Hui Xu. "Electrospun Polystyrene/Graphene Nanofiber Film as Novel Adsorbent of Thin Film Microextraction for Extraction of Aldehydes In Human Exhaled Breath Condensates." *Analytica Chimica Acta* 878, (2015): 102–8.

Jafari, Zahra, and Mohammad Reza Hadjmohammadi. "In Situ Growth of Zeolitic Imidazole Framework-8 on Woven Cotton Yarn for the Thin Film Microextraction of Quercetin in Human Plasma and Food Samples." *Analytica Chimica Acta* 1131, (2020): 45–55.

Jiang, Ruifen, and Janusz Pawliszyn. "Thin-Film Microextraction Offers Another Geometry for Solid-Phase Microextraction." *TrAC Trends in Analytical Chemistry*. 39, (2012): 245–53.

Karimi, Shima, Zahra Talebpour, and Noushin Adib. "Sorptive Thin Film Microextraction Followed by Direct Solid State Spectrofluorimetry: A Simple, Rapid and Sensitive Method for Determination of Carvedilol in Human Plasma." *Analytica Chimica Acta* 924, (2016): 45–52.

Li, Dan, Juan Zou, Pei-Shan Cai, Chao-Mei Xiong, and Jin-Lan Ruan. "Preparation of Magnetic Ods-Pan Thin-films for Microextraction of Quetiapine and Clozapine in Plasma and Urine Samples Followed by HPLC-UV Detection." *Journal of Pharmaceutical and Biomedical Analysis* 125, (2016): 319–28.

Liu, Feilong, and Hui Xu. "Development of a Novel Polystyrene/Metal-Organic Framework-199 Electrospun Nanofiber Adsorbent for Thin Film Microextraction of Aldehydes in Human Urine." *Talanta* 162, (2017): 261–7.

Lončarević, Davor, and Željko Čupić. 2019. "The Perspective of Using Nanocatalysts in the Environmental Requirements and Energy Needs of Industry." In *Industrial Applications of Nanomaterials*, edited by Sabu Thomas, Yves Grohens and Yasir Beeran Pottathara. France: Elsevier.

Luque-Garcia, Jose L., and Thomas A. Neubert. "Sample Preparation for Serum/Plasma Profiling and Biomarker Identification by Mass Spectrometry." *Journal of Chromatography A* 1153, (2007): 259–76.

Mirnaghi, Fatemeh S., Yong Chen, Leonard M. Sidisky, and Janusz Pawliszyn. "Optimization of the Coating Procedure for High-throughput 96-blade Solid Phase Microextraction System Coupled with LC-MS/MS for Analysis of Complex Samples." *Analytical Chemistry* 83, (2011): 6018–25.

Mirnaghi, Fatemeh S., Dietmar Hein, and Janusz Pawliszyn. "Thin-Film Microextraction Coupled to Mass Spectrometry and Liquid Chromatography Mass-Spectrometry." *Chromatographia* 76, (2013): 1215–23.

Mirnaghi, Fatemeh S., and Janusz Pawliszyn. "Reusable Solid-Phase Microextraction Coating for Direct Immersion Whole-blood Analysis and Extracted Blood Spot Sampling Coupled with Liquid Chromatography-Tandem Mass Spectrometry and Direct Analysis in Real Time-tandem Mass Spectrometry." *Analytical Chemistry* 84, (2012): 8301–9.

Mirnaghi, Fatemeh S., Maria Rowena N. Monton, and Janusz Pawliszyn. "Thin-FIlm Octadecyl-Silica Glass Coating for Automated 96-blade Solid-Phase Microextraction Coupled with Liquid Chromatography-Tandem Mass Spectrometry for Analysis of Benzodiazepines." *Journal of Chromatography A* 1246, (2012): 2–8.

Mishra, Abhilasha, Neha Bhatt, and A. K. Bajpai. 2019. "Nanostructured Superhydrophobic Coatings for Solar Panel Applications." In *Nanomaterias-Based Coatings.*, edited by Phuong Nguyen Tri, Sami Rtimi and Claudiane M. Ouellet Plamondon. Canada: Elsevier.

Mohammadzadeh, Alireza, Seyedeh Kiana Naghib Zadeh, Mohammad Hassan Saidi, Mahdi Sharifzadeh. 2019. "Mechanical Engineering of Solid Oxide Fuel Cell Systems: Geometric Design, Mechanical Configuration, and Thermal Analysis." In *Design and Operation of Solid Oxide Fuel Cells*, edited by Mahdi Sharifzadeh, 85–130. London: Academic Press.

Montesano, Camilla., Manuel Sergi, Sara Odoardi, Maria Chiara Simeoni, Dario Compagnone, and Roberta Curini. "A µ-SPE Procedure for the Determination of Cannabinoids and Their Metabolites in Urine by LC-MS/MS." *Journal of Pharmaceutical and Biomedical Analysis* 91, (2014): 169–75.

Moradi, Ebrahim, Homeira Ebrahimzadeh, and Zahra Mehrani. "ElecTrospun Acrylonitrile Butadiene Styrene Nanofiber Film as an Efficient Nanosorbent for Head Space Thin Film Microextraction of Polycyclic Aromatic Hydrocarbons From Water And Urine Samples." *Talanta* 205, (2019): 120080.

Morés, Lucas, Adriana Neves Dias, andEduardo Carasek. "Development of a High-throughput Method Based on Thin-filmmicroextraction Using a 96-Well Plate System with a Cork Coatingfor the Extraction of Emerging Contaminants in River Watersamples." *Journal of Separation Science* 41, (2018): 697–703.

Niu, Zongliang, Weiwei Zhang, Chunwei Yu, Jun Zhang, and Yingying Wen. "Recent Advances in Biological Sample Preparation Methods Coupled with Chromatography, Spectrometry and Electrochemistry Analysis Techniques." *TrAC Trends in Analytical Chemistry* 102, (2018): 123–46.

Nozoe, Takuma, Shigemi Goda, Roman Selyanchyn, Tao Wanga, Kohji Nakazawaa, Takeshi Hiranoa, Hidetaka Matsuic, and Seung-Woo Leea. "In Vitro Detection of Small Molecule Metabolites Excreted From Cancer Cells Using A Tenax Ta Thin-film Microextraction Device." *Journal of Chromatography B* 991, (2015): 99–107.

Olcer, Yekta Arya, Marcos Tascon, Ahmet E. Eroglu, and Ezel Boyaci. "Thin-Film Microextraction: Towards Faster and More Sensitive Microextraction." *TrAC Trends in Analytical Chemistry* 113, (2019): 93–101.

Pawliszyn, Janusz.2012. *Handbook of Solid Phase Microextraction*. Ontario: Chemical Industry Press.

Qin, Zhipei, Leslie Bragg, Gangfeng Ouyang, Janusz Pawliszyn. "Comparison of Thin-film Microextraction and Stir Bar Sorptive Extraction for the Analysis of Polycyclic Aromatic Hydrocarbons in Aqueous Samples with Controlled Agitation Conditions." *Journal of Chromatography A* 1196–1197, (2008): 89–95.

Reddy, A. Babul, G. Siva Mohan Reddy, Veluri Sivanjineyulu, J. Jayaramudu, Kokkarachedu Varaprasad, and E. Rotimi Sadiku. 2016. "Hydrophobic/Hydrophilic Nanostructured Polymer Blends." In *Design and Applications of Nanostructured Polymer Blends and Nanocomposite Systems*, edited by Sabu Thomas, Robert Shanks and Sarathchandran Chandrasekharakurup. India: William Andrew.

Reyes-Garcés, Nathaly, Barbara Bojko, and Janusz Pawliszyn. "High Throughput Quantification of Prohibited in Plasma Using Thin Film Solid Phase Microextraction." *Journal of Chromatography A* 1374, (2014): 40–49.

RiaziKermani, Farhad. "Optimization of Solid Phase Microextraction for Determination of Disinfection By-products in Water." *UWSpace*, (2012).

Ríos-Gómez, Julia, Rafael Lucena, and Soledad Cárdenas. "Paper Supported Polystyrene Membranes for Thin Film Microextraction." *Microchemical Journal*(2017): 90–95.

Sahoo, S. K., B. Manoharan, N. Sivakumar. Introduction – Why Perovskite and Perovskite Solar Cells? In *Perovskite Photovoltaics*. India: Academic Press. 2018.

Saraji, Mohammad, and Bahman Farajmand. "Chemically Modified Cellulose Paper as Thin Film Microextraction Phase." *Journal of Chromatography A* 1314, (2013): 24–30.

Simões, Rodrigo Almeida, Pierina Sueli Bonato, Fatemeh S. Mirnaghi, Barbara Bojko, and Janusz Pawliszyn. "Bioanalytical Method for *In Vitro* Metabolism Study of Repaglinide Using 96-blade Thin-film Solid-Phase Microextraction and LC-MS/MS." *Bioanalysis* 7, (2015): 65–77.

Wilcockson, John B., and Frank A. P. Gobas. "Thin-Film Solid-Phase Extraction to Measure Fugacities of Organic Chemicals with Low Volatility in Biological Samples." *Environmental Science and Technology* 35, (2001): 1425–31.

Zargar, Tahereh, Taghi Khamyamian, and Mohammad T. Jafari. "Immobilized Aptamer Paper Spray Ionization Source Ion Mobility Spectrometry." *Journal of Pharmaceutical and Biomedical Analysis* 132, (2017): 232–7.

7 Application of Single Drop Microextraction in Analytical Toxicology

Archana Jain[1], Manju Gupta[2] and Krishna K. Verma[1]
[1]Department of Chemistry, Rani Durgavati University, Madhya Pradesh, India
[2]Department of Chemistry, St. Aloysius College (Autonomous), Madhya Pradesh, India

CONTENTS

7.1 Introduction 131
7.2 Strategy of Microextraction 133
    7.2.1 Complexity of Biological Samples 133
    7.2.2 Single Drop Microextraction 133
    7.2.3 Drop Protection 133
7.3 Modes of Extraction and Applications 135
    7.3.1 Direct Immersion SDME 135
    7.3.2 Headspace SDME 136
    7.3.3 Three-Phase SDME 137
7.4 Recent Advances 138
    7.4.1 Newer Solvents 138
    7.4.2 Automation 139
7.5 Conclusions 140
Acknowledgement 140
References 141

7.1 Introduction

Miniaturized approaches for extraction of analytes using solvents are collectively grouped under liquid-phase microextraction (LPME) or solvent microextraction, and the growth in activities in this field began when a drop of organic solvent of a few microliters size was employed for extraction, the technique being termed single drop microextraction (SDME). The innovative format of SDME, consisting of a drop suspended at the needle tip positioned in the aqueous sample solution, has been the source of a variety of SDME modes and a collection of LPME techniques integrating different principles (Tang et al. 2018). Initial experiments involved aspirating the gaseous sample over a collector drop hanging at the tip of a silica capillary, and diverting the drop to the automated analysis system (Liu and Dasgupta 1995). Later, the extraction was performed into a 1 μL drop of n-octane suspended from a microsyringe needle tip being placed in a stirred aqueous sample (direct immersion SDME, DI-SDME) (Jeannot and Cantwell 1997) or the headspace of the sample (headspace-SDME, HS-SDME) (Theis et al. 2001), and after extraction analyzed by gas chromatography (GC). Two modifications were suggested to overcome the low pre-concentration factor of analytes observed in SDME. First was a three-phase extraction mode (liquid-liquid-liquid microextraction, LLLME), where 30 μL of n-octane layer confined inside a PTFE ring was placed over the donor aqueous sample and in turn accommodated a 1 μL aqueous acceptor drop, supported by a microsyringe (Ma and Cantwell 1999). The other modification was continuous-flow microextraction, in which the aqueous sample solution was continuously circulated over the organic acceptor drop held inside an extraction chamber, and after a given period of extraction, the drop was withdrawn with a GC syringe for analysis (Liu and Lee 2000). Still another route to optimize pre-concentration, called directly suspended droplet microextraction, involved the addition of water immiscible and low density solvent, such as n-butyl acetate or 1-octanol, to the aqueous sample, which was stirred using a magnetic bar to make a vortex. The organic solvent as a self-stable drop was pulled to the bottom of the conical area and kept rotating while in contact with the sample. After a pre-determined time, the extract was picked up with a microsyringe for analysis (Yangcheng et al. 2006). A simple modification to this method (solidification of floating organic drop microextraction) used a much smaller volume of a solvent, such as 1-undecanol, wiith a melting point in the range 10–30°C, which was solidified by placing the extraction vial in an ice bath after extraction. The frozen extract was gathered with a micro-spatula, allowed to melt in a micro-vial, and, thereafter, a portion of extract was injected into the chromatograph (Zanjani et al. 2007). Attempts to optimize extraction efficiency in a shorter time by increasing the surface area of the drop resulted in a compound drop technique, called bubble-in-drop, in which an air bubble was intentionally housed in the solvent drop (Williams et al. 2011, 2014). The extraction efficiency was greatly and quickly enhanced by mass transfer under the electric field in SDME (Song and Yang 2019) and in LLLME (Raterink et al. 2013; He et al. 2021).

LPME procedures are rapid and convenient to conduct, and they constitute an excellent alternative to classical solvent extraction that uses large volumes of toxic organic solvent, and where the final extract also necessitates a pre-concentration, usually by solvent evaporation before analysis. Thus, micro-volume extraction methods, particularly SDME, result in extract volume that can be directly used in analysis by an instrumental method. Calibration in SDME requires extraction of a series of standards under the condition as will be used for the sample, and refereeing the signal for the sample with the calibration graph. Both polar and non-polar substances can be extracted using an appropriate mode of SDME, taking adequate measures to enhance extraction, chromatography, and detection. There is freedom from analytes carryover since a renewed drop of solvent is employed for each sample. Other advantages are utilizing the whole single drop of extract in the final analysis to optimize sensitivity, and full automation of the extraction-analysis process.

SDME finds its application as an in-line extraction method for capillary electrophoresis (CE), where the analytes are extracted from a sample donor to the acceptor drop hanging at the inlet end of the capillary, simulating all possible modes of conducting SDME. As the sample volume in CE is only in the nanoliter range, extensive work in this area has been performed for automating and improving the reproducibility of on-line SDME-CE (ALOthman et al. 2012). SDME-CE coupling has the advantage of handling small volume samples, which could be diluted and analytes extracted into nanoliter-size droplets, thus increasing total pre-concentration and improving the limit of detection. An electro-extraction SDME has been coupled to CE and on-line mass spectrometric detection for sensitive analysis of metabolites (Oedit et al. 2021). A sensitive and precise capillary zone electrophoresis method for homocysteine thiolactone in urine has been developed utilizing extraction in chloroform by SDME and UV-detection (Purgat et al. 2020). In these methods, SDME plays an important role in increasing sensitivity since extraneous matter such as proteins and salts are excluded during extraction.

SDME is a widely accepted technique constituting a major step of sample preparation for trace analysis by a large number of analytical methods, and demonstrating an ample account of creative ideas to advance extraction and analysis. Recently, a short historical account of SDME and its automation and newer developments was reviewed by Kokosa (2015), Tang et al. (2018), and Jain and Verma (2011, 2020); modes of SDME were reviewed by Mogaddam et al. (2019), Tegladza et al. (2020), and Przyjazny (2019); chemical reactions in LPME were reviewed by Basheer et al. (2019); and applications for biomolecules were reviewed by Kailasa et al. (2021).

7.2 Strategy of Microextraction

7.2.1 Complexity of Biological Samples

Biological samples for analysis of drugs and toxic substances have complicated matrices and are often not available in large quantities, making a sharp distinction from environmental water analysis. The object of sample preparation is to remove interfering sample matrix substances, enrich the target analytes to optimize signal-to-noise ratio in the final analytical method, and phase transfer so that the extract is compatible with the instrumental method (Reddy et al. 2019). SDME is typically not taken to a steady state to allow a reasonable extraction, and thus it is not an exhaustive process. Use of a few microliters volume of drops for SDME results in a highly reduced ratio of extraction drop-to-sample volume; therefore, due to kinetic control, a high pre-concentration of analytes is achieved before any significant diffusion of matrix substances can occur. This is further supported by careful selection of the solvent for a favourable clean-up of extracted analytes. The kinetic control on SDME distinguishes this technique from one or another mode of dispersive liquid-liquid microextraction (DLLME), where attainment of a large enrichment factor in the shortest time is the outcome of a large surface area of extraction solvent droplets (Jain and Singh 2016). Thus, the sensitivity in SDME is principally due to low background noise in the final analysis, a situation favourable for biological sample analysis. Rapidity, low operational cost, utilization of general laboratory equipment, immensely reduced sample size, and consumption of extremely low volumes of extraction solvents make SDME a popular sample handling technique. Still other features of merit comprise applicability to analytes of diverse natures, ease of conducting derivatization in parity with separation and detection of analytes, diverse modes of extraction, use of the whole extract in the final analysis to gain optimum sensitivity, and easy full automation of methods.

SDME methods based on different modes have been reviewed for sample preparation in bioanalytical methods (Bitas and Samanidou 2020; Hansen et al. 2020; He and Concheiro-Guidan 2019). Sample complexity, method applicability, and quality of results affect the performance and rapidity of these modes. Thus, a particular mode cannot be selected in isolation, but due consideration should be given to the nature of analyte and the matrix, and the operations necessary before extraction, such as derivatization. These additional steps may affect the rapidity of the total method and the analytical precision.

7.2.2 Single Drop Microextraction

SDME is a miniaturized version of liquid-liquid extraction, replacing the separatory funnel with the common laboratory syringe and sample vials, and it is capable of working in a variety of modes to suit the nature of the analyte and the analytical problem (Figure 7.1). Only microliter volumes of extraction solvent are required, and due to high pre-concentration factors, the sample size is also small. The choice of low boiling and low viscosity organic solvent is due to extract compatibility with gas chromatography (GC). However, solvent volatility and drop instability are common problems. A variety of newer solvents mostly avoid shortcomings of volatile solvents, and they also enable final analysis by high-performance liquid chromatography (HPLC). Since SDME is performed with solvent drop freely hanging at the needle tip, drop dislodgement is still a vexing problem. This imposes a limit on sample stirring rates, a process commonly used to enhance the rate of extraction. Nevertheless, SDME is focused on achieving high enrichment factors and on mitigating sample matrix interferences.

7.2.3 Drop Protection

Extraction drop solvent evaporation, miscibility with water, high sample stirring rates, and temperatures of extraction are common reasons for accidental dislodgement of solvent drop of extraction. Modified needle tip with increased cross-section served to increase the adhesion force to stabilize the drop (Figure 7.2). A silicone ring provided stability for 45 min to a 5 μL hexane drop in DI-SDME at a

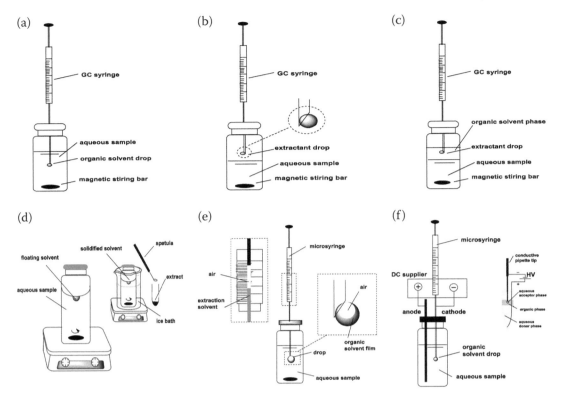

FIGURE 7.1 Modes of SDME. (A) Direct immersion SDME, (B) headspace SDME, (C) liquid-liquid-liquid microextraction, (D) directly suspended droplet microextraction (inset, solidification of floating organic drop microextraction), (E) bubble-in-drop microextraction, and (F) electroenhanced SDME (inset, electroenhanced liquid-liquid-liquid microextraction). Reproduced with permission from Elsevier, and American Chemical Society.

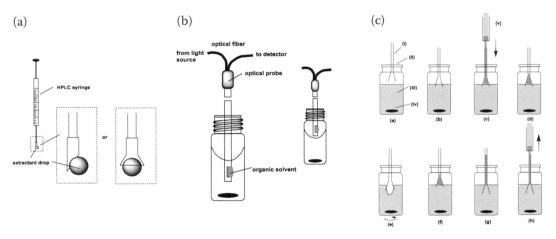

FIGURE 7.2 Solvent drop protection. (A) PTFE sleeve or funnel attached to microsyringe tip, (B) optical probe as solvent microdrop holder in direct immersion SDME (inset, optical probe in headspace SDME), and (C) bell-shaped extraction device assisted liquid-liquid microextraction. Reproduced with permission from Royal Society of Chemistry, American Chemical Society, and Elsevier.

stirring rate of 200 rpm (Fernances et al. 2012). Similarly, a large cross-section needle allowed extraction for up to 80 min with a stirring rate of 1,700 rpm using a solvent drop of 0.9 µL (Ahmadi et al. 2006). Other protective devices used were a plastic membrane on a wire holder accommodating 15 µL of toluene drop for HS-SDME (Ma and Ma 2017); copper fibre fixed to an SPME holder, coated with highly porous copper foam and in turn impregnated with organic solvent (Saraji et al. 2016); and a bell-shaped mesh holding 15 µL of deep eutectic solvent (DES) of choline chloride and oxalic acid (Mehravar et al. 2021). For the HPLC method, where a blunt-end needle is required for sample introduction, the needle was fitted with an angle-cut PTFE sleeve holding either 7 µL of 1-butanol (Pillai et al. 2009) or 20 µL of ionic liquid (IL) (Wen et al. 2013). An assembly consisting of a quartz capillary, funnel cap, and microsyringe was developed to support microextraction with bubble-in-drop (Xie et al. 2014). This assembly provided flexibility in accommodating air bubbles of different sizes and use of low-density organic solvents.

The sleeve attached to the needle of the sampling syringe could terminate in a small polytetrafluoroethylene (PTFE) funnel to hold a larger volume of extraction solvent, 3.5–20 µL; for a longer extraction period, 20–40 min; and a high stirring rate, 1,000–1,200 rpm (Tian et al. 2014; Sharma et al. 2011; Wang et al. 2012). The micro-funnel has been demonstrated to hold 400 µL of toluene as extraction solvent for 90 min (Saleh et al. 2014), a 30 µL volume of mixed solvent of decanoic acid and tetrabutylammonium hydroxide for 60 min (Lopez-Jimenez et al. 2008), and 12 µL drop of IL for 25 min at 80°C (He et al. 2012). In an interesting study on the selection of sleeve material for producing a non-polar solvent drop in DI-SDME, PTFE was observed to cause solvent spreading over the capillary tip due to the prominence of adhesive forces over cohesive forces. The situation was contrary with glass capillary, which produced a spherical drop, and the cohesive forces allowed formation of a stable bigger drop (7 µL) of IL, which was retracted fully after extraction (Nunes et al. 2021).

Systems in which the microsyringe was not used to hold the extraction drop employed an optical probe with an optical window to house 40 µL of the extraction solvent in DI-SDME (Zaruba et al. 2017) and HS-SDME modes (Zaruba et al. 2016). Still another system used a bell-shaped device, placed with an organic solvent lighter than water, adjusted on the surface of the liquid sample. The organic solvent formed a vortex in the aqueous sample on stirring the sample to begin the extraction, and it returned to its original position inside the bell on switching off the stirrer. The bell was lowered into the sample to raise the extract in the upper narrow tube of the bell to allow its collection by a microsyringe (Cabala and Bursova 2012).

7.3 Modes of Extraction and Applications

7.3.1 Direct Immersion SDME

In direct immersion SDME (DI-SDME), the solvent drop is kept immersed in the liquid sample for extraction of target analytes under pre-optimized conditions. The general method involves taking 0.5–3 µL of organic solvent in a microsyringe with a bevel tip needle that is pierced through the septum of a sample vial containing 1–5 mL of the test sample. The needle is immersed in the sample, keeping well below the meniscus, and a single drop of solvent is carefully formed, dangled at the needle tip. The sample is magnetically stirred at a low rate, typically 150–300 rpm, and after the given period of extraction, usually 15–30 min, the solvent drop is withdrawn into the syringe and utilized in analysis. Traditionally, DI-SDME makes use of low boiling water insoluble solvent for extraction of partially volatile substances, and GC is an analytical method of choice. There are a number of experimental parameters that could affect extraction and need to be optimized. Such variables include volumes of the sample and the extraction solvent, the nature of solvent, the presence of salt, temperature, stirring rate, and period of extraction (Kokosa 2015; Tegladza et al. 2020). The most important of these variables are the selection of the solvent for drop formation and sample stirring rates. Salts usually increase the extraction, but in many reports, the effect has been opposite.

Microextraction methods, including DI-SDME, have been reviewed to determine amphetamines as an example of the importance of illicit drugs in biological samples (Chalavi et al. 2019). Perchlorate is an iodine inhibitor in the thyroid gland, and its trace analysis is important to avoid incidences of increased perchlorate concentration in the human body. Ion-pair formation with a cationic surfactant, SDME with methyl isobuty ketone, and analysis by attenuated total reflectance spectroscopy have been used to determine perchlorate in human breast milk and urine (Chandrawanshi et al. 2018). Bubble-in-drop SDME using a mixed solvent system was demonstrated to have advantages over conventional SDME and SPE on application to growth hormones in the urine of farmed animals (George 2015). The utility of bubble-in-drop was further verified, and attainment of enrichment factors between 536 and 1097 was reported in the analysis of carbamate pesticides (Chullasat et al. 2020).

Lipid droplets are energy reservoir organelles that have a crucial role in lipid metabolism. Their undue intracellular accumulation is related to obesity, diabetes, steatosis, etc. A novel technique based on in-tip solvent microextraction was developed to separate phosphatidylcholines and triglycerides. A single lipid droplet was sucked into a nanotip that was subsequently filled with an organic solvent suitable for lipid extraction, and the extract was subjected to nano-electrospray ionization-mass spectrometry (Zhao et al. 2019). Metabolite concentration variations in single cells are significant for exploring the dynamic regulation of important biological processes, such as cell development and differentiation. A quantitative method for single-cell metabolites, glucose-phosphate has been proposed by combining a microwell array with droplet microextraction mass spectrometry (Feng et al. 2019). The extraction efficiency is greatly enhanced by acceleration of analyte mass transfer by the electric field (Song and Yang 2019). This principle has been utilized in the analysis of amphetamines in human urine.

7.3.2 Headspace SDME

In HS-SDME the solvent drop supported by the syringe needle tip is placed in the headspace of the aqueous sample to extract volatile or semi-volatile analytes that have emerged in the air space in the sample vial (Mogaddam et al. 2019). Thus, HS-SDME consists of a liquid-air-liquid system, but there is a greater choice in the physical state of the sample. Two factors, sample stirring and temperature, are vital factors since both high stirring rates and moderately higher temperatures can be used during extraction to promote mass transfer of volatile analytes to the headspace. The higher extraction recoveries in HS-SDME are due to larger diffusion coefficients of analytes in gaseous phases, and the effect is further augmented by bigger solvent drop due to rapid mass transfer. Relative to DI-SDME, there is a wider choice of solvents since the drop is not in contact with the aqueous sample; the only prerequisite is on fair involatility at the condition of extraction. The extract is free from involatile substances, such as salts, high boiling organics, proteins, etc., and particulate matter, making the whole extraction process convenient and sensitive in the final analytical procedure. HS-SDME in a vacuum has received recent attention in accelerating the extraction kinetics of analytes, and also enabling conduction of the sampling at ambient temperatures (Psillakis 2020).

To determine captopril in human serum, a drop of colloidal solution of Au nanoparticles was placed in the headspace of a sample to act as both acceptor/labelling agent for the thiol group of captopril. Next, the drop was injected in a glass microchip and detected by microchip-photothermal lens microscopy (Abbasi-Ahd et al. 2017). A temperature gradient HS-SDME method was found selective and sensitive sensor for ammonia based on the blue-emitting Ag nanocluster as fluorescence probe. A sample solution of ammonium salt was added over solid sodium hydroxide when the high temperature generated assisted ammonia vapours with contacting the silver nanocluster droplet in the headspace to decrease the fluorescence of the sensor (Dong et al. 2017). Owing to confinement of electrons and holes, CdSe/ZnS quantum dots have unique optical properties, and the luminescence was quenched on contact with a volatile species. A number of volatile species have been assayed by microvolume spectrofluorimetry (Costas-Mora et al. 2011). A sensitive detection of hydrogen sulphide in biosamples was done on contact of its vapours with Ag-Au core-shell nanoprism in HS-SDME. The analysis was completed by a smartphone camera and colour measuring software (Tang et al. 2019). A mixed reagent drop of Au nanoparticles and Tollens reagent (Ag ammonia complex) was used in HS-SDME as a sensor for formaldehyde in chicken and octopus flesh, utilizing redox reaction of the latter

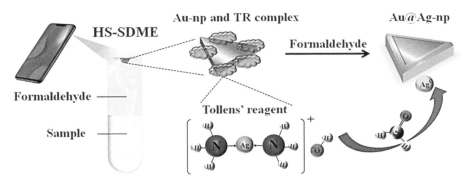

FIGURE 7.3 Schematic of the headspace-SDME smartphone nanocolorimetry for formaldehyde detection based on the reduction of Ag^+ (Tollens reagent) and coating of Ag on Au nanoparticles. Reproduced with permission from Elsevier.

with the Tollens reagent forming reddish orange Au@Ag nanoparticles (Figure 7.3). A smartphone camera was used for spectrophotometry (Qi et al. 2020). A drop of phosphoric acid was employed for the headspace absorption of ammonia and subjected to indophenol red/blue species formation by LPME, and colour measurement by micro-spectrophotometry (Jain et al. 2021). An automated flow-batch system was developed for headspace absorption of ammonia and its on-drop conductometric measurement. Absorption in boric acid provided a lower conductivity background and a wider linear range (Jiang et al. 2021).

7.3.3 Three-Phase SDME

This technique is very successful in extracting acidic (phenol, carboxylic acids) or basic (amines) substances from their aqueous solution (Kokosa 2015). Following the sequence of operations for attaining unionized forms of target compounds by the addition of acids to acidic substances, and bases to basic substances, and their extraction into a lower density organic solvent now floating over the aqueous sample phase, an aqueous drop of base or acid at the tip of a syringe is produced in the organic layer for back extraction of the acid or base, respectively. Amphetamines (He and Kang 2006) and azithromycin (Ebrahimzadeh et al. 2010) in urine have been analyzed by first extraction in n-hexane and n-octane, respectively, and SDME in phosphoric acid. The technique has been extended to other types of substances by involving different principles, such as complex formation (Costas-Mora et al. 2013). A homemade vial with a narrow neck to accommodate the intermediate organic layer made placement of an aqueous drop convenient, and it also allowed stirring of the aqueous sample with high rates (Bagheri et al. 2008). To hold the anorganic-aqueous compound droplet, a coupling microdevice was designed to produce droplets of different sizes, varying the volume ratio of the organic phase to the aqueous phase (Jahan et al. 2015). By using a 1.2 μL toluene-aqueous compound droplet (volume ratio 0.2:1), a 350 to 1,712 fold enrichment of statins was achieved within 4 min.

For application in CE, a drop of an acceptor phase covered with an organic layer was hung at the inlet tip of a separation capillary. By adjusting the pH of the aqueous sample, analytes in the neutral form were extracted into the organic layer, and then back-extracted into the acceptor phase. Application of this method was demonstrated for ionic arsenic by employing the carrier-mediated counter-transport using $CH_3(C_8H_{17})_3N^+Cl^-$ (Aliquat 336) in the organic layer (Cheng et al. 2013).

Electro-driven extraction involves active migration of charged analytes in an applied electric field, and the extraction completes in 2.5 min to 33.3 min. An online three-phase electroextraction setup has been developed (Figure 7.4) in which the extraction unit is coupled to a mass spectrometer by using a switching valve, syringe pump, and HPLC pump (He et al. 2021). The setup was applied to propranolol, amitriptyline, bupivacaine, and oxeladin in human urine and plasma samples. Type and composition of the organic phase and of the acceptor phase, and the extraction voltage and time were optimized. An ultrafast extraction within 30 s and enrichment factors in the range 105–569 were obtained. This online setup has great potential for high-throughput sample analysis.

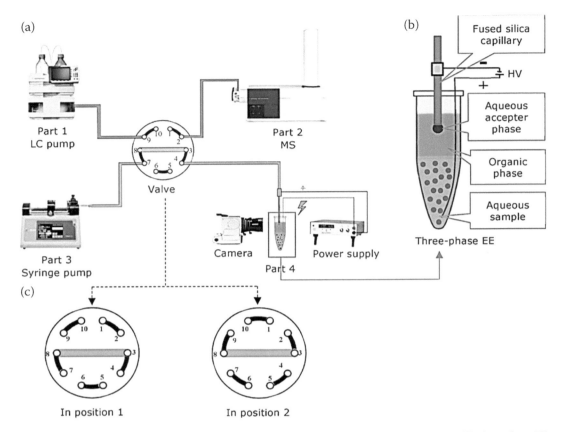

FIGURE 7.4 (A) The schematic diagram of the online three-phase electroextraction (EE) setup, (B) three-phase EE process inside the Eppendorf tube, and (C) the switching valve positions: Position 1, extraction process, and Position 2, flow injection transfer of extract to mass spectrometer. Reproduced with permission from Elsevier.

To avoid the inherent inconveniences in employing the lengthy amplification process, a magnetic three-phase SDME approach was developed for the quantification of nucleic acids in human serum. By integrating a fast, magnetic three-phase SDME and formation of hyperbranched DNA/Fe_3O_4 networks, triggered by nucleic acids, a highly sensitive method for nucleic acid detection was developed (Tang et al. 2020). A layer of dodecane was layered over the aqueous solution. A droplet of 3,3′,5,5′-tetramethylbenzidine and hydrogen peroxide affixed to the end of a magnetic bar was lowered into the organic phase. The DNA/Fe_3O_4 networks were then rapidly attracted to the reagent droplet and catalyzed the colour reaction. Micro-spectrophotometric detection was used to measure the colour in the drop. The networks were separated and enriched within 6 s, producing highly sensitive signals for the quantification of nucleic acids. The method has potential for application to other biomolecules.

7.4 Recent Advances

7.4.1 Newer Solvents

SDME does not seriously violate the principles of green chemistry since it requires only a few microliters volume of organic solvent (Carasek et al. 2021). Nonetheless, there is the requirement of avoiding toxic solvents as far as possible or their replacement by innocuous alternatives (Plotka-Wasylka et al. 2017). Different modes of extraction by SDME, and other methods of LPME, have specific solvent requirements (Kokosa 2019). There is a constant trend of search for solvents of adequate physico-chemical properties, such as boiling and melting points, density, 1-octanol/water partition constant, viscosity, and surface tension, which can provide suitable drop stability, increased extraction, compatibility with final analytical technique, and safety. Forensic toxicology needs sensitive and multi-analyte analyses on peripheral blood, urine, and gastric fluid. Such samples contain abundant amounts of blood proteins or phospholipids, which may produce viscous extracts that are difficult to handle and likely to vitiate the detector performance, especially mass ionization. Removal of such extraneous matter, and satisfactory extraction of multi-analytes of wider chemical nature, make solvent selection a challenging task.

Ionic liquids (ILs) have been found to be excellent alternatives to organic solvents, owing to their low volatility and flammability, excellent thermal stability, and ease of synthesis to specific applications. ILs are non-molecular ionic compounds consisting of a huge asymmetric organic cation (e.g., alkylated imidazolium, pyrrolidinium, or phosphonium) and a small organic or inorganic anion,(e.g., bis(trifluoromethylsulphonyl)imide or hexafluorophosphate) (Berthod et al. 2018). ILs allow the formation of a bigger and stable solvent drop in SDME, and working at elevated temperatures for longer periods gives better extraction and precision. Applications of ILs as extraction solvents for a large number of forensic drugs in biological samples have been reviewed (De Boeck et al. 2019).

An interesting class of ILs, called magnetic ionic liquids (MILs), has emerged as extraction solvents in SDME. MILs are produced by incorporating a paramagnetic component in the IL structure (Trujillo-Rodriguez et al. 2017; Mafra et al. 2019). Their property of exhibiting a strong response to the external magnetic field has been used to simplify the extraction process and minimize sources of errors. They enable the use of larger volumes of extraction drops, and the collection of extract using an external magnetic rod.

Deep eutectic solvents (DESs) are analogues of ILs comprising a hydrogen-bond acceptor, e.g., quaternary ammonium or phosphonium ion with a halide ion, and a hydrogen-bond donor, such as amine, carbohydrate, alcohol, or carboxylic acid, to form a eutectic mixture that has a much lower melting point than that of each constituent compound. Contrary to ILs, the biodegradable nature of many DESs makes them greener solvents (Zhang et al. 2012; Cunha and Fernandes, 2018). DESs are liquid at ambient temperature and are composed of two or more safe components that are capable of self association through hydrogen bonding. DESs exhibit physico-chemical properties identical to those of conventional ILs, but they are much cheaper and convenient to synthesize in a laboratory. Besides extraction capabilities, DESs also find newer applications in analytical chemistry that include chromatographic separation, electrochemical analysis, synthesis, and modification of sorption materials (Shishov et al. 2020).

Both hydrophilic and hydrophobic DESs, made from water-soluble and insoluble components, respectively, are in use. The former are water soluble, and phase separation is carried out either by the addition of high salt concentration to produce salting-out effect, or an emulsifier to form a cloudy solution. DES of choline chloride and 4-chlorophenol (other tried phenol and ethylene glycol as proton donor) has been used in HS-SDME for the extraction of triazole fungicides from juices (Abolghasemi et al. 2020), and of menthol and phenylacetic acid for SDME-SFOD of pesticides in human saliva and exhaled breath condensate (Jouyban et al. 2019); in both methods, the extract was analyzed by direct injection into GC-MS. Some other workers reported dilution of DES extract with ethanol (Triaux et al. 2020) or hexane (Mehravar et al. 2021) before injection into GC-MS, ostensibly due to high viscosity and involatility of DES.

7.4.2 Automation

Though most of the SDME methods perform manually carried out operations, semi or fully automated SDME systems offer a number of advantages, such as reduction in sample and reagents consumption, economy of time, minimization of errors, improving sensitivity and precision, and minimization of waste. In addition to miniaturization of LPME, interest in automation of SDME led to ample innovations, and such efforts have been reviewed (Kocurova et al. 2013). Two examples employed sequential injection manifold for metal ion complex formation and extraction. The autosampler arm of the atomic absorption spectrometer was housed with a sampling capillary needle containing organic solvent to form a drop into the aqueous sample. After extraction, the autosampler arm was directed toward the graphite tube for injection (Pena et al. 2008). In another example, the solvent drop was suspended at the capillary tip of the home-designed flow-cell, where the complex solution flows around the drop continuously. Thereafter, the drop was retracted into the holding coil and delivered to a graphite tube (Anthemidis and Adam 2009). In a dynamic in-syringe LPME approach, a software-controlled autosampler allowed sample extraction, analytes derivatization, and extract injection by a fully automatic method for GC-MS (Lee and Lee 2011). With the use of an autosampler, the entire procedure entailing 1-octanol drop formation with air bubble inside for HS-SDME and GC-MS of extraction was performed automatically for nitro musks (Guo et al. 2016). A compound drop of the basic aqueous phase covered with a thin film of the organic phase was formed by controlled back-and-forth pressures at the tip of the capillary by two different commercial CE instruments to perform automatic LLLME. Analytes from the stirred acidified donor phase diffused through the organic film into the basic acceptor phase to attain a pre-concentration factor of 2000 within 10 min (Choi et al. 2009).

In a lab-in-syringe experiment, the syringe was used to concoct a size-adaptable reaction chamber and to perform a series of protocols, including aspirating the liquid sample and base, mixing to evolve ammonia, and collecting ammonia in the headspace at reduced pressure. The syringe piston was drilled with a hole to form a drop of bromothymol blue in the headspace. The on-drop colour measurements were made by fibre optics (Sramkova et al. 2016). Two more such techniques used an air-bubble stabilized solvent drop of either dithizone in DI-SDME into the aqueous sample of lead(II) (Sramkova et al. 2018), or acidified dichromate in HS-SDME of ethanol in a wine sample (Sramkova et al. 2014). An innovative 96-well format used a set of magnetic pins to stabilize the MIL drops for the extraction of parabens, bisphenol A, and triclocarban (Mafra et al. 2019). The system has the advantage of high sample throughput, and suitability for full automation.

Among lab-automation strategies, adoption of dedicated robots for analytical purposes has received widespread importance. An innovative platform has been assembled to hyphenate online sample microextraction techniques with instrumental techniques. This robot is programmed to automated sample clean-up, pre-concentration of analytes by LPME, syringe collection of microextract, and its delivery to interfaced online analytical systems. A lab-made multipurpose autosampler was used in robotic-assisted microextraction using large solvent drops in DI-SDME (Cabal et al. 2019; Medina et al. 2019). The system demonstrated its performance in a dynamic and static large-drop based microextraction in an automated manner with minimal requirements of hardware and software. The synergic interaction between the use of solvent large drops and the automated dynamic mode of extraction was claimed to provide the best extraction efficiencies.

7.5 Conclusions

This chapter provided a brief introduction to the modes of SDME and their application to a variety of analytes, including biomolecules in their real matrix. The emphasis was placed on the creative ideas that have been developed over a period of time, mostly in the last ten years, to provide solutions to analytical problems in sample preparation that are otherwise cumbersome by other means. Nanoparticle-enhanced SDME, and the use of ILs and DESs as solvents for drop formation, have greatly improved extraction efficiencies. Certain examples of automation in SDME were cited, but activities in this area should increase to handle biomaterials safely and generate results rapidly and precisely. Medina 2019

Acknowledgement

The authors are thankful to the Department of Science and Technology, Government of India, New Delhi, for financial support under the Women Scientists Scheme to A.J. (project number SR/WOS-B/658/2016) and M.G. (project number (SR/WOS-A/CS-7/2018).

REFERENCES

Abbasi-Ahd, Atefeh, Nadr Shokoufi, and Kazem Kargosha. "Headspace Single Drop Microextraction Coupled to Microchip-Photothermal Lens Microscopy for Highly Sensitive Determination of Captopril in Human Serum and Pharmaceuticals." *Microchimica Acta* 184 (2017): 2403–9.

Abolghasemi, Mir Mahdi, Marzieh Piryaei, and Roghayeh Moghtader Imani. "Deep Eutectic Solvents as Extraction Phase in Headspace Single Drop Microextraction for Determination of Pesticides in Fruit Juice and Vegetable Samples." *Microchemical Journal* 158 (2020): 105041.

Ahmadi, Fardin, Yaghoub Assadi, S. Mohammad-Reza Milani Hosseini, and Mohammad Rezaee. "Determination of Organophosphorus Pesticides in Water Samples by Single Drop Microextraction and Gas Chromatography-Flame Photometric Detector." *Journal of Chromatography A* 1101 (2006): 307–12.

ALOthman, Zeid A., Mohammed Dawod, Jihye Kim, and Doo Soo Chung. "Single-Drop Microextraction as a Powerful Pretreatment Tool for Capillary Electrophoresis: A Review." *Analytica Chimica Acta* 739 (2012): 14–24.

Anthemidis, Aristidis N., and Ibrahim S.I. Adam. "Development of On-Line Single-Drop Micro-extraction Sequential Injection System for Electrothermal Atomic Absorption Spectrometric Determination of Trace Metals." *Analytica Chimica Acta* 632 (2009): 216–20.

Bagheri, Habib, Faezeh Khalilian, Esmaeil Babanezhad, Ali Es-haghi, and Mohammad-Reza Rouini. "Modified Solvent Microextraction with Back Extraction Combined with Liquid Chromatography-Fluorescence Detection for the Determination of Citalopram in Human Plasma." *Analytica Chimica Acta* 610 (2008): 211–16.

Basheer, Chanbasha, Muhammad Kamran, Muhammad Ashraf, and Hian Kee Lee. "Enhancing Liquid-Phase Microextraction Efficiency Through Chemical Reactions." *Trends in Analytical Chemistry* 118 (2019): 426–33.

Berthod, Alain, Maria Jose Ruiz-Angel, and Samuel Carda-Broch. "Recent Advances on Ionic Liquid Uses in Separation Techniques." *Journal of Chromatography A* 1559 (2018): 2–16.

Bitas, Dimitrios, and Victoria Samanidou. "Biomedical Applications." In *Liquid-Phase Extraction*, edited by Colin F. Poole, pp. 683–723. Elsevier, 2020.

Cabal, Luis Felipe Rodriguez, Deyber Arley Vargas Medina, Adriel Martins Lima, Fernando Mauro Lancas, and Alvaro Jose Santos-Neto. "Robotic-assisted Dynamic Large Drop Microextraction." *Journal of Chromatography A* 1608 (2019): 460416.

Cabala, Radomir, and Miroslava Bursova. "Bell-shaped Extraction Device Assisted Liquid-liquid Microextraction Technique and Its Optimization Using Response-Surface Methodology." *Journal of Chromatography A* 1230 (2012): 24–9.

Carasek, Eduardo, Gabrieli Berhardi, Diogo Morelli, and Josias Merib. "Sustainable Green Solvents for Microextraction Techniques: Recent Developments and Applications." *Journal of Chromatography* 1640 (2021): 461944.

Chalavi, Soheila, Sakine Asadi, Saeed Nojavan, and Ali Reza Fakhari. "Recent Advances in Microextraction Procedures for Determination of Amphetamines in Biological Samples." *Bioanalysis* 11 (2019): 437–60.

Chandrawanshi, Swati, Santosh K. Verma, and Manas K. Deb. "Collective Ion-Pair Single-Drop Microextraction Attenuated Total Reflectance Fourier Transform Infrared Spectroscopic Determination of Perchlorate in Bioenvironmental Samples." *Journal of AOAC International* 101 (2018): 1145–55.

Cheng, Khley, Kihwan Choi, Jihye Kim, Hye Sung, and Doo Soo Chung. "Sensitive Arsenic Analysis by Carrier-Mediated Counter-Transport Single Drop Microextraction Coupled to Capillary Electrophoresis." *Microchemical Journal* 106 (2013): 220–5.

Choi, Kihwan, Su Ju Kim, Yoo Gon Jin, Yong Oh Jang, Jin-Soo Kim, and Doo Soo Chung. "Single Drop Microextraction Using Commercial Capillary Electrophoresis Instruments." *Analytical Chemistry* 81 (2009): 225–30.

Chullasat, Kochaporn, Zhenzhen Huang, Opas Bunkoed, Proespichaya Kanatharana, and Hian Kee Lee. "Bubble-in-Drop Microextraction of Carbamate Pesticides Followed by Gas Chromatography-Mass Spectrometric Analysis." *Microchemical Journal* 155 (2020): 104666.

Costas-Mora, Isabel, Vanesa Romero, Francisco Pena-Pereira, Isela Lavilla, and Carlos Bendicho. "Quantum Dot-Based Headspace Single-Drop Microextraction Technique for Optical Sensing of Volatile Species." *Analytical Chemistry* 83 (2011): 2388–93.

Costas-Mora, Isabel, Vanesa Romero, Isela Lavilla, and Carlos Bendicho. "Solid-State Chemiluminescence Assay for Ultrasensitive Detection of Antimony Using On-Vial Immobilization of CdSe Quantum Dots Combined with Liquid-Liquid-Liquid Microextraction." *Analytica Chimica Acta* 788 (2013): 114–21.

Cunha, Sara C., and Jose O. Fernandes. "Extraction Techniques with Deep Eutectic Solvents." *Trends in Analytical Chemistry* 105 (2018): 225–30.

De Boeck, Marieke, Wim Dehaen, Jan Tytgat, and Eva Cuypers. "Microextractions in Forensic Toxicology: The Potential Role of Ionic Liquids." *Trends in Analytical Chemistry* 111 (2019): 73–84.

Dong, Jiang Xue, Zhong Feng Gao, Ying Zhang, Bang Lin Li, Nian Bing Li, and Hong Qun Luo. "A Selective and Sensitive Optical Sensor for Dissolved Ammonia Detection Via Agglomeration of Fluorescent Ag Nanoclusters and Temperature Gradient Headspace Single Drop Microextraction." *Biosensors and Bioelectronics* 91 (2017): 155–61.

Ebrahimzadeh, Homeira, Yadollah Yamini, Katayoun Mahdavi Ara, Fahimeh Kamarei, and Farahnaz Khalighi-Sigaroodi. "Determination of Azithromycin in Biological Samples by LLLME Combined with LC." *Chromatographia* 72 (2010): 731–35.

Feng, Jiaxin, Xiaochao Zhang, Liang Huang, Huan Yao, Chengdui Yang, Xiaoxiao Ma, Sichun Zhang, and Xinrong Zhang. "Quantitation of Glucose-Phosphate in Single Cells by Microwell-Based Nanoliter Droplet Microextraction and Mass Spectrometry." *Analytical Chemistry* 91 (2019): 5613–20.

Fernances, Virginia C., Viswanathan Subramanian, Nuno Mateus, Valentina F. Domingues, and Cristina Deleruc-Matos. "The Development and Optimization of a Modified Single-Drop Microextraction Method for Organochlorine Pesticides Determination by Gas Chromatography-Tandem Mass Spectrometry." *Microchimica Acta* 178 (2012): 195–202.

George, Mosotho J., Ljiljana Marjanovic, and D. Bradley G. Williams. "Picogram-Level Quantification of Some Growth Hormones in Bovine Urine Using Mixed-Solvent Bubble-in-Drop Single Drop Microextraction." *Talanta* 144 (2015): 445–50.

Guo, Liang, Nurliyana Binte Nawi, and Hian Kee Lee. "Fully Automated Headspace Bubble-in-Drop Microextraction." *Analytical Chemistry* 88 (2016): 8409–14.

Hansen, Frederik, Elisabeth Leere Oiestad, and Stig Pedersen-Bjergaard. "Bioanalysis of Pharmaceuticals Using Liquid-Phase Microextraction Combined with Liquid Chromatography-Mass Spectrometry." *Journal of Pharmaceutical and Biomedical Analysis* 189 (2020): 113446.

He, Xiaowen, Fucheng Zhang, and Ye Jiang. "An Improved Ionic Liquid-Based Headspace Single-Drop Microextraction-Liquid Chromatography Method for the Analysis of Camphor and Trans-anethole in Compound Liquorice Tablets." *Journal of Chromatographic Science* 50 (2012): 457–63.

He, Yi, and Marta Concheiro-Guidan. "Microextraction Sample Preparation Techniques in Forensic Analytical Toxicology." *Biomedical Chromatography* 33 (2019): e4444.

He, Yupeng, Paul Miggiels, Bert Wouters, Nicolas Drouin, Faisa Guled, Thomas Hankemeier, and Petrus W. Lindenburg. "A High-Throughput, Ultrafast, and Online Three-Phase Electroextraction Method for Analysis of Trace Level Pharmaceuticals." *Analytica Chimica Acta* 1149 (2021): 338204.

He, Yi and Youn-Jung Kang. "Single Drop Liquid-Liquid-Liquid Microextraction of Methamphetamine and Amphetamine in Urine." *Journal of Chromatography A* 1133 (2006): 35–40.

Jahan, Sharmin, Haiyang Xie, Ran Zhong, Jian Yan, Hua Xiao, Liuyin Fan, and Chengxi Cao. "A Highly Efficient Three-Phase Single Drop Microextraction Technique for Sample Preconcentration." *Analyst* 140 (2015): 3193–200.

Jain, Archana, and Krishna K. Verma. "Recent Advances in Applications of Single-Drop Microextraction: A Review." *Analytica Chimica Acta* 706 (2011): 37–65.

Jain, Archana, and Krishna K. Verma. "Single-Drop Microextraction." In *Liquid-Phase Extraction*, edited by Colin F. Poole, pp. 439–72. Elsevier, 2020.

Jain, Archana, Soumitra Soni, and Krishna K. Verma. "Combined Liquid Phase Microextraction and Fiber-Optics-Based Cuvetteless Micro-spectrophotometry for Sensitive Determination of Ammonia in Water and Food Samples by the Indophenol Reaction." *Food Chemistry* 340 (2021): 128156.

Jain, Rajeev, and Ritu Singh. "Applications of Dispersive Liquid-Liquid Micro-extraction in Forensic Toxicology." *Trends in Analytical Chemistry* 75 (2016): 227–37.

Jeannot, Michael A., and Frederick F. Cantwell. "Mass Transfer Characteristics of Solvent Extraction into a Single Drop at the Tip of a Syringe Needle." *Analytical Chemistry* 69 (1997): 235–9.

Jiang, Yongrong, Xuezhi Dong, Yuzhe Li, Yan Li, Ying Liang, and Min Zhang. "An Environmentally-Benign Flow-Batch System for Headspace Single-Drop Microextraction and On-Drop Conductometric Detecting Ammonium." *Talanta* 224 (2021): 121849.

Jouyban, Abolghasem, Mir Ali Farajzadeh, Maryam Khubnasabjafari, Vahid Jouyban-Gharamaleki, and Mohammad Reza Afshar Mogaddam. "Development of Deep Eutectic Solvent Based Solidification of Organic Droplets-Liquid Phase Microextraction; Application to Determination of Some Pesticides in Farmers Saliva and Exhaled Breath Condensate Samples." *Analytical Methods* 11 (2019): 1530–40.

Kailasa, Suresh Kumar, Janardhan Reddy Koduru, Tae Jung Park, Rakesh Kumar Singhal, and Hui-Fen Wu. "Applications of Single-Drop Microextraction in Analytical Chemistry: A Review." *Trends in Environmental Analytical Chemistry* 29 (2021): e00113.

Kocurova, Livia, Ioseph S. Baloch, and Vasil Andruch. "Solvent Microextraction: A Review of Efforts at Automation." *Microchemical Journal* 110 (2013): 599–607.

Kokosa, John M. "Recent Trends in Using Single-Drop Microextraction and Related Techniques in Green Analytical Methods." *Trends in Analytical Chemistry* 71 (2015): 194–204.

Kokosa, John M. "Selecting an Extraction Solvent for a Greener Liquid Phase Microextraction (LPME) Mode-Based Analytical Method." *Trends in Analytical Chemistry* 118 (2019): 238–47.

Lee, Jingyi, and Hian Kee Lee. "Fully Automated Dynamic In-Syringe Liquid-Phase Microextraction and On-Column Derivatization of Carbamate Pesticides with Gas Chromatography/Mass Spectrometric Analysis." *Analytical Chemistry* 83 (2011): 6856–61.

Liu, Shaorong, and Purnendu K. Dasgupta. "Liquid Droplet. A Renewable Gas Sampling Interface." *Analytical Chemistry* 67 (1995): 2042–9.

Liu, Wuping, and Hian Kee Lee. "Continuous-Flow Microextraction Exceeding 1000-Fold Concentration of Dilute Analytes." *Analytical Chemistry* 72 (2000): 4462–7.

Lopez-Jimenez, Francisco Jose, Soledad Rubio, and Dolores Perez-Bendito. "Single-Drop Coacervative Microextraction of Organic Compounds Prior to Liquid Chromatography: Theoretical and Practical Considerations." *Journal of Chromatography A* 1195 (2008): 25–33.

Ma Xiao, and Jun Ma. "Determination of Trace Amounts of Chlorobenzenes in Water Using Membrane-Supported Headspace Single-Drop Microextraction and Gas Chromatography-Mass Spectrometry." *Journal of Analytical Chemistry* 72 (2017): 890–6.

Ma, Minhui, and Frederick F. Cantwell. "Solvent Microextraction with Simultaneous Back-Extraction for Sample Cleanup and Preconcentration: Preconcentration into a Single Microdrop." *Analytical Chemistry* 71 (1999): 388–93.

Mafra, Gabriela, Augusto A. Vieira, Josias Merib, Jared L. Anderson, and Eduardo Carasek. "Single Drop Microextraction in a 96-Well Plate Format: A Step Toward Automated and High-Throughput Analysis." *Analytica Chimica Acta* 1063 (2019): 159–66.

Medina, Deyber Arley Vargas, Luis Felipe Rodriguez Cabal, Fernando Mauro Lancas, and Alvaro Jose Santos-Neto. "Sample Treatment Platform for Automated Integration of Microextraction Techniques and Liquid Chromatography Analysis." *HardwareX* 6 (2019): e00056.

Medina, Deyber Arley Vargas, Luis Felipe Rodriguez Cabal, Guilherme Miola Titato, Fernando Mauro Lancas, and Alvaro Jose Santos-Neto. "Automated Online Coupling of Robot-Assisted Single Drop Microextraction and Liquid Chromatography." *Journal of Chromatography* 1595 (2019): 66–72.

Mehravar, Amir, Alireza Feizbakhsh, Amir Hosein Mohsen Sarafi, Elaheh Konoz, and Hakim Faraji. "Validation of Chemometric-Assisted Single-Drop Microextraction Based on Sustainable Solvents to Analyze Polyaromatic Hydrocarbons in Water Samples." *Analytical Methods* 13 (2021): 242–9.

Mogaddam, Mohammad Reza Afshar, Ali Mohebbi, Azar Pazhohan, Fariba Khodadadeian, and Mir Ali Farajzadeh. "Headspace Mode of Liquid Phase Microextraction: A Review." *Trends in Analytical Chemistry* 110 (2019): 8–14.

Nunes, Leane Santos, Maria Gracas Andrade Korn, and Valfredo Azevedo Lemos. "A Novel Direct-Immersion Single-Drop Microextraction Combined with Digital Colorimetry Applied to the Determination of Vanadium in Water." *Talanta* 224 (2021): 121893.

Oedit, Amar, Thomas Hankemeier, and Peter W. Lindenburg. "On-Line Coupling of Two-Phase Microextraction to Capillary Electrophoresis-Mass Spectrometry for Metabolomics Analyses." *Microchemical Journal* 162 (2021): 105741.

Pena, Francisco, Isela Lavilla, and Carlos Bendicho. "Immersed Single-Drop Microextraction Interfaced with Sequential Injection Analysis for Determination of Cr(VI) in Natural Waters by Electrothermal-Atomic Absorption Spectrometry." *Spectrochimica Acta B* 63 (2008): 498–503.

Pillai, Aradhana K.K.V., Khileshwari Gautam, Archana Jain, and Krishna K. Verma. "Headspace In-Drop Derivatization of Carbonyl Compounds for Their Analysis by High-Performance Liquid Chromatography-Diode Array Detection." *Analytica Chimica Acta* 632 (2009): 208–15.

Plotka-Wasylka, Justyna, Malgorzata Rutkowska, Katarzyna Owczarek, Marek Tobiszewski, and Jacek Namiesnik. "Extraction with Environmentally Friendly Solvents." *Trends in Analytical Chemistry* 91 (2017): 12–25.

Przyjazny, Andrzej. *Liquid-Phase Microextraction. Encyclopedia of Analytical Science*, 3rd edn, pp. 52–62. Amsterdam: Elsevier, 2019.

Psillakis, Elefteria. "The Effect of Vacuum: An Emerging Experimental Parameter to Consider During Headspace Microextraction Sampling." *Analytical and Bioanalytical Chemistry* 412 (2020): 5989–97.

Purgat, Krystian, Patrycja Olejarz, Izabella Koska, Rafal Glowacki, and Pawel Kubalczyk. "Determination of Homocysteine Thiolactone in Human Urine by Capillary Zone Electrophoresis and Single Drop Microextraction." *Analytical Biochemistry* 596 (2020): 113640.

Qi, Tong, Mengyuan Xu, Yao Yao, Wenhui Chen, Mengchan Xu, Sheng Tang, Wei Shen, Dezhao Kong, Xingwei Cai, Haiwei Shi, and Hian Kee Lee. "Gold Nanoprism/Tollens' Reagent Complex as Plasmonic Sensor in Headspace Single-Drop Microextraction for Colorimetric Detection of Formaldehyde in Food Samples Using Smartphone Readout." *Talanta* 220 (2020): 121388.

Raterink, Robert-Jan, Peter W. Lindenburg, Rob J. Wreeken, and Thomas Hankemeier. "Three-Phase Electroextraction: A New (Online) Sample Purification and Enrichment Method for Bioanalysis." *Analytical Chemistry* 85 (2013): 7762–8.

Reddy, Gangireddy Navitha, Aarati Dilip Zagade, and Pinaki Sengupta. "Current Direction and Advances in Analytical Sample Extraction Techniques for Drugs with Special Emphasis on Bioanalysis." *Bioanalysis* 11 (2019): 313–32.

Saleh, Abolfazl, Neda Sheijooni Fumani, and Saeideh Molaei. "Microfunnel-Supported Liquid-Phase Microextraction: Application to Extraction and Determination of Irgarol 1051 and Diuron in Persian Gulf Seawater Samples." *Journal of Chromatography A* 1356 (2014): 32–7.

Saraji, Mohammad, Milad Ghani, Behzad Rezei, and Maryam Mokhtarianpour. "Highly Porous Nanostructured Copper Foam Fiber Impregnated with an Organic Solvent for Headspace Liquid-Phase Microextraction." *Journal of Chromatography A* 1469 (2016): 25–34.

Sharma, Nisha, Archana Jain, Vandana Kumari Singh, and Krishna K. Verma. "Solid-Phase Extraction Combined with Headspace Single-Drop Microextraction of Chlorophenols as Their Methyl Ethers and Analysis by High-Performance Liquid Chromatography-Diode Array Detection." *Talanta* 83 (2011): 994–9.

Shishov, Andrey, Aleksei Pochivalov, Lawrence Nugbienyo, Vasil Andruch, and Andrey Bulatov. "Deep Eutectic Solvents Are Not Only Effective Extractants." *Trends in Analytical Chemistry* 129 (2020): 115956.

Song, Aiying, and Jing Yang. "Efficient Determination of Amphetamine and Methylamphetamine in Human Urine Using Electro-Enhanced Single-Drop Microextraction with In-Drop Derivatization and Gas Chromatography." *Analytica Chimica Acta* 1045 (2019): 162–8.

Sramkova, Ivana H., Burkhard Horskotte, Katerina Fikarova, Hana Sklenarova, and Petr Solich. "Direct-Immersion Single-Drop Microextraction and In-Drop Stirring Microextraction for the Determination of Nanomolar Concentrations of Lead Using Automated Lab-in-Syringe Technique." *Talanta* 184 (2018): 162–72.

Sramkova, Ivana, Burkhard Horstkotte, Hana Sklenarova, Petr Solich, and Spas D. Kolev. "A Novel Approach to Lab-in-Syringe Headspace Single-Drop Microextraction and On-Drop Sensing of Ammonia." *Analytica Chimica Acta* 934 (2016): 132–44.

Sramkova, Ivana, Burkhard Horstkotte, Petr Solich, and Hana Sklenarova. "Automated In-Syringe Single-Drop Headspace Microextraction Applied to the Determination of Ethanol in Wine Samples." *Analytica Chimica Acta* 828 (2014): 53–60.

Tang, Sheng, Tong Qi, Prince Dim Ansah, Juliette Chancellevie Nalouzebi Fouemina, Wei Shen, Chanbasha Basheer, and Hian Kee Lee. "Single-Drop Microextraction." *Trends in Analytical Chemistry* 108 (2018): 306–13.

Tang, Sheng, Tong Qi, Dasha Xia, Mengchan Xu, Mengyuan Xu, Anni Zhu, Wei Shen, and Hian Kee Lee. "Smartphone Nano-colorimetric Determination of Hydrogen Sulfide in Biosamples After Silver-Gold Core-Shell Nanoprism-Based Headspace Single-Drop Microextraction." *Analytical Chemistry* 91 (2019): 5888–95.

Tang, Sheng, Tong Qi, Yao Yao, Liangxiu Tang, Wenhui Chen, Tianyu Chen, Wei Shen, Dezhao Kong, Hai-Wei Shi, Tianlong Liu, and Hian Kee Lee. "Magnetic Three-Phase Single-Drop Microextraction for Rapid Amplification of the Signals of DNA and MicroRNA Analysis." *Analytical Chemistry* 92 (2020): 12290–6.

Tegladza, Isaac Delove, Tong Qi, Tianyu Chen, Kingdom Alorku, Sheng Tang, Wei Tang, Wei Shen, Dezhao Kong, Aihua Yuan, Jianfeng Liu, and Hian Kee Lee. "Direct Immersion Single Drop Microextraction of Semi-volatile Organic Compounds in Environmental Samples: A Review." *Journal of Hazardous Materials* 393 (2020): 122403.

Theis, Aaron L., Adam J. Waldack, Susan M. Hansen, and Michael A. Jeannot. "Headspace Solvent Microextraction." *Analytical Chemistry* 73 (2001): 5651–4.

Tian, Fei, Weijie Liu, Hansun Fang, Min An, and Shunshan Duan. "Determination of Six Organophosphorus Pesticides in Water by Single-Drop Microextraction Coupled with GC-NPD." *Chromatographia* 77 (2014): 487–92.

Triaux, Zelie, Hugues Petitjean, Eric Marchioni, Maria Boltoeva, and Christophe Marcic. "Deep Eutectic Solvent-Based Headspace Single Drop Microextraction for Quantification of Terpenes in Spices." *Analytical and Bioanalytical Chemistry* 412 (2020): 933–48.

Trujillo-Rodriguez, Maria J., Veronica Pino, and Jared L. Anderson. "Magnetic Ionic Liquids as Extraction Solvents in Vacuum Headspace Single-Drop Microextraction." *Talanta* 172 (2017): 86–94.

Wang, Xiuhong, Jing Cheng, Xiangfang Wang, Min Wu, and Min Cheng. "Development of an Improved Single-Drop Microextraction Method and Its Application for the Analysis of Carbamate and Organophosphorus Pesticides in Water Samples." *Analyst* 137 (2012): 5339–45.

Wen, Xiaodong, Qingwen Deng, Jiwei Wang, Shengchun Yang, and Xia Zhao. "A New Coupling of Ionic Liquid Based Single Drop Microextraction with Tungsten Coil Electrothermal Atomic Absorption Spectrometry." *Spectrochimica Acta A* 105 (2013): 320–5.

Williams, D. Bradley G., Mosotho J. George, and Ljiljana Marjanovic. "Rapid Detection of Atrazine and Metolachlor in Farm Soils: Gas Chromatography-Mass Spectrometry-Based Analysis Using the Bubble-in-Drop Single Drop Microextraction Enrichment Method." *Journal of Agricultural and Food Chemistry* 62 (2014): 7676–81.

Williams, D. Bradley G., Mosotho J. George, Riaan Meyer, and Ljiljana Marjanovic. "Bubbles in Solvent Microextraction: The Influence of Intentionally Introduced Bubbles on Extraction Efficiency." *Analytical Chemistry* 83 (2011): 6713–16.

Xie, Hai-Yang, Jian Yan, Sharmin Jahan, Ran Zhong, Liu-Yin Fan, Hua Xiao, Xin-Qiao Jin, and Cheng-Xi Cao. "A New Strategy for Highly Efficient Single-Drop Microextraction with a Liquid–Gas Compound Pendant Drop." *Analyst* 139 (2014): 2545–50.

Yangcheng, Lu, Lin Quan, Luo Guangsheng, and Dai Youyuan. "Directly Suspended Droplet Microextraction." *Analytica Chimica Acta* 566 (2006): 259–64.

Zanjani, Mohammad Reza Khalili, Yadollah Yamini, Shahab Shariati, and Jan Ake Jonsson. "A New Liquid-Phase Microextraction Method Based on Solidification of Floating Organic Drop." *Analytica Chimica Acta* 585 (2007): 286–93.

Zaruba, Serhii, Andriy B. Vishnikin, Jana Skrlikova, Alina Diuzheva, Ivana Ozimanicova, Kiril Gavazov, and Vasil Andruch. "A Two-in-One Device for Online Monitoring of Direct Immersion Single-Drop Microextraction: An Optical Probe as Both Microdrop Holder and Measuring Cell." *Royal Society of Chemistry Advances* 7 (2017): 29421–7.

Zaruba, Serhii, Andriy B. Vishnikin, Jana Skrlikova, and Vasil Andruch. "Using an Optical Probe as the Microdrop Holder in Headspace Single Drop Microextraction: Determination of Sulfite in Food Samples." *Analytical Chemistry* 88 (2016): 10296–300.

Zhang, Qinghua, Karine De Oliveira Vigier, Sebastien Royer, and Francois Jerome. "Deep Eutectic Solvents: Syntheses, Properties and Applications." *Chemical Society Reviews* 41 (2012): 7108–46.

Zhao, Yaoyao, Zhen Chen, Yue Wu, Takayuki Tsukui, Xiaoxiao Ma, Xinrong Zhang, Hitoshi Chiba, and Shu-Ping Hui. "Separating and Profiling Phosphatidylcholines and Triglycerides from Single Cellular Lipid Droplet by In-Tip Solvent Microextraction Mass Spectrometry." *Analytical Chemistry* 91 (2019): 4466–71.

8

Applications of Liquid-Phase Microextraction in Analytical Toxicology

María Ramos-Payán[1], Samira Dowlatshah[2], and Mohammad Saraji[2]
[1]Department of Analytical Chemistry, Faculty of Chemistry, University of Seville, Seville, Spain
[2]Department of Chemistry, Isfahan University of Technology, Iran

CONTENTS

8.1 Introduction ... 147
8.2 Liquid-Phase Microextraction Configurations ... 148
    8.2.1 Single Drop Microextraction (SDME) ... 149
    8.2.2 Dispersive Liquid-Liquid Microextraction (DLLME) ... 149
    8.2.3 Hollow-Fibre Liquid-Phase Microextraction (HF-LPME) ... 150
    8.2.4 Microfluidic LPME ... 150
8.3 Applications of LPME in Toxicology ... 150
    8.3.1 Analysis of Amphetamines ... 153
    8.3.2 Analysis of Sedative and Hypnotic Drugs ... 153
    8.3.3 Analysis of Opium Alkaloids, Opiates, and Other Alkaloids ... 153
    8.3.4 Analysis of Cannabinoids ... 153
    8.3.5 Analysis of Antidepressant Drugs ... 153
    8.3.6 Analysis of Hallucinogens ... 154
    8.3.7 Comparison between DLLME Methods ... 154
8.4 Conclusions ... 154
References ... 154

8.1 Introduction

Due to the trace and ultratrace amounts of analytes in complex matrices, sample preparation is still required prior to quantification in analytical toxicology, even with the high sensitivity and selectivity of modern analytical instruments. In addition, sample preparation plays a critical role in the accuracy of analytical methods applied for toxicology since its operation accounts for one-third of the errors in a whole analytical process (Majors 1991). The traditional sample preparation methods are liquid-liquid extraction (LLE) and solid-phase extraction (SPE). Although these methods have been applied in analytical toxicology, they entail time-consuming and complicated steps and the consumption of large amounts of chemicals and reagents (at least a few millilitres, in the case of SPE via minicartridges); therefore, they are far away from the 'green chemistry' concept. As a result, trends have gained momentum toward developing environmentally friendly, low cost, simple and fast sample preparation techniques that consume negligible or minimum amounts of organic solvent. The various efforts in this area evolved into the invention of a solvent-minimized sample preparation procedure as a new form of LLE method, namely liquid-phase microextraction (LPME). Considering the impossibility of solvent elimination, LPME has significantly reduced the volumes employed in the procedure so that it demands only several microliters or microdrops of organic solvent to concentrate target compounds in different matrices. Moreover, the method provides unique advantages rather than traditional ones, such as rapidity, simplicity of operation, low cost, high recovery, or high enrichment factor. Nowadays, a good number of LPME methods in different formats are available, categorized into single drop microextraction (SDME), hollow-fibre LPME (HF-LPME), and dispersive liquid-liquid microextraction (DLLME) (Sharifi et al. 2016). Among these formats, HF-LPME and DLLME have received great interest because of their individual benefits.

LPME was introduced in 1996 by Dasgupta and Cantwell not only to facilitate the automation of this technique but also to lower the required volume of extractant and sample (Liu and Dasgupta 1996). Since the procedure provided a high enrichment factor, it could be of crucial importance when a very small volume of the sample was available (Lucena et al. 2009). In a first format, the principle of the method was based on compound distribution between several microliters of two immiscible liquid phases, termed donor and acceptor phase. The donor phase was an aqueous containing the analytes of interest, and the acceptor phase was a water-immiscible solvent so that subject substances were transferred from sample solution to acceptor based on passive diffusion. In this format, the acceptor phase could be suspended above the sample for headspace sampling (HS-SDME) or could be directly immersed in the donor phase to perform a direct immersion SDME procedure (Liu and Dasgupta 1995). In this regard, a microdrop of a water-immiscible organic phase was immersed in a large amount of sample solution. Although the method was efficient and significantly reduced the organic volume, the microdrop was not stable. In order to enhance the stability of the technique, a new method termed HF-LPME was presented in 1999, which was based on a supported liquid membrane with two sampling modes (two- and three-phase) (Pedersen-Bjergaard and Rasmussen 1999). In the two-phase mode, the analytes of interest were extracted from the sample solution through the membrane into an organic phase, while in the three-phase mode, the analytes were extracted from the sample (aqueous solution) to the acceptor phase (aqueous solution) across the supported liquid membrane (SLM). Three phases can also be configured as carrier-assisted LPME. That variant consisted of adding some carriers (e.g., Di(2-etilhexil)ftalato and Tris(2-ethylhexyl)phosphat) to the composition of the SLM in order to improve migration of the analytes through the SLM into the acceptor solution. In the contact region of the liquid membrane and the acceptor solution, the analytes are released from the ion-pair complex into the acceptor solution.

The effort for the development of different modes of the LPME method has never ceased, which led to the introduction of the DLLME method in 2006 (Rezaee et al. 2006). In this mode, the acceptor phase (a mixture of extracting solvent and disperse) is injected into the donor phase containing the analytes to form a cloudy solution (Nuhu et al. 2011; Barroso et al. 2012). Recently, LPME was miniaturized in microfluidic systems to benefit not only the automation of this technique but also to require a lower volume of extractant and sample volume as well as to provide more extraction efficiency. In these systems, two liquid streams are flowed from two inlets into a microchannel in parallel, acting as a donor and acceptor phase that can be separated in most cases by a membrane in which the extractant is impregnated. This approach has seen tremendous strides over the age of designing downscaled sample preparation methods in different fields.

Nowadays, different modes of LPME have intensively been applied to the separation and quantification of various compounds in different fields, such as in analytical toxicology. Here, different LPME configurations and their main applications in the extraction and pre-concentration of toxic analytes from complex matrices are reviewed and discussed.

8.2 Liquid-Phase Microextraction Configurations

As mentioned above, LPME has developed different configurations: LLLME, SDME, HF-LPME, and DLLME. Figure 8.1 shows the general scheme of LPME for analyte extraction from the donor to the acceptor phase. Figures 8.1A and 8.1B represent two- and three-phase configurations, respectively.

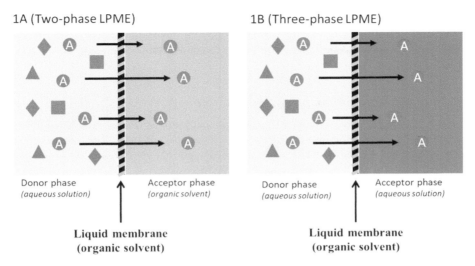

FIGURE 8.1 General scheme for two (1A) and three-phase (1B) configuration of LPME.

8.2.1 Single Drop Microextraction (SDME)

The LPME method, in which a single drop is the extraction medium, is named single drop microextraction (SDME), and it was introduced by Liu in 1995 (Liu and Dasgupta 1995). The SDME procedure is implemented in two different modes, termed direct immersion SDME and headspace SDME. They are based on the distribution of a partition between the analyte in the aqueous sample and a microdroplet of extraction solvent. In this method, a microliter organic solvent droplet, as an extracting solvent, is suspended from a microsyringe to insert into the liquid samples (DI-SDME) or expose to the headspace of the samples containing the analytes (HS-SDME). After extraction, the microdroplet is withdrawn into the microsyringe and coupled to the analytical instruments for further analysis. On the other hand, the method is classified into two- and three-phase appraches. In the two-phase approach, the target compounds are directly extracted from the aqueous sample solution to the organic solvent microdroplet, such as direct immersion SDME and continuous flow SDME. In this mode, the droplet can be disturbed by suspended particles or impurities in the sample solution as well as the droplet suffers from instability. In the three-phase approach, the compounds are extracted by an organic microdroplet or headspace and then back-extracted into an aqueous microdroplet to have headspace SDME and the drop-to-drop SDME method, respectively. This method is more appropriate for analysis of volatile compounds in complex matrices as well as the droplet is more stable and is not influenced by impurities in the sample solution (Kokosa 2015). In the SDME method, some influenced parameters should be considered to obtain desirable results, such as relatively high boiling point or relatively low vapour pressure, density, high viscosity, and compatibility with chromatographic instruments (Tang et al. 2018).

8.2.2 Dispersive Liquid-Liquid Microextraction (DLLME)

DLLME was presented by Assadi and co-workers in 2006 (Rezaee et al. 2006) as a novel sample pre-treatment technique in which a ternary component solvent system, including disperser solvent, extraction solvent (AP), and aqueous phase sample (DP) containing the compounds, was utilized to implement the procedure. In brief, a proper mixture of disperser and organic solvent is prepared to inject into the sample solution in which a cloudy solution is formed to enrich the analytes of interest. In this method, the most effective factors are the physical properties of disperser and organic solvent; for instance, their density should be considered. To improve the DLLME method surfactant solution, ionic liquids have been used as the organic solvent (Han et al. 2012; Trujillo-Rodríguez et al. 2019). Moreover, magnetic nanoparticles have been applied as dispersers due to their unique advantages, such as low vapour pressure, high viscosity, good thermal stability, miscibility with water or organic solvent, and greater use of larger, reproducible extracting volume. In another approach, dispersion solvent was eliminated to enhance the partition coefficient of the compounds into the extraction solvent (An et al. 2017).

8.2.3 Hollow-Fibre Liquid-Phase Microextraction (HF-LPME)

In order to enhance the stability of the SDME technique and keep the droplet from being influenced by impurities in the sample solution, HF-LPME was presented in 1999 (Pedersen-Bjergaard and Rasmussen 1999). The method is based on a supported liquid membrane with two sampling modes (two- and three-phase). In the two-phase mode, the organic solvent is immobilized in the pores and inserted in the hollow-fibre lumen. As the result, the analytes of interest are distributed from the sample solution into the organic phase by passive diffusion. In the three-phase mode, which is utilized to improve the performance of the two-phase mode, three solvents are utilized for extraction of analytes of interest, in which the transferred analytes (from the sample phase to the organic phase) are back-extracted by an aqueous acceptor phase. The extractant organic solvent and aqueous acceptor phase are filled in the wall pores of the lumen and in the hollow-fibre lumen, respectively. The extracted analytes by the HF-LPME method can be detected by different types of analytical instruments. The extraction solvent plays a significant role in achieving a good selectivity and high enrichment factor. This solvent should provide an appropriate affinity toward analytes and possesses similar polarity to the hollow fibre, and it should have no reaction with any of the compounds in the sample solution. Hollow fibre is another crucial factor to achieve the optimum conditions due to its participation in the concentration of the analyte. In this regard, propylene is considered to be the most effective fibre to enrich the analytes of interests (Sharifi et al. 2016; Zhao and Lee 2002).

8.2.4 Microfluidic LPME

Microfluidic liquid-liquid device has seen tremendous strides over the age of designing downscaled sample preparation methods (Xu and Xie 2017). In this approach, two liquid streams are flowed from two inlets into a microchannel in parallel. The initial experiments in a microfluidic liquid-liquid system were performed in 2000 by Sato (Sato et al. 2000). Subsequently, scientists carried out the studies concerning microfluidic extraction between two immiscible liquids for preparation and separation of compounds into various application areas (Shen et al. 2013).

The two liquid-phase microfluidic system creates a large interface area and short diffusion distance between the fluid phases in which the subject substances are transferred from one phase to another phase. In another approach, named three liquid phases or more multiplier, analytes are extracted from an aqueous sample through an organic phase into the other acceptor phase. In fact, the transferred analytes are back-extracted based on the diffusivity difference (Tetala et al. 2009).

8.3 Applications of LPME in Toxicology

Since LPMEs are fast, affordable, selective, and relatively solvent-free sample-preparation methods, they are of great significance to quantify compounds, especially in toxic determination (Sharifi et al. 2016; Jain and Singh 2016; He and Concheiro-Guisan 2019). The most relevant applications of LPME methods in analytical toxicology reported in the scientific literature are described in the following sections. The methods are applied to extract drugs of abuse, hallucinogens, illicit drugs, cannabinoids, narcotic substances, etc. For instance, amphetamines, amphetamine-type stimulants, ketamine, cocaine, analogues, lysergic acid diethylamide, buprenorphine, methadone and fentanyl, benzodiazepines, and Z-compounds have been extracted from urine, blood, plasma, and serum. Table 8.1 also lists other applications where LPME is used in toxicological applications and where it is also observed that the most widely used technique is DLLME.

TABLE 8.1

Applications of LPMEs in analytical toxicology

Analyte	Matrix	Method	EF	LOD (µg/L)	Ref.
AP and MA	Urine	DLLME-HPLC	56, 48	2, 3	(Saber Tehrani et al. 2012)
AP and MA	Urine	DLLME-HPLC	58.5, 62.4	8, 2	(Ahmadi-Jouibari et al. 2014)
MA and MDMA	Urine	DLLME-GC-FID	427, 285	2, 18	(Djozan et al. 2012)
MDMA, LSD, and PCP	Urine	DLLME-CE-UV	—	1, 4.41, 4.52	(Airado-Rodríguez et al. 2012)
APs, opiates, and cocaine	Urine	DLLME-CE-UV	—	0.1–10	(Kohler et al. 2013)
MA, MDMA, ketamine, and heroin	Seized forensic samples	DLLME-CE-UV	545–611	0.08–0.2	(Meng et al. 2011)
BZD	Plasma	DLLME-HPLC	—	1.7–10.6	(Fernández et al. 2013)
BZD	Urine and water	DLLME-UPLC	—	0.6–6.2	(Fernández et al. 2014)
BZD	Urine	DLLME-GC-FID	—	0.02–0.05	(Ghobadi et al. 2014)
7-aminoflunitrazepam	Urine	DLLME-LC-MS	20	0.025	(Melwanki et al. 2009)
Chlordiazepoxide	Water, urine, plasma, and chlordiazepoxide tablets	DLLME-HPLC	—	0.5	(Khodadoust and Ghaedi 2013)
Barbituric acid	Urine, serum, and tablet	DLLME-UV/VIS	30	2	(Zarei and Gholamian 2011)
Opium alkaloids	Urine	DLLME-HPLC-UV	63–104.5	0.2–10	(Shamsipur Mojtaba and Fattahi 2011)
Opium alkaloids	Plasma	DLLME-HPLC-UV	110–165	0.5–5	(Ahmadi-Jouibari et al. 2013)
Fentanyl, alfentanil, and sufentanil	Urine and plasma	DLLME-HPLC	275–325	0.4–1.9	(Saraji et al. 2011)
Fantanyl	Urine	DLLME-GC-MS	—	1	(Gardner et al. 2015)
Methadone	Urine, plasma, saliva, and sweat	DLLME-HPLC-PDA	98–100	0.22–25	(Ranjbari et al. 2012)
Amphetamine-type stimulants, cathinones, phenethylamines, and ketamine analogues	Blood and urine	DLLME-GC-MS	—	2–50	(Mercieca et al. 2018)
Cannabinoids	Urine	DLLME-HPLC-UV	190–292	0.1–0.5	(Moradi et al. 2011)
Imipramine, desipramine, amitriptyline, nortriptyline, and clomipramine	Urine	DLLME-GC-MS	—	0.2–0.5	(Ito et al. 2011)

(Continued)

TABLE 8.1 (Continued)
Applications of LPMEs in analytical toxicology

Analyte	Matrix	Method	EF	LOD (µg/L)	Ref.
Methamphetamine	Urine	DLLME-HPLC-UV/Vis	–	10	(Ge and Lee 2013)
Imipramine and trimipramine	Urine	DLLME-HPLC-UV	161–186	0.6	(Shamsipur and Mirmohammadi 2014)
Amitryptiline, trimipramine, and doxepine	Plasma and urine	DLLME-EME-GC-FID	383–1065	0.25–15	(Seidi et al. 2013)
Duloxetine	Plasma	DLLME-HPLC-FLD	98	2.5	(Suh et al. 2013)
Ketamine and its main metabolites	Urine	HF-LPME- GC-MS	–	>1	(de Bairros et al. 2014)
THC-COOH	Urine	HF-LPME- GC-MS	–	1.5	(De Souza Eller et al. 2014)
Amphetamine-type stimulants	Hair	HF-LPME- GC-MS	40	0.02–0.04	(Pantaleão et al. 2012)
Recreational drugs	Blood	DLLME-GC-MS	19–24	10	(Lin et al. 2017)
Amphetamine-type stimulants	Urine	DLLME-GC-MS	–	0.05–0.1	(Cunha et al. 2016)
Cocaine's major adulterants	Urine	DLLME-HPLC-PDA	–	–	(Sena et al. 2017)
Drugs of abuse	Blood	DLLME-UPLC-MS/MS	–	0.05–2	(Fisichella et al. 2015)
Methadone	Serum	DLLME-HPLC-UV	134	3.34	(Taheri et al. 2015)
Benzodiazepines and benzodiazepine-like hypnotics	Blood	DLLME-LC-MS/MS	–	0.03–4.74	(De Boeck et al. 2017)

Abbreviations: 3,4-methylenedioxymethamphetamine (MDMA), amphetamine (AP), amphetamines (APs), methamphetamine (MA), 3,4-methylenedioxyamphetamine (MDA), amphetamine-type stimulants (ATS), phencyclidine (PCP), lysergic acid diethylamide (LSD), benzodiazepines (BZD).

8.3.1 Analysis of Amphetamines

In 2006, He and Kang reported the extraction of amphetamines in urine samples using three-phase SDLPME (single drop liquid-liquid-liquid microextraction) (He and Kang 2006). Headspace two-phase single drop LPME was suggested by He et al. in 2007 (He et al. 2007), in which sub-µg/L level detection limits could be achieved with HPLC-UV detection due to the high enrichment capability of the method. One of the most efficient LPMEs is provided by the combination between DLLME and SPE, in which extraction of amphetamines from urine and plasma samples showed increasing recovery compared with previous studies (Mashayekhi et al. 2014). In this study, 10 mL of sample solution containing amphetamines was initially pre-concentrated by C18 SPE cartridge and eluted by 2 mL of acetone MeCN. In the next step, MeCN was applied as disperser solvent for the DLLME method. The combined method was coupled to gas chromatography with flame ionization detector (GC-FID) for separation and detection of the amphetamines. Limits of detection (LODs) were in the range of 0.05–7 µg/L, and recoveries were in the ranges of 94–105% with 1.64% of relative standard deviation (RSD).

8.3.2 Analysis of Sedative and Hypnotic Drugs

For extraction and pre-concentration of sedative and hypnotic drugs, the proposed methods have been capable of providing superior merits for extraction of barbituric acid in water and biological samples (Zarei and Gholamian 2011), with LOD of 0.002 µg/mL and recoveries in the ranges of 94–105%.

8.3.3 Analysis of Opium Alkaloids, Opiates, and Other Alkaloids

In order to propose a method for the determination of nicotine, as the principal alkaloid of tobacco, and its metabolites such as cotinine, a dispersive liquid-liquid microextraction method based on solidification of floating organic drop (DLLME-SFO) was coupled to HPLC-UV. In this method, addition of no disperser solvent was found to be more effective for proper extraction efficiency. Instead, manual shaking was applied to form the extraction solvent emulsion. In addition, a binary extraction solvent was utilized to extract both nicotine and cotinine due to their different polarity. The obtained LOD was 0.002 µg/mL for both compounds, with spiking recoveries in the range of 72–105% (Wang et al. 2014).

8.3.4 Analysis of Cannabinoids

Hollow-fibre LPME-GC–MS/MS has been applied for the determination of cannabinoids in human hair. In this approach, the analytes were extracted in 20 min. LODs and extraction efficiencies between 0.5–15 pg/mg and 4.4 to 8.9% were obtained, respectively (Emídio et al. 2010). Among the methods developed for analysis of cannabinoids, a method combining surfactant-assisted and dispersive liquid-liquid microextraction (SA-DLLME) coupled to HPLC-UV has provided the best results (Moradi et al. 2011). The extraction time was about 15 min for the determination of cannabinoids in urine samples. In this work, cationic, anionic, and non-ionic surfactant as disperser solvent and toluene, 1-octanol, and 1-dodecanol as extraction solvent were screened using a one-variable-at-a-time (OVAT) approach. The proposed SA-DLLME-HPLC-UV method enjoyed reasonable analytical parameters, such as good RSDs in the range of 0.1–0.5 µg/L, with an enrichment factor over the range of 190–292.

8.3.5 Analysis of Antidepressant Drugs

Jafari et al. reported three-phase-HF-LPME coupled with electrospray ionization-ion mobility spectrometry (ESI-IMS) for the simultaneous determination of trimipramine and desipramine, as antidepressant drugs, in urine and plasma samples. RSDs in the range of 5–6 µg/L, with spiking recoveries over the range of 92–97% (Jafari et al. 2011), were obtained. In the latter study, using a combination of DLLME and electromembrane extraction (EME), coupled with GC-FID, provided the analysis of amitryptiline, trimipramine, and doxepine in urine and plasma samples (Seidi et al. 2013). A hollow fibre was filled with acceptor solution and used to extract the analytes of interest. The fibre was dipped in the sample solution while an electrical field was applied to transfer the analytes.

8.3.6 Analysis of Hallucinogens

For the analysis of lysergic acid diethylamide, phencyclidine, and 3,4-methylenedioxymethamphetamine (MDMA), as hallucinogens, a combined DLLME method with capillary zone electrophoresis (CZE) and UV detection was developed by Airado-Rodríguez et al. (Airado-Rodríguez et al. 2012). The LODs were found to be in the range of 1/4.5 ng mL for all the three analytes.

8.3.7 Comparison between DLLME Methods

Table 8.1 summarizes other DLLME applications in analytical toxicology. As can be seen, the highest enrichment factors over the ranges of 545–611 and 383–1065 were obtained for the determination of methamphetamine (MA), MDMA, ketamine, heroin (Meng et al. 2011), and amitryptiline, and trimipramine and doxepine (Seidi et al. 2013), respectively. Good enrichment factors between 100 and 300 were obtained for the determination of MA and MDMA (Djozan et al. 2012), opium alkaloids (Shamsipur Mojtaba and Fattahi 2011; Ahmadi-Jouibari et al. 2013), fentanyl, alfentanil, and sufentanil (Saraji et al. 2011), methadone (Ranjbari et al. 2012; Taheri et al. 2015), cannabinoids (Moradi et al. 2011), and imipramine and trimipramine (Shamsipur and Mirmohammadi 2014). The lowest LODs over the ranges of 0.02–0.04 and 0.0–0.05 were obtained for amphetamine-type stimulants (Pantaleão et al. 2012) and benzodiazepines (Ghobadi et al. 2014), respectively.

8.4 Conclusions

In this chapter, the procedures most used for the extraction and determination of toxicological compounds using LPME were described. The LPME technique that is most used is DLLME, offering very high enrichment factors for some compounds, such as MDMA, MA, ketamine, heroin, fentanyl, alfentanil, amitryptiline, trimipramine, doxepine, sufentanil, and opium alkaloids. These compounds have been determined by different instrumental techniques, such as HPLC, CE, and GC. In some cases, hollow-fibre has been used for hair samples; however, the rest have been successfully applied to urine, plasma, and blood samples, mainly. Thus, DLLME has been shown to be a well-established technique for the analysis of compounds of this nature, offering low detection limits that allow its quantification in real samples at expected doping concentrations. The well-known hollow fibre has proven to be a good option for the extraction of compounds of a different nature that offer excellent clean-up, so it would be interesting to investigate in more detail the use of this technique for urine and plasma samples. The only drawback of this technique is that it is necessary to apply it directly to liquid samples, so hair samples would be excluded, unless a stage prior to this extraction is carried out.

REFERENCES

Ahmadi-Jouibari, Toraj, Nazir Fattahi, and Mojtaba Shamsipur. "Rapid Extraction and Determination of Amphetamines in Human Urine Samples Using Dispersive Liquid-Liquid Microextraction and Solidification of Floating Organic Drop Followed by High Performance Liquid Chromatography." *Journal of Pharmaceutical and Biomedical Analysis* 94 (2014): 145–51.

Ahmadi-Jouibari, Toraj, Nazir Fattahi, Mojtaba Shamsipur, and Meghdad Pirsaheb. "Dispersive Liquid-Liquid Microextraction Followed by High-Performance Liquid Chromatography-Ultraviolet Detection to Determination of Opium Alkaloids in Human Plasma." *Journal of Pharmaceutical and Biomedical Analysis* 85 (2013): 14–20.

Airado-Rodríguez, Diego, Carmen Cruces-Blanco, and Ana M. García-Campaña. "Dispersive Liquid-Liquid Microextraction Prior to Field-Amplified Sample Injection for the Sensitive Analysis of

3,4-Methylenedioxymethamphetamine, Phencyclidine and Lysergic Acid Diethylamide by Capillary Electrophoresis in Human Urine." *Journal of Chromatography A* 1267 (2012): 189–97.

An, Jiwoo, María J. Trujillo-Rodríguez, Verónica Pino, and Jared L. Anderson. "Non-conventional Solvents in Liquid Phase Microextraction and Aqueous Biphasic Systems." *Journal of Chromatography A* 1500 (2017): 1–23.

Barroso, Mário, Ivo Moreno, Beatriz Da Fonseca, João António Queiroz, and Eugenia Gallardo. "Role of Microextraction Sampling Procedures in Forensic Toxicology." *Bioanalysis* 4 (2012): 1805–26.

Cunha, Ricardo Leal, Wilson Araujo Lopes, and Pedro Afonso P. Pereira. "Determination of Free (Unconjugated) Amphetamine-Type Stimulants in Urine Samples by Dispersive Liquid-Liquid Microextraction and Gas Chromatography Coupled to Mass Spectrometry (DLLME-GC-MS)." *Microchemical Journal* 125 (2016): 230–5.

de Bairros, André Valle, Rafael Lanaro, Rafael Menck de Almeida, and Mauricio Yonamine. "Determination of Ketamine, Norketamine and Dehydronorketamine in Urine by Hollow-Fiber Liquid-Phase Microextraction Using an Essential Oil as Supported Liquid Membrane." *Forensic Science International* 243 (2014): 47–54.

De Boeck, Marieke, Sophie Missotten, Wim Dehaen, Jan Tytgat, and Eva Cuypers. "Development and Validation of a Fast Ionic Liquid-Based Dispersive Liquid–Liquid Microextraction Procedure Combined with LC–MS/MS Analysis for the Quantification of Benzodiazepines and Benzodiazepine-Like Hypnotics in Whole Blood." *Forensic Science International* 274 (2017): 44–54.

De Souza Eller, Sarah Carobini Werner, Luma Gonçalves Flaiban, Beatriz-Aparecida-Passos-Bismara Paranhos, José Luiz Da Costa, Felipe Rebello Lourenço, and Mauricio Yonamine. "Analysis of 11-Nor-9-carboxy-δ9-tetrahydrocannabinol in Urine Samples by Hollow Fiber-Liquid Phase Microextraction and Gas Chromatography-Mass Spectrometry in Consideration of Measurement Uncertainty." *Forensic Toxicology* 32 (2014): 282–91.

Djozan, Djavanshir, Mir Ali Farajzadeh, Saeed Mohammad Sorouraddin, and Tahmineh Baheri. "Molecularly Imprinted-Solid Phase Extraction Combined with Simultaneous Derivatization and Dispersive Liquid-Liquid Microextraction for Selective Extraction and Preconcentration of Methamphetamine and Ecstasy from Urine Samples Followed by Gas Chromatography." *Journal of Chromatography A* 1248 (2012): 24–31.

Emídio, Elissandro-Soares, Vanessa de Menezes Prata, Fernando-Jose-Malagueno de Santana, and Haroldo-Silveria Dórea. "Hollow Fiber-Based Liquid Phase Microextraction with Factorial Design Optimization and Gas Chromatography-Tandem Mass Spectrometry for Determination of Cannabinoids in Human Hair." *Journal of Chromatography B Analytical Technologies in the Biomedical and Life Sciences* 878 (2010): 2175–83.

Fernández, Purificación, Cristina González, M. Teresa Pena, Antonia M. Carro, Rosa A. Lorenzo. "A Rapid Ultrasound-Assisted Dispersive Liquid-Liquid Microextraction Followed by Ultra-performance Liquid Chromatography for the Simultaneous Determination of Seven Benzodiazepines in Human Plasma Samples." *Analytica Chimica Acta* 767 (2013): 88–96.

Fernández, Purificación, Maria Regenjo, A.M. Fernández, Rosa A. Lorenzo, and Antonia M. Carro. "Optimization of Ultrasound-Assisted Dispersive Liquid-Liquid Microextraction for Ultra Performance Liquid Chromatography Determination of Benzodiazepines in Urine and Hospital Wastewater." *Analytical Methods* 6 (2014): 8239–46.

Fisichella, Marco, Sara Odoardi, and Sabina Strano-Rossi. "High-Throughput Dispersive Liquid/liquid Microextraction (DLLME) Method for the Rapid Determination of Drugs of Abuse, Benzodiazepines and Other Psychotropic Medications in Blood Samples by Liquid Chromatography-Tandem Mass Spectrometry (LC-MS/MS) and Application to Forensic Cases." *Microchemical Journal* 123 (2015): 33–41.

Gardner, Michael A., Sheena Sampsel, Werner W. Jenkins, and Janel E. Owens. "Analysis of Fentanyl in Urine by DLLME-GC-MS." *Journal of Analytical Toxicology* 39 (2015): 118–25.

Ge, Dandan, and Hian Kee Lee. "Ionic Liquid Based Dispersive Liquid-Liquid Microextraction Coupled with Micro-solid Phase Extraction of Antidepressant Drugs from Environmental Water Samples." *Journal of Chromatography A* 1317 (2013): 217–22.

Ghobadi, Masoomeh, Yadollah Yamini, and Behnam Ebrahimpour. "SPE Coupled with Dispersive Liquid-Liquid Microextraction Followed by GC with Flame Ionization Detection for the Determination of Ultra-trace Amounts of Benzodiazepines." *Journal of Separation Science* 37 (2014): 287–94.

Han, Dandan, Baokun Tang, Yu Ri Lee, and Kyung Ho Row. "Application of Ionic Liquid in Liquid Phase Microextraction Technology." *Journal of Separation Science* 35 (2012): 2949–61.

He, Yi, and Marta Concheiro-Guisan. "Microextraction Sample Preparation Techniques in Forensic Analytical Toxicology." *Biomedical Chromatography* 33 (2019): 1–2.

He, Yi, and Youn-Jung Kang. "Single Drop Liquid-Liquid-Liquid Microextraction of Methamphetamine and Amphetamine in Urine." *Journal of Chromatography A* 1133 (2006): 35–40.

He, Yi, Angelica Vargas, and Youn-Jung Kang. "Headspace Liquid-Phase Microextraction of Methamphetamine and Amphetamine in Urine by an Aqueous Drop." *Analytica Chimica Acta* 589 (2007): 225–30.

Ito, Rie, Masaru Ushiro, Yuki Takahashi, Koichi Saito, Tetsuo Ookubo, Yusuke Iwasaki, and Hiroyuki Nakazawa. "Improvement and Validation the Method Using Dispersive Liquid-Liquid Microextraction with In Situ Derivatization Followed by Gas Chromatography-Mass Spectrometry for Determination of Tricyclic Antidepressants in Human Urine Samples." *Journal of Chromatography B Analytical Technologies in the Biomedical and Life Sciences* 879 (2011): 3714–20.

Jafari, Mohammad Taghi, Mohammad Saraji, and Hossein Sherafatmand. "Electrospray Ionization-Ion Mobility Spectrometry as a Detection System for Three-Phase Hollow Fiber Microextraction Technique and Simultaneous Determination of Trimipramine and Desipramine in Urine and Plasma Samples." *Analytical and Bioanalytical Chemistry* 399 (2011): 3555–64.

Jain, Rajeev, and Ritu Singh. "Applications of Dispersive Liquid-Liquid Micro-extraction in Forensic Toxicology." *TrAC - Trends in Analytical Chemistry* 75 (2016): 227–37.

Khodadoust, Saeid, and Mehrorang Ghaedi. "Optimization of Dispersive Liquid-Liquid Microextraction with Central Composite Design for Preconcentration of Chlordiazepoxide Drug and Its Determination by HPLC-UV." *Journal of Separation Science* 36 (2013): 1734–42.

Kohler, Isabelle, Julie Schappler, T. Sierro, and Serge Rudaz. "Dispersive Liquid-Liquid Microextraction Combined with Capillary Electrophoresis and Time-of-Flight Mass Spectrometry for Urine Analysis." *Journal of Pharmaceutical and Biomedical Analysis* 73 (2013): 82–9.

Kokosa, John M. "Recent Trends in Using Single-Drop Microextraction and Related Techniques in Green Analytical Methods." *TrAC - Trends in Analytical Chemistry* 71 (2015): 194–204.

Lin, Zebin, Jiaolun Li, Xinyu Zhang, Meihong Qiu, Zhibin Huang, and Yulan Rao. "Ultrasound-Assisted Dispersive Liquid-Liquid Microextraction for the Determination of Seven Recreational Drugs in Human Whole Blood Using Gas Chromatography–Mass Spectrometry." *Journal of Chromatography B Analytical Technologies in the Biomedical and Life Sciences* 1046 (2017): 177–84.

Liu, Hanghui, and Purnendu K. Dasgupta. "Analytical Chemistry in a Drop. Solvent Extraction in a Microdrop." *Analytical Chemistry* 68 (1996): 1817–21.

Liu, Shaorong, and Purnendu K. Dasgupta. "Liquid Droplet. A Renewable Gas Sampling Interface." 1995. https://pubs.acs.org/sharingguidelines (accessed December 7, 2020).

Lucena, Rafael, Marta Cruz-Vera, Soledad Cáárdenas, and Miguel Valcáárcel. "Liquid-Phase Microextraction in Bioanalytical Sample Preparation." *Bioanalysis* 1 (2009): 135–49.

Majors, Ronald E. "An Overview of Sample Preparation." *LC GC* 9, no. 1 (1991): 16–20.

Mashayekhi, Hossein Ali, Mohammad Rezaee, and Faezeh Khalilian. "Solid-Phase Extraction Followed by Dispersive Liquid-Liquid Microextraction for the Sensitive Determination of Ecstasy Compounds and Amphetamines in Biological Samples." *Bulletin of the Chemical Society of Ethiopia* 28 (2014): 339–48.

Melwanki, Mahaveer B., Wei-Shan Chen, Hsin-Yu Bai, Tzuen-Yeuan Lin, and Ming-Ren Fuh. "Determination of 7-Aminoflunitrazepam in Urine by Dispersive Liquid-Liquid Microextraction with Liquid Chromatography-Electrospray-Tandem Mass Spectrometry." *Talanta* 78 (2009): 618–22.

Meng, Liang, Bin Wang, Feng Luo, Guijun Shen, Zhengqiao Wang, and Ming Guo. "Application of Dispersive Liquid-Liquid Microextraction and CE with UV Detection for the Chiral Separation and Determination of the Multiple Illicit Drugs on Forensic Samples." *Forensic Science International* 209 (2011): 42–7.

Mercieca, Gilbert, Sara Odoardi, Marisa Cassar, and Sabina Strano Rossi. "Rapid and Simple Procedure for the Determination of Cathinones, Amphetamine-Like Stimulants and Other New Psychoactive Substances in Blood and Urine by GC–MS." *Journal of Pharmaceutical and Biomedical Analysis* 149 (2018): 494–501.

Moradi, Morteza, Yadollah Yamini, and Tahmineh Baheri. "Analysis of Abuse Drugs in Urine Using Surfactant-Assisted Dispersive Liquid-Liquid Microextraction." *Journal of Separation Science* 34 (2011): 1722–9.

Moradi, Morteza, Yadollah Yamini, and Tahmineh Baheri. "Analysis of Abuse Drugs in Urine Using Surfactant-Assisted Dispersive Liquid-Liquid Microextraction." *Journal of Separation Science* 34 (2011): 1722–9.

Nuhu, Abdulmumin A., Chanbasha Basheer, and Bahruddin Saad. "Liquid-Phase and Dispersive Liquid-Liquid Microextraction Techniques with Derivatization: Recent Applications in Bioanalysis." *Journal of Chromatography B Analytical Technologies in the Biomedical and Life Sciences* 879 (2011): 1180–8.

Pantaleão, Lorena do Nascimento, Beatriz Aparecida Passos Bismara Paranhos, and Mauricio Yonamine. "Hollow-Fiber Liquid-Phase Microextraction of Amphetamine-Type Stimulants in Human Hair Samples." *Journal of Chromatography A* 1254 (2012): 1–7.

Pedersen-Bjergaard, Stig, and Knut Einar Rasmussen. "Liquid-Liquid-Liquid Microextraction for Sample Preparation of Biological Fluids Prior to Capillary Electrophoresis." *Analytical Chemistry* 71 (1999): 2650–6.

Ranjbari, Elias, Ali-asghar Golbabanezhad-Azizi, and Mohammad Reza Hadjmohammadi. "Preconcentration of Trace Amounts of Methadone in Human Urine, Plasma, Saliva and Sweat Samples Using Dispersive Liquid-Liquid Microextraction Followed by High Performance Liquid Chromatography." *Talanta* 94 (2012): 116–22.

Rezaee, Mohammad, Yaghoub Assadi, Mohammad-Reza Milani Hosseini, Elham Aghaee, Fardin Ahmadi, and Sana Berijani. "Determination of Organic Compounds in Water Using Dispersive Liquid-Liquid Microextraction." *Journal of Chromatography A* 1116 (2006): 1–9.

Saber Tehrani, Mohammad, Mohammad Hadi Givianrad, and Nasibeh Mahoor. "Surfactant-Assisted Dispersive Liquid-Liquid Microextraction Followed by High-Performance Liquid Chromatography for Determination of Amphetamine and Methamphetamine in Urine Samples." *Analytical Methods* 4 (2012): 1357–64.

Saraji, Mohammad, Malihe Khalili Boroujeni, Ali Akbar, and Hajialiakbari Bidgoli. "Comparison of Dispersive Liquid-Liquid Microextraction and Hollow Fiber Liquid-Liquid-Liquid Microextraction for the Determination of Fentanyl, Alfentanil, and Sufentanil in Water Synthesis Carbon and Silica Mesoporous Materials and LDHs. Fabrication New Class of MOFs and COFs. View Project Radiolytical Synthesis of Bimetallic Nanoparticles View Project." *Artic. Analytical and Bioanalytical Chemistry* 400 (2011): 2149–58.

Sato, Kiyoshi, Manabu Tokeshi, Tsuguo Sawada, and Takehiko Kitamori. "Molecular Transport between Two Phases in a Microchannel." *Analytical Sciences* 16 (2000): 455–6.

Seidi, Shahram, Yadollah Yamini, and Maryam Rezazadeh. "Combination of Electromembrane Extraction with Dispersive Liquid-Liquid Microextraction Followed by Gas Chromatographic Analysis as a Fast and Sensitive Technique for Determination of Tricyclic Antidepressants." *Journal of Chromatography B Analytical Technologies in the Biomedical and Life Sciences* 913–914 (2013): 138–46.

Sena, Laís Cristina Santana, Humberto Reis Matos, Haroldo Silveira Dórea, Maria Fernanda Pimentel, Danielle Cristine Almeida Silva de Santana, and Fernando José Malagueño de Santana. "Dispersive Liquid-Liquid Microextraction Based on Solidification of Floating Organic Drop and High-Performance Liquid Chromatography to the Analysis of Cocaine's Major Adulterants in Human Urine." *Toxicology* 376 (2017): 102–12.

Shamsipur, Mojtaba, and Mehrosadat Mirmohammadi. "High Performance Liquid Chromatographic Determination of Ultratraces of Two Tricyclic Antidepressant Drugs Imipramine and Trimipramine in Urine Samples After Their Dispersive Liquid-Liquid Microextraction Coupled with Response Surface Optimization." *Journal of Pharmaceutical and Biomedical Analysis* 100 (2014): 271–8.

Shamsipur, Mojtaba, and Nazir Fattahi. "Extraction and Determination of Opium Alkaloids in Urine Samples Using Dispersive Liquid-Liquid Microextraction Followed by High-Performance Liquid Chromatography." *Journal of Chromatography B Analytical Technologies in the Biomedical and Life Sciences* 879 (2011): 2978–83.

Sharifi, Vahid, Ali Abbasi, and Anahita Nosrati. "Application of Hollow Fiber Liquid Phase Microextraction and Dispersive Liquid-Liquid Microextraction Techniques in Analytical Toxicology." *Journal of Food and Drug Analysis* 24 (2016): 264–76.

Shen, Yao, Teris A. Van Beek, Han Zuilhof, and Bo Chen. "Hyphenation of Optimized Microfluidic Sample Preparation with Nano Liquid Chromatography for Faster and Greener Alkaloid Analysis." *Analytica Chimica Acta* 797 (2013): 50–6.

Suh, Joon-Hyuk, Yun-Young Lee, Hee-Joo Lee, Myunghee Kang, Yeoun Hur, Sun-Neo Lee, Dong-Hyug Yang, and Sang-Beom Han. "Dispersive Liquid-Liquid Microextraction Based on Solidification of Floating Organic Droplets Followed by High Performance Liquid Chromatography for the Determination of Duloxetine in Human Plasma." *Journal of Pharmaceutical and Biomedical Analysis* 75 (2013): 214–19.

Taheri, Salman, Fahimeh Jalali, Nazir Fattahi, Ronak Jalili, and Gholamreza Bahrami. "Sensitive Determination of Methadone in Human Serum and Urine by Dispersive Liquid-Liquid Microextraction Based on the Solidification of a Floating Organic Droplet Followed by HPLC-UV." *Journal of Separation Science* 38 (2015): 3545–51.

Tang, Sheng, Tong Qi, Prince Dim Ansah, Juliette Chancellevie Nalouzebi Fouemina, Wei Shen, Chanbasha Basheer, and Hian Kee Lee. "Single-Drop Microextraction." *TrAC - Trends in Analytical Chemistry* 108 (2018): 306–13.

Tetala, Kishore K.R., Jan W. Swarts, Bo Chen, Anja E.M. Janssen, and Teris A. Van Beek. "A Three-Phase Microfluidic Chip for Rapid Sample Clean-Up of Alkaloids from Plant Extracts." *Lab on a Chip* 9 (2009): 2085–92.

Trujillo-Rodríguez, Maria J., He Nan, Marcelino Varona, Miranda N. Emaus, Israel D. Souza, and Jared L. Anderson. "Advances of Ionic Liquids in Analytical Chemistry." *Analytical Chemistry* 91 (2019): 505–31.

Wang, Xu, Yu Wang, Xiaoli Zou, and Yun Cao. "Improved Dispersive Liquid–Liquid Microextraction Based on the Solidification of Floating Organic Droplet Method with a Binary Mixed Solvent Applied for Determination of Nicotine and Cotinine in Urine." *Analytical Methods* 6 (2014): 2384–9.

Xu, Cong, and Tingliang Xie. "Review of Microfluidic Liquid-Liquid Extractors." *Industrial & Engineering Chemistry Research* 56 (2017): 7593–622.

Zarei, Ali-Reza, and Forouzan Gholamian. "Development of a Dispersive Liquid-Liquid Microextraction Method for Spectrophotometric Determination of Barbituric Acid in Pharmaceutical Formulation and Biological Samples." *Analytical Biochemistry* 412 (2011): 224–8.

Zhao, Limian, and Hian Kee Lee. "Liquid-Phase Microextraction Combined with Hollow Fiber as a Sample Preparation Technique Prior to Gas Chromatography/Mass Spectrometry." *Analytical Chemistry* 74 (2002): 2486–92.

9
Dispersive Liquid-Liquid Microextraction and Its Variants

Rakesh Roshan Jha[1,2], Rabindra Singh Thakur[3], and Rajeev Jain[4]

[1]*Centre of Analytical Bioscience, School of Pharmacy, University of Nottingham, UK*
[2]*Analytical Chemistry Laboratory, Regulatory Toxicology Group, CSIR-Indian Institute of Toxicology Research (CSIR-IITR), Vishvigyan Bhawan, UP, India*
[3]*Academy Council of Scientific and Innovative Research (AcSIR), CSIR-IITR Campus, UP, India*
[4]*Forensic Toxicology Division, Central Forensic Science Laboratory, India*

CONTENTS

9.1 Introduction 159
9.2 Variants of DLLME 162
    9.2.1 Based on Extraction Solvents 162
        9.2.1.1 Ionic Liquid-Based DLLME (IL-DLLME) 162
        9.2.1.2 Low-Density Solvent-Based DLLME (LDS-DLLME, DLLME-SFO) 162
    9.2.2 Based on Dispersion Solvents 163
        9.2.2.1 Auxiliary Dispersion Solvents 163
        9.2.2.2 Surfactant as Dispersion Solvent 163
    9.2.3 Assistance-Based Modification of DLLME 163
        9.2.3.1 Solid-Phase Extraction DLLME (SPE-DLLME) 163
        9.2.3.2 Molecularly Imprinted SPE-DLLME (MISPE-DLLME) 164
        9.2.3.3 Ultrasound-Assisted DLLME (UA-DLLME) 164
        9.2.3.4 Salt-Assisted Liquid-Liquid Extraction DLLME (SALLE-DLLME) 164
        9.2.3.5 Miscellaneous Modifications in DLLME 165
9.3 Applications of DLLME and Its Variants 165
    9.3.1 Analysis of Urine 165
    9.3.2 Analysis of Blood, Plasma, and Serum 166
    9.3.3 Analysis of Tissue and Viscera 166
    9.3.4 Analysis of Saliva 166
    9.3.5 Miscellaneous Applications of DLLME 167
9.4 Conclusion and Future Trends 167
References 167

9.1 Introduction

Analytical chemistry deals with identification and quantification of various analytes of different origin, whereas toxicology is the study of adverse effects of any chemical entities on living organisms.

Analytical toxicology covers the qualitative and quantitative estimation of chemical toxicants as well as small biological molecules to find research answers for the effect of toxicants on biological systems (Kenkel 2002; Rushton 1997).

Analytical scientists deal with various types of chemicals, poisons, drugs, pesticides, and other analytes in the routine analysis of complex matrices, such as saliva, urine, serum, plasma, blood, hair, and vitreous humour (Manousi and Samanidou 2021). This complexity in matrices makes the analysis of these chemical substances more tedious. Similar problems are frequently observed in forensic science, where every case's uniqueness and the ambiguous analyte existence make it very hard to standardize specific analytical procedures (Manousi and Samanidou 2021). The complexity of matrices, trace amounts of analytes of interest, and limited sample quantities can complicate sample preparation, which is the most important step of analytical method development (Kumari et al. 2015; Sharma et al. 2018).

Sample preparation comprises conversion of the analyte of interest in a detectable and quantitative state from its complex matrices with the best possible sensitivity and the least interferences. For optimum analytical method development, sample preparation should have the following criteria:

- Require a smaller amount of the sample
- Be less time-consuming
- Have the selectivity to extract the target analyte(s) from the sample
- Employ the least/no amount of toxic solvents
- Give efficient recovery of the analyte with reproducibility
- Give a clean extract of analyte without matrix interferences and impurities
- Be suitable for the derivatizing steps
- Be suitable to couple with various analytical instruments

There are several analytical techniques available to date from traditional extraction methods, like liquid-liquid extraction (LLE), Soxhlet extraction, and solid-phase extraction (SPE), to modern miniaturized extraction methods, such as liquid-liquid microextraction (LLME), dispersive liquid-liquid microextraction (DLLME), solid-phase microextraction (SPME), and single droplet microextraction (SDME). LLE and SPE are the finest available analytical techniques, and even today they are preferred for most routine analysis by reputed referral laboratories (Jha et al. 2017, 2018). However, modern analytical techniques have the upper hand on traditional available analytical extraction techniques as there are some limitations associated with them, such as the following:

- They require a large quantity of the sample.
- They require a large volume of toxic organic solvents for the extraction of analytes.
- They are environmentally unfriendly.
- They are tedious and time-consuming.
- They are multi-step extraction procedures.
- They are costly.
- They have low enrichment of analyte from the sample matrices.
- They require clean-up before analysis on an instrument.

The need for multiple time extraction affects the reproducibility of the results (Jha et al. 2018). Emulsion formation and matrix interference also often create obstacles in the preparation of samples. Such disadvantages make analytes more vulnerable to lose during sample preparation steps, which affects the reproducibility and extraction efficiency of the analytical method. The clean-up step becomes necessary after LLE to reduce the matrix effect. Another extraction technique that has long been used in forensic laboratories is SPE (Mudiam et al. 2014). In this technique, the extraction, as well as the clean-up, can be performed simultaneously. This extraction technique is based on the affinity of the analyte between the solid and liquid phases. SPE requires a smaller amount of extraction solvent in comparison to LLE. It involves a cartridge filled with solid packing material acting as a stationary phase, which sometimes has an affinity specific to individual analytes. Commercial SPE cartridges are now also available with small amounts of stationary phases, which require only microliters of solvent to extract analytes from samples. There are some crucial step of multi-step SPE processes such as pre-conditioning, adsorption, elution, and pre-concentration of the final extract. Clogging can cause trouble in the case of real sample handling while performing SPE (Balinova et al. 2007; Jha et al. 2018).

In the 1990s, Arthur and Pawliszyn revolutionized the sample preparation procedures by inventing SPME (Arthur et al. 1990). SPME is a solvent-less microextraction process consisting of a fibre coated with a specific polymeric stationary phase on the surface, on which analytes are adsorbed or absorbed. Microextraction techniques overcome the drawback of conventional extraction. SPME is an expensive extraction technique, and the fibres used for extraction are very delicate and fragile, requiring special maintenance and protection. As the fibre lifespan is short, SPME fibres need periodic replacement. Besides this, the sample carryover is also a significant drawback of SPME (Ulrich 2000).

The primary emphasis of microextraction techniques is on decreasing the number of steps involved in sample preparation, thus reducing time consumption and minimizing the use of toxic organic solvents. Miniaturization, economical operation, coupling capability with a broad range of analytical instruments with high enrichment factors, and better extraction efficiency are the remarkable benefits of microextraction techniques (Kataoka 2010). Consequently, Rezaee et al. implemented the LLE method's miniaturization in 2006, which was termed dispersive liquid-liquid microextraction (DLLME) (Rezaee et al. 2006).

This microextraction comprises the three-component system (extraction phase, dispersion phase, and aqueous phase). The conventional DLLME employs the extraction solvent with a higher density than water, and they are toxic organic solvents (e.g., chloroform, carbon tetrachloride, trichloroethylene). The dispersion phase has suitable solubility with the aqueous and organic phases, which increases the interaction between the two phases. The solvent premix of the dispersion solvent with the extraction solvent is speedily injected into the aqueous phase, which results in a cloudy solution (Jha et al. 2018) and, after centrifugation, is ready for instrumental analysis (Figure 9.1).

Several modifications in terms of dispersion solvent, extraction solvent, and mode of dispersion have been done with the first coined DLLME by Rezaee et al. in 2006, and these are the variants of DLLME. There are several advantages of DLLME and its variants, which have been widely applied over the years for the analysis of a wide range of analytes from environmental, biological, and food matrices. In this

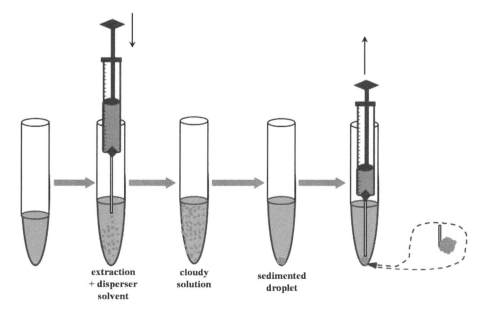

FIGURE 9.1 Generalized scheme of dispersive liquid-liquid microextraction (DLLME) (reproduced with permission from Jain and Singh 2016).

chapter, we focus on the different variants of DLLME that have been developed, their advantages, and their application for the analysis of drugs and poisons in biological matrices such as blood, plasma, urine, hair, nail, vitreous humour, and tissue.

9.2 Variants of DLLME

9.2.1 Based on Extraction Solvents

Currently used extraction solvents in DLLME are toxic, with a limited ability to extract different analytes with a range of polarities. Thus, it is necessary to look for other available solvents for DLLME. In order to extend DLLME's application scope on the basis of extraction solvents, researchers have concentrated on the use of low-density and polar organic solvents or new eco-friendly solvents, such as ionic liquids (ILs).

9.2.1.1 Ionic Liquid-Based DLLME (IL-DLLME)

ILs, which are generally known as green solvents, are a group of organic salts in liquid state at room temperature. ILs have some specific physicochemical properties, such as insignificant variable viscosity, high thermal stability, and vapour pressure. Liu et al. (2009) first introduced the use of 1-hexyl-3-methylimidazolium hexafluorophosphate ([C6MIM][PF6]) as an extraction solvent with the disperser solvent (methanol), similar to the conventional DLLME method. 1-butyl-3-methylimidazolium hexafluorophosphate ([BMIM][PF6]) has been used to extract benzodiazepines from blood samples (De Boeck et al. 2017, 2018). Except for organic compounds, ILs also demonstrated strong extractability as neutral or charged complexes for metal ions. Arsenic was extracted from urine and whole blood samples (Shirkhanloo et al. 2011) by using [BMIM][PF6] IL. Ultrasound was applied to increase the interaction of analytes with IL, and further extraction efficiency can be enhanced by controlling temperature. Thus, ultrasound-enhanced temperature-controlled ionic liquid dispersive liquid-liquid microextraction (UETC-IL-DLLME) was applied for triazole pesticide in a plasma sample (Li et al. 2013). Many applications focused on the IL-DLLME technique were carried out in the years that followed, emphasizing metal ions (Shirkhanloo et al. 2011), pesticides (Li et al. 2013), and benzodiazepines (De Boeck et al. 2017, 2018) in the biological sample.

9.2.1.2 Low-Density Solvent-Based DLLME (LDS-DLLME, DLLME-SFO)

In conventional DLLME, the extraction phase collection is tedious because it sediments in the lower portion of the tube. Therefore, the introduction of low-density solvents makes this task more comfortable and increases the range of solvents used for extraction (Jha et al. 2017). These extraction solvents (hexane, toluene, xylene, chloroform) remain over the aqueous phase, easily collected by the needle. The benefit of using LDS-DLLME is that any matrix part that remains at the bottom of the extraction vessel will be sediment after centrifugation, while the extraction solvent will remain floating on the surface and results in a cleaner extract, which can be recovered easily (Mudiam et al. 2014).

To recover the low-density solvent layer, Xu et al. proposed a modified method that employs low-density solvents such as 1-dodecanol, 2-dodecanol, hexadecanol, and 1-undecanol with a melting point less than room temperature (Xu et al. 2009). A droplet of extraction solvent tends to float over the surface due to its low density. The sample vial is moved into an ice bath for some time, making it easier to solidify the floating organic droplet due to its lower melting point below room temperature. The solidified droplet is then melted, which is subjected to instrumental analysis. Several solvents meeting these criteria, such as 1-undecanol (Jha et al. 2017; Saber Tehrani et al. 2012), were approved in DLLME-SFO (Dispersive liquid-liquid microextraction-solidification of organic droplet) and used for certain organic compounds in complex samples, such as urine and plasma (Suh et al. 2013).

9.2.2 Based on Dispersion Solvents

9.2.2.1 Auxiliary Dispersion Solvents

The disperser solvent's miscibility in both the extraction solvent and the aqueous phase are criteria for its selection in DLLME. Methanol, acetone, acetonitrile, and ethanol are commonly used as the disperser solvents in this process, which have been confirmed in the literature multiple times. Based on the polarity, analytes can show good solubility in disperser solvents, especially when the compounds are more polar, increasing the partition of the analytes with extraction solvent droplets and leading to an increase in the efficiency of extraction.

9.2.2.2 Surfactant as Dispersion Solvent

For the first time, due to the amphipathic nature, a surfactant was explored to act as the disperser solvent. This process is termed surfactant-assisted dispersive liquid-liquid micro-extraction (SA-DLLME). Herein, the surfactants could contribute significantly to the dispersion of extraction solvents into the aqueous phase, leading to a significant decrease in the interfacial tension between the two phases (Moradi et al. 2010), showing a better effect than the solvents commonly used. The authors also argued that based on pH, the surfactant could form ion pairs with target analytes, which would most likely increase extraction efficiency. Some analytical benefits, such as low costs, fast handling, and lack of toxic effects, have already been demonstrated by surfactants. Consequently, the versatility of surfactants as efficient disperser solvents was shown by several other surfactants, such as cetyl trimethyl ammonium bromide (CTAB) (Behbahani et al. 2013), sodium dodecyl sulphate (SDS) (Saber Tehrani et al. 2012), and tetradecyl trimethyl ammonium bromide (TTAB) (Moradi et al. 2011). Hence, for the use of disperser solvents in DLLME, this is considered a new choice these days.

9.2.3 Assistance-Based Modification of DLLME

Due to the complex nature of the matrix and the trace-level presence of analytes, sometimes, for optimum extraction, DLLME needs to be coupled with other extraction techniques. This combination of DLLME with other methods could increase pre-concentration of the analyte, decrease matrix interferences, and increase sensitivity of analytes in some cases. Thus, several studies have concentrated on DLLME coalitions with more methods of purification or extraction, exploring consistency with multiple samples. To date, large combinations associated with DLLME have been recorded.

9.2.3.1 Solid-Phase Extraction DLLME (SPE-DLLME)

Initially, the DLLME technique was applied to the study based on the simplest sample matrices, primarily water. It can be concluded that the primary purpose of DLLME is to maximize sensitivity at the cost of selectivity. The combination of SPE and DLLME techniques for the isolation and pre-concentration of chlorophenols (CPs) in complex matrices has also been studied (Fattahi et al. 2007). In this work, the eluent from SPE was served as a disperser solvent, and together with the extraction solvent, it was rapidly injected into the additional aqueous phase (water) for the DLLME process, which enriched the analyte in the extraction phase and lowered the detection limit (Quigley et al. 2016). The combination of SPE with DLLME increased the method's applicability and adaptability to different sample matrixes. Furthermore, the same solvent, operating as two actors, also proved the viability of such a combination in the two operations. In complex samples, the improved SPE-DLLME approach was subsequently widely adopted and applied to extract different analytes, such as benzodiazepines (Ghobadi et al. 2014), amphetamines (Mashayekhi et al. 2014), and cocaine and its metabolites (Martins et al. 2017).

9.2.3.2 Molecularly Imprinted SPE-DLLME (MISPE-DLLME)

Furthermore, a significantly higher degree of specificity and sensitivity was asserted by combining molecularly imprinted polymers (MIP) with SPE-DLLME. The critical advantage of MIP compared to SPE is the specificity of MIP toward the target analyte molecules, which is due to its analyte-specific template formed during their synthesis. By precipitation polymerization, MIP was synthesized using the targeted analyte as a template. The sample was loaded on an MI-SPE cartridge, the analyte was bound with polymer, and then it was eluted with a suitable solvent (methanol). This methanol was then proceeded further for the DLLME procedure (Peñuela-Pinto et al. 2017). Sometimes the extract from MISPE was combined with butyl-chloroformate (BCF), which acted as an extraction solvent and a derivatizing reagent. High-density BCF settled on the sediment phase after centrifugation, which was injected into GC-FID. The MISPE-DLLME with simultaneous derivatization was used for MA and MDMA analysis in a urine sample (Djozan et al. 2012).

9.2.3.3 Ultrasound-Assisted DLLME (UA-DLLME)

Another modification in the DLLME procedure was the use of ultrasonic waves to enhance extraction efficiency. Ultrasonication was used in UA-DLLME to facilitate emulsion formation, thus increasing the extraction efficiency by speeding up the mass transfer process between the extraction solvent and the aqueous phase (Jain et al. 2013). Ultrasonication helped to attain the equilibrium very easily and helped to reduce extraction time. The surface-to-volume ratio of the extracting drops was improved by ultrasonic wave (Fernández et al. 2014). While extracting benzodiazepines from urine, the ultrasonic wave potentially induced the intermolecular interaction cleavage of benzodiazepines from the matrix (Fernández et al. 2013). The UA-DLLME method was applied for a psychoactive substance in urine (Jain et al. 2013; Reddy Mudiam et al. 2012), blood (Chen et al. 2017), tissue (brain) (Mudiam et al. 2014), wastewater (Fernández et al. 2014), saliva (Shekari et al. 2020), etc.

Using a lower-density solvent as an extraction solvent and ultrasound energy to assist in emulsification without any dispersive solvent enhanced the extraction efficiency with ease of collecting the extraction phase. Also, the low-density solvent used in extraction was easily obtained after de-emulsification, which reduced the influence of a complex sample matrix and was ideal for biological samples (Meng et al. 2015). The UA-LDS-DLLME was applied to extract a psychoactive substance from urine (Meng et al. 2011, 2015Meng 2015) and blood (Meng et al. 2015). It has also been shown that the pre-treatment approach is efficient in eliminating the influence of complex biological sample matrices.

9.2.3.4 Salt-Assisted Liquid-Liquid Extraction DLLME (SALLE-DLLME)

The DLLME procedure is not enough when dealing with semi-solid or solid complex matrices, such as tissue or viscera. On performing the conventional DLLME, the extraction phase could not be separated, and the extracts thus obtained were contaminated. When DLLME was performed, no sedimentary drop occurred, and sometimes the analyte was used to solubilize again in the sample. In such cases, salt addition reduced the solubility of the analytes in the sample solution and simultaneously enhanced the analytes' distribution in the organic phase, contributing to an improvement in extraction efficiency. Besides, the homogeneous solution obtained during extraction from the complex matrix was broken down by dissolving a sufficient amount of salt as the phase separation agent (Mohebbi et al. 2018). In SALLE after the salt addition, water-soluble sample components stayed in the aqueous process after centrifugation, giving a cleaner extract of the extraction phase. SALLE-DLLME has been successfully applied to human organs (kidney, liver, brain, heart, lung, spleen, abdominal fat) (Pastor-Belda et al. 2019). In addition to SALLE, Ali et al. used dispersive solid-phase extraction as a clean-up step before performing DLLME to extract tricyclic antidepressant drugs (Mohebbi et al. 2018).

9.2.3.5 Miscellaneous Modifications in DLLME

An innovative microextraction technique called electro-membrane extraction (EME) was recently introduced by Pedersen-Bjergaard and Rasmussen (Pedersen-Bjergaard et al. 2006). An electrical potential was used in the EME, enabling the analytes' extraction through the hollow-fibre membrane. EME can also extract analytes without sample pre-treatment, thereby removing the resulting problems due to this step. EME's drawback is its incompatibility with the gas chromatographic method. The current technique eliminates the need for relatively high-cost SPE cartridges and tedious extraction steps, especially solvent evaporation. EME-DLLME allows DLLME to be quickly applied to complex matrices, removes the disadvantage of the EME technique, improves sensitivity due to analytes' aggregation in significantly fewer micro volumes of the extraction solvent, and provides high sample clean-up. A new pre-treatment approach for the extraction of tricyclic antidepressants (TCA) from biological matrices is being developed by integrating the benefits of EME with DLLME (Seidi et al. 2013).

In the solid-based DLLME extraction method, a mixture of butyl chloroformate (derivatizing reagent) with 1,1,2,2-tetrachloroethane (extraction solvent) is added to a sugar cube, which is then inserted into an aqueous sample having the analytes and a catalyst, such as 3-methylpyridine. The extractant and derivatization agent are slowly released into the aqueous sample as fine droplets during the dissolving of the sugar cube by shaking. The resulting cloudy solution is centrifuged, and the sedimentary phase is processed for instrumental analysis (Farajzadeh et al. 2015).

9.3 Applications of DLLME and Its Variants

9.3.1 Analysis of Urine

Urine is a complex matrix and is one of the major pieces of evidence in forensic science. Drug use and toxicant metabolite were analyzed in a urine sample (Meng et al. 2015). Hallucinogens are drugs that change an individual's perception and mood, without activating or preventing brain activity. Lysergic acid diethylamide (LSD), phencyclidine (PCP), and 34-methylenedioxymethamphetamine (MDMA) are the most used hallucinogens in the world. Rodríguez et al. applied the DLLME process with capillary zone electrophoresis (CE) and UV detection to study LSD, PCP, and MDMA in human urine samples. Diluted urine samples were made alkaline with ammonia and subjected to DLLME using acetonitrile (disperser solvent) and dibromomethane (extraction solvent). Amphetamines are known as synthetic stimulants of the central nervous system, including amphetamine (AP), methamphetamine (MA), 3,4-methylenedioxyamphetamine (MDA), and MDMA. According to the World Drug Report, 2019, amphetamine-type and prescription stimulants (excluding ecstasy) were the third most widely used illegal drugs in 2017, with an estimated 29 million users (United Nations Office on Drugs and Crime UNODC 2018).

Antidepressants, such as TCAs, are the class of psychoactive medications used to treat major depressive disorders. However, an overdose of TCA can result in arrhythmia, hypertension, and death in some cases. A DLLME-high-performance liquid chromatography mobile phase (HPLC) with ultraviolet (UV) detection method for the extraction and determination of psychoactive drugs such as thioridazine, clomipramine, and amitryptiline was developed in urine samples (Xiong et al. 2009). The rapid injection of acetonitrile and carbon tetrachloride (CCl4) extracted these drugs into the urine sample led to a cloudy solution. The DLLME-HPLC-UV method was used to evaluate two TCA drugs, i.e., imipramine and trimipramine, in urine samples. The author reported that the proposed method could assess the target analyte concentration in urine samples after 5 hours (Shamsipur et al. 2014).

The DLLME-SFO technique was recently reported using 1-undecanol as an extraction solvent for MA and AP in urine samples in which acetonitrile was used as a disperser solvent. Using HPLC-UV, the isolation and identification of analytes was performed.

9.3.2 Analysis of Blood, Plasma, and Serum

The DLLME-HPLC procedure was applied for various samples, such as plasma, urine, water, and chlordiazepoxide tablets to determine chlordiazepoxide, a BZD. The chloroform and methanol were injected into the aqueous sample, accompanied by centrifugation. Dilution of plasma and urine samples minimized the matrix effect (Khodadoust et al. 2013).

Seven BZDs (alprazolam, bromazepam, lormetazepam, diazepam, clonazepam, lorazepam, and tetrazepam) were extracted from the plasma sample using UA-DLLME. Methanol deproteinizes the plasma sample and works as a disperser solvent with chloroform as an extraction solvent. After centrifugation, the supernatant with chloroform was injected into ultrapure water (pH 9). The mixture was then subjected to ultrasound, accompanied by centrifugation, and samples were analyzed by ultra performance liquid chromatography (Fernández et al. 2013).

DLLME with SFO was applied to extract and pre-concentrate the opium alkaloids in human plasma. The sample's protein precipitation was done using 15% zinc sulphate-acetonitrile solution (50:40, v/v). The solution was made alkaline by sodium chloride (NaCl) in which 1-undecanol was injected with acetone to extract the alkaloids. After centrifugation, the sample was put in an ice bath to solidify the floating organic droplet. This SFO drop was melted and proceeded for HPLC analysis (Ahmadi-Jouibari et al. 2013).

Duloxetine, a medication used in the treatment of depressive disorders, was extracted using DLLME-SFO. In this process, the analyte was extracted using 1-undecanol as an extraction solvent, from the plasma sample. The sample matrix was first deproteinized using zinc sulphate and acetonitrile; in this process, acetonitrile was acting as a dispersant itself. The extract was further analyzed by using HPLC with fluorescence detection (Suh et al. 2013).

9.3.3 Analysis of Tissue and Viscera

One of the most common suicide strategies in developing countries is self-poisoning with pesticides. For the analysis of pesticides in matrices of toxicological value, such as skin, blood, and urine, highly sensitive and rapid analytical methods are required. A low density-DLLME method coupled with gas chromatography-electron capture detection (GC-ECD) has been developed to analyze cypermethrin in tissue and blood samples of rats treated with cypermethrin. Tissue samples (brain, liver, and kidney) were first homogenized in acetone and then centrifuged. The supernatant acetone was used as a solvent disperser and was quickly injected into ultra-pure water in conjunction with n-hexane for pre-concentration of cypermethrin in n-hexane. Blood samples were mixed with water and subjected to a similar process of DLLME (Mudiam et al. 2012).

The primary metabolites and biomarkers for exposure to pyrethroid pesticides were 3-phenoxybenzoic acid (3-PBA), and 4-phenoxy-3-hydroxybenzoic acid (OH-PBA) was determined in the rat brain by UA-DLLME. Methyl chloroformate (MCF) acted as a derivatizing reagent in the process, a single stage derivatization cum extraction method was developed and combined with large-volume injection-gas chromatography-tandem mass spectrometry (LVI-GC-MS/MS) for pyrethroid metabolite analysis (Mudiam et al. 2014).

The MISPE-DLLME method for evaluating 3-PBA in rat liver and blood samples has been reported to extract 3-PBA selectively. MIP was synthesized having a 3-PBA binding site, and the eluent obtained after MISPE was subjected to DLLME, followed by injection pot silylation (IPS) inside the injection port and further proceeded for gas chromatography-tandem mass spectrometry (GC-MS/MS) analysis (Mudiam et al. 2014).

9.3.4 Analysis of Saliva

Methadone is a synthetic medication used in the treatment of dependency on opiates. A DLLME system coupled to HPLC-UV for the pre-concentration and analysis of methadone was reported in four matrices (human urine, plasma, saliva, and sweat). Methanol and chloroform were used as dispersers and extraction solvents, respectively, for the DLLME extraction of methadone in samples. After centrifugation, the sediment phase obtained was evaporated and makeup in methanol for analysis by HPLC. Before DLLME, the pH was set to 10 to hold methadone all together in its molecular shape. This method showed greater sensitivity compared to conventional approaches, such as SPE and LLE (Ranjbari et al. 2012).

9.3.5 Miscellaneous Applications of DLLME

DLLME-CE-UV was applied for determining multiple illicit substances (MDMA, MA, opium, and ketamine) in forensic samples, such as kraft paper and banknotes silver paper, as well as in plastic bags. The samples were soaked in acetic acid and filtered; this filtrate was made alkaline by sodium hydroxide (NaOH). The isopropyl alcohol (IPA) and chloroform were quickly injected into this filtrate to form the cloud solution. The effect of pH on amines and heroin was different; as a result, amphetamines and ketamine displayed an improved recovery, while recovery of heroin was decreased. The method showed better analysis speed, and it was possible to isolate the entire target analytes within 10 min (Meng et al. 2011).

The UA-DLLME and DLLME method was applied to the beverages for the extraction of benzodiazepines. Dichloromethane and acetonitrile were rapidly injected into the sample, and the extract was further processed for the HPLC-UV analysis (Piergiovanni et al. 2018).

9.4 Conclusion and Future Trends

Elaboration of the DLLME technique focuses on the standard innovations, developments, and various implementations in different fields. Exhaustive attempts have been made to expand the applications to various analytes with more complex biological matrices than aqueous samples with substantial achievements. In terms of internal technique modifications (seeking other usable extraction or disperser solvents) and the combination of DLLME with other techniques, all the methods discussed in this study provide both advantages and disadvantages. In most of the works, extensive DLLME research is concerned with extending the variety of extraction solvents that are used to improve the method's extraction efficiency. This technique also will be of immense use in the routine analysis of chemical and biological compounds in various scopes of work.

REFERENCES

Ahmadi-Jouibari, Toraj et al. "Dispersive Liquid-Liquid Microextraction Followed by High-Performance Liquid Chromatography-Ultraviolet Detection to Determination of Opium Alkaloids in Human Plasma." *Journal of Pharmaceutical and Biomedical Analysis* 85 (2013): 14–20.

Arthur, C.L. et al. "Solid Phase Microextraction with Thermal Desorption Using Fused Silica Optical Fibers." *Analytical Chemistry* 62 (1990): 2145–48.

Behbahani, Mohammad et al. "Application of Surfactant Assisted Dispersive Liquid-Liquid Microextraction as an Efficient Sample Treatment Technique for Preconcentration and Trace Detection of Zonisamide and Carbamazepine in Urine and Plasma Samples." *Journal of Chromatography A* 1308 (2013): 25–31.

Balinova, Anna et al. "Solid-Phase Extraction on Sorbents of Different Retention Mechanisms Followed by Determination by Gas Chromatography-Mass Spectrometric and Gas Chromatography-Electron Capture Detection of Pesticide Residues in Crops." *Journal of Chromatography A* 1150 (2007): 136–44.

Chen, X. et al. "Ultrasound-Assisted Low-Density Solvent Dispersive Liquid–Liquid Microextraction for the Simultaneous Determination of 12 New Antidepressants and 2 Antipsychotics in Whole Blood by Gas Chromatography–Mass Spectrometry." *Journal of Pharmaceutical and Biomedical Analysis* 142 (2017): 19–27.

De Boeck, Marieke et al. "Ionic Liquid-Based Liquid–Liquid Microextraction for Benzodiazepine Analysis in Postmortem Blood Samples." *Journal of Forensic Sciences* 63 (2018): 1875–9.

De Boeck, Marieke et al. "Development and Validation of a Fast Ionic Liquid-Based Dispersive Liquid–Liquid Microextraction Procedure Combined with LC–MS/MS Analysis for the Quantification of Benzodiazepines and Benzodiazepine-Like Hypnotics in Whole Blood." *Forensic Science International* 274 (2017): 44–54.

Djozan, Djavanshir et al. "Molecularly Imprinted-Solid Phase Extraction Combined with Simultaneous Derivatization and Dispersive Liquid-Liquid Microextraction for Selective Extraction and Preconcentration of Methamphetamine and Ecstasy from Urine Samples Followed by Gas Chromatography." *Journal of Chromatography A* 1248 (2012): 24–31.

Farajzadeh, Mir Ali et al. "Simultaneous Derivatization and Solid-Based Disperser Liquid-Liquid Microextraction for Extraction and Preconcentration of Some Antidepressants and an Antiarrhythmic Agent in Urine and Plasma Samples Followed by GC-FID." *Journal of Chromatography B Analytical Technologies in the Biomedical and Life Sciences* 983–984 (2015): 55–61.

Fattahi, N. et al. "Solid-Phase Extraction Combined with Dispersive Liquid-Liquid Microextraction-Ultra Preconcentration of Chlorophenols in Aqueous Samples." *Journal of Chromatography A* 1169 (2007): 63–9.

Fernández, Purificación et al. "A Rapid Ultrasound-Assisted Dispersive Liquid-Liquid Microextraction Followed by Ultra-performance Liquid Chromatography for the Simultaneous Determination of Seven Benzodiazepines in Human Plasma Samples." *Analytica Chimica Acta* 767 (2013): 88–96.

Fernández, Purificación et al. "Optimization of Ultrasound-Assisted Dispersive Liquid-Liquid Microextraction for Ultra Performance Liquid Chromatography Determination of Benzodiazepines in Urine and Hospital Wastewater." *Analytical Methods* 6 (2014): 8239–46.

Ghobadi, M. et al. "SPE Coupled with Dispersive Liquid-Liquid Microextraction Followed by GC with Flame Ionization Detection for the Determination of Ultra-trace Amounts of Benzodiazepines." *Journal of Separation Science* 37 (2014): 287–94.

Jain, Rajeev et al. "Ultrasound Assisted Dispersive Liquid-liquid Microextraction Followed by Injector Port Silylation: A Novel Method for Rapid Determination of Quinine in Urine by GC-MS." *Bioanalysis* 5 (2013): 2277–86.

Jain, Rajeev and Ritu Singh. "Applications of Dispersive Liquid–Liquid Micro-Extraction in Forensic Toxicology." *TrAC Trends in Analytical Chemistry* 75 (2016): 227–237.

Jha, Rakesh Roshan et al. "Ionic Liquid Based Ultrasound Assisted Dispersive Liquid-Liquid Microextraction for Simultaneous Determination of 15 Neurotransmitters in Rat Brain, Plasma and Cell Samples." *Analytica Chimica Acta* 1005 (2018): 43–53.

Jha, Rakesh Roshan et al. "Ultrasound-Assisted Emulsification Microextraction Based on a Solidified Floating Organic Droplet for the Rapid Determination of 19 Antibiotics as Environmental Pollutants in Hospital Drainage and Gomti River Water." *Journal of Separation Science* 40 (2017): 2694–702.

Jha, Rakesh Roshan et al. "Dispersion-Assisted Quick and Simultaneous Extraction of 30 Pesticides from Alcoholic and Non-alcoholic Drinks with the Aid of Experimental Design." *Journal of Separation Science* 41 (2018): 1625–34.

Kataoka, Hiroyuki "Recent Developments and Applications of Microextraction Techniques in Drug Analysis." *Analytical and Bioanalytical Chemistry* 396 (2010): 339–64.

Kenkel, John *Analytical Chemistry for Technicians*, 2002.

Khodadoust, Saeid et al. "Optimization of Dispersive Liquid-Liquid Microextraction with Central Composite Design for Preconcentration of Chlordiazepoxide Drug and Its Determination by HPLC-UV." *Journal of Separation Science* 36 (2013): 1734–42.

Kumari, Rupender et al. "Fast Agitated Directly Suspended Droplet Microextraction Technique for the Rapid Analysis of Eighteen Organophosphorus Pesticides in Human Blood." *Journal of Chromatography A* 1377 (2015): 27–34.

Liu, Yubo et al. "Determination of Four Heterocyclic Insecticides by Ionic Liquid Dispersive Liquid-Liquid Microextraction in Water Samples." *Journal of Chromatography A* 1216 (2009): 885–91.

Li, Yubo et al. "Determination of Triazole Pesticides in Rat Blood by the Combination of Ultrasound-Enhanced Temperature-Controlled Ionic Liquid Dispersive Liquid-Liquid Microextraction Coupled to High-Performance Liquid Chromatography." *Analytical Methods* 5 (2013): 2241–8.

Manousi, N. and V. Samanidou "Green Sample Preparation of Alternative Biosamples in Forensic Toxicology." In *Sustainable Chemistry and Pharmacy* 20 (2021, May 1): 100388.

Martins, A.F. et al. "Occurrence of Cocaine and Metabolites in Hospital Effluent - A Risk Evaluation and Development of a HPLC Method Using DLLME." *Chemosphere* 170 (2017): 176–82.

Mashayekhi, Hossein Ali et al. "Solid-Phase Extraction Followed by Dispersive Liquid-Liquid Microextraction for the Sensitive Determination of Ecstasy Compounds and Amphetamines in Biological Samples." *Bulletin of the Chemical Society of Ethiopia* 28 (2014): 339–48.

Meng, Liang et al. "Application of Dispersive Liquid-Liquid Microextraction and CE with UV Detection for the Chiral Separation and Determination of the Multiple Illicit Drugs on Forensic Samples." *Forensic Science International* 209 (2011): 42–7.

Meng, Liang et al. "Ultrasound-Assisted Low-Density Solvent Dispersive Liquid-Liquid Extraction for the Determination of Amphetamines in Biological Samples Using Gas Chromatography-Mass Spectrometry." *Journal of Forensic Science and Medicine* 1 (2015): 114–18.

Meng, Liang et al. "Ultrasound-Assisted Low-Density Solvent Dispersive Liquid-Liquid Microextraction for the Determination of Eight Drugs in Biological Samples by Gas Chromatography-Triple Quadrupole Mass Spectrometry." *Chinese Journal of Chromatography (Se Pu)* 33 (2015): 304–8.

Mohebbi, Ali et al. "Determination of Tricyclic Antidepressants in Human Urine Samples by the Three–Step Sample Pretreatment Followed by HPLC–UV Analysis: An Efficient Analytical Method for Further Pharmacokinetic and Forensic Studies." *EXCLI Journal* 17 (2018): 952–63.

Moradi, Morteza et al. "Analysis of Abuse Drugs in Urine Using Surfactant-Assisted Dispersive Liquid-Liquid Microextraction." *Journal of Separation Science* 34 (2011): 1722–9.

Moradi, Morteza et al. "Application of Surfactant Assisted Dispersive Liquid-Liquid Microextraction for Sample Preparation of Chlorophenols in Water Samples." *Talanta* 82 (2010): 1864–9.

Mudiam, Mohana Krishna Reddy et al. "Development of Ultrasound-Assisted Dispersive Liquid-Liquid Microextraction-Large Volume Injection-Gas Chromatography-Tandem Mass Spectrometry Method for Determination of Pyrethroid Metabolites in Brain of Cypermethrin-Treated Rats." *Forensic Toxicology* 32 (2014): 19–29.

Mudiam, Mohana Krishna Reddy et al. "Low Density Solvent Based Dispersive Liquid-Liquid Microextraction with Gas Chromatography-Electron Capture Detection for the Determination of Cypermethrin in Tissues and Blood of Cypermethrin Treated Rats." *Journal of Chromatography B Analytical Technologies in the Biomedical and Life Sciences* 895–896 (2012): 65–70.

Mudiam, Mohana Krishna Reddy et al. "Molecularly Imprinted Polymer Coupled with Dispersive Liquid-Liquid Microextraction and Injector Port Silylation: A Novel Approach for the Determination of 3-Phenoxybenzoic Acid in Complex Biological Samples Using Gas Chromatography-Tandem Mass Spectrometry." *Journal of Chromatography B Analytical Technologies in the Biomedical and Life Sciences* 945–946 (2014): 23–30.

Pastor-Belda, Marta et al. "Bioaccumulation of Polycyclic Aromatic Hydrocarbons for Forensic Assessment Using Gas Chromatography-Mass Spectrometry." *Chem. Res. Toxicol.* 32 (2019): 1680–8.

Peñuela-Pinto, Otilia et al. "Selective Determination of Clenbuterol Residues in Urine by Molecular Imprinted Polymer—Ion Mobility Spectrometry." *Microchemical Journal* 134 (2017): 62–7.

Piergiovanni, Maurizio et al. "Determination of Benzodiazepines in Beverages Using Green Extraction Methods and Capillary HPLC-UV Detection." *Journal of Pharmaceutical and Biomedical Analysis* 154 (2018): 492–500.

Pedersen-Bjergaard, Stig et al. "Electrokinetic Migration Across Artificial Liquid Membranes: New Concept for Rapid Sample Preparation of Biological Fluids." *Journal of Chromatography A* 1109, no. 2 (2006): 183–90.

Quigley, Andrew et al. "Dispersive Liquid-Liquid Microextraction in the Analysis of Milk and Dairy Products: A Review." *Journal of Chemistry* 2016 (2016): 12.

Ranjbari, Elias et al. "Preconcentration of Trace Amounts of Methadone in Human Urine, Plasma, Saliva and Sweat Samples Using Dispersive Liquid-Liquid Microextraction Followed by High Performance Liquid Chromatography." *Talanta* 94 (2012): 116–22.

Reddy Mudiam, Mohana Krishna et al. "Optimization of UA-DLLME by Experimental Design Methodologies for the Simultaneous Determination of Endosulfan and Its Metabolites in Soil and Urine Samples by GC-MS." *Analytical Methods* 4 (2012): 3855–63.

Rezaee, Mohammad et al. "Determination of Organic Compounds in Water Using Dispersive Liquid-Liquid Microextraction." *Journal of Chromatography A* 1116 (2006): 1–9.

Rushton, B. "Basic Analytical Toxicology." *Journal of Clinical Pathology* 50, no. 2 (1997): 177.

Saber Tehrani, Mohammad et al. "Surfactant-Assisted Dispersive Liquid-Liquid Microextraction Followed by High-Performance Liquid Chromatography for Determination of Amphetamine and Methamphetamine in Urine Samples." *Analytical Methods* 4 (2012): 1357–64.

Shamsipur, Mojtaba et al. "High Performance Liquid Chromatographic Determination of Ultra Traces of Two Tricyclic Antidepressant Drugs Imipramine and Trimipramine in Urine Samples After Their Dispersive

Liquid-Liquid Microextraction Coupled with Response Surface Optimization." *Journal of Pharmaceutical and Biomedical Analysis* 100 (2014): 271–8.

Sharma, Divya et al. "Benzene Induced Resistance in Exposed Drosophila Melanogaster: Outcome of Improved Detoxification and Gene Modulation." *Chemosphere* 201 (2018): 144–58.

Seidi, Shahram et al. "Combination of Electromembrane Extraction with Dispersive Liquid-Liquid Microextraction Followed by Gas Chromatographic Analysis as a Fast and Sensitive Technique for Determination of Tricyclic Antidepressants." *Journal of Chromatography B Analytical Technologies in the Biomedical and Life Sciences* 913–914 (2013): 138–46.

Shekari, Ahmad et al. "Validation and Optimization of Ultrasound-Assisted Dispersive Liquid-Liquid Microextraction as a Preparation Method for Detection of Methadone in Saliva with Gas Chromatography-Mass Spectrometry Technique." *Advanced Pharmaceutical Bulletin* 10 (2020): 329–33.

Shirkhanloo, Hamid et al. "Ultra-trace Arsenic Determination in Urine and Whole Blood Samples by Flow Injection-Hydride Generation Atomic Absorption Spectrometry After Preconcentration and Speciation Based on Dispersive Liquid-Liquid Microextraction." *Bulletin of the Korean Chemical Society* 32 (2011): 3923–7.

Suh, J.H. et al. "Dispersive Liquid-Liquid Microextraction Based on Solidification of Floating Organic Droplets Followed by High Performance Liquid Chromatography for the Determination of Duloxetine in Human Plasma." *Journal of Pharmaceutical and Biomedical Analysis* 75 (2013): 214–19.

Ulrich, S. "Solid-Phase Microextraction in Biomedical Analysis." *Journal of Chromatography A* 902 (2000): 167–94.

United Nations Office on Drugs and Crime (UNODC). World Drug Report 2018, 2018.

Xiong, C. et al. "Extraction and Determination of Some Psychotropic Drugs in Urine Samples Using Dispersive Liquid-Liquid Microextraction Followed by High-Performance Liquid Chromatography." *Journal of Pharmaceutical and Biomedical Analysis* 49 (2009): 572–8.

Xu, H. et al. "A Novel Dispersive Liquid-Liquid Microextraction Based on Solidification of Floating Organic Droplet Method for Determination of Polycyclic Aromatic Hydrocarbons in Aqueous Samples." *Analytica Chimica Acta* 636 (2009): 28–33.

10
Electromembrane Extraction in Analytical Toxicology

Marothu Vamsi Krishna, Kantamaneni Padmalatha, and Gorrepati Madhavi
Vijaya Institute of Pharmaceutical Sciences for Women, Vijayawada, AP, India

CONTENTS

10.1	Introduction	172
10.2	Principle, EME Setup, and Procedure	172
10.3	Parameters Influencing Flux of Analytes across SLM	173
	10.3.1 Composition (Organic Solvent) of SLM, Viscosity, and Thickness	173
	10.3.2 Extraction Voltage and Time	174
	10.3.3 pH of Donor and Acceptor Phases	174
	10.3.4 Volume of Sample and Acceptor Solution	175
	10.3.5 Agitation/Stirring Speed	175
	10.3.6 Presence of Salt/Salt Effect	175
	10.3.7 Temperature	175
10.4	Technical Developments in EME	175
	10.4.1 On-Chip EME	175
	10.4.2 Low-Voltage EME	176
	10.4.3 Drop-to-Drop EME	176
	10.4.4 Pulsed EME	177
	10.4.5 EME Followed by Low-Density Solvent-Based Ultrasound-Assisted Emulsification Micro-Extraction (EME-LDS-USAEME)	178
	10.4.6 Parallel EME (Pa-EME)	178
	10.4.7 EME Using Free Liquid Membranes (FLMs)	178
	10.4.8 Gel-Electromembrane Extraction (G-EME)	178
10.5	Applications	179
	10.5.1 Extraction of Thebaine	179
	10.5.2 Extraction of Six Basic Drugs	179
	10.5.3 Extraction of Mebendazole	179
	10.5.4 Extraction of Nalmefene and Naltrexone	180
	10.5.5 Extraction of Citalopram, Loperamide, Methadone, and Sertraline from Dried Blood Spots	180
	10.5.6 Extraction of Six Basic Drugs of Abuse	180
	10.5.7 Extraction of Amphetamine-Type Stimulants from Human Urine	180
	10.5.8 Extraction of Lithium from Human Body Fluids	180
	10.5.9 Extraction of Heavy Metal Cations	181
	10.5.10 Extraction of Nerve Agent Degradation Products	181
	10.5.11 Extraction of Chlorophenols from Sea Water	181
10.6	Conclusion	181
References		181

DOI: 10.1201/9781003128298-10

10.1 Introduction

Biological samples, environmental samples, and pharmaceutical products are very complex and often contain interfering components like acids, bases, salts, metals, proteins, peptides, and several organic compounds with similar chemical properties to that of analytes. Thus, sample preparation is an important step required for extraction of preferred components from complex materials and for analyte enrichment. Sensitivity and analyte enrichment are enhanced by the extraction process. Liquid-liquid extraction (LLE) and solid-phase extraction (SPE) are traditionally used techniques for sample preparation, but LLE requires a great amount of organic solvents, is time consuming, and is a tedious process. SPE is a costly process and requires evaporation of the eluent after extraction.

An important feature of modern analytical chemistry is efficient and effective sample preparation. Quality of the analytical results will be affected by interfering compounds present in the sample matrix and poor sample preparation techniques. The focus of recent sample preparation techniques is miniaturization and novel modifications to conventional methods (LLE or SPE) to meet the needs of current analytical methods. Small sample size, high throughput process, potential for automation and online coupling, less use of hazardous organic solvents, cost-effectiveness, and user-friendly equipment are the common requirements of modern sample preparation techniques. Hollow-fibre liquid-phase microextraction (HF-LPME), developed in the mid-1990s, meets some of these requirements. It is simple, is rapid, and needs microliters of solvent. In this technique, pores of HFs are impregnated with the organic solvent and form a supported liquid membrane (SLM). The thickness of the SLM and amount of organic solvent in the SLM are specified by thickness and by the porosity and pore size of the HF, correspondingly. A very small amount of acceptor solution is filled into the lumen of the HF, and the system is positioned in the donor (sample) solution for extraction. Proper mixing of the sample solution is commonly required throughout extraction. Once the extraction is completed, the acceptor solution will be withdrawn and directly injected into analytical systems, such as gas chromatography (GC), liquid chromatography (LC), and capillary electrophoresis (CE).

The basic principle involved in the HF-LPME is passive diffusion of analytes from the donor solution (sample solution) through SLM into an acceptor solution. Distribution ratios of analytes between different aqueous and organic solvents determine their flux through the SLM. Generally, HF-LPME requires long extraction time to attain an equilibrium level. Extraction times in the range of 15–45 min are required to get maximum recoveries based on the sample volumes. It is reported that very large sample volumes require extraction times up to 2 h. Several parameters influence the extraction speed in HF-LPME, such as analyte distribution coefficient between sample solution and organic solvent in the SLM, distribution coefficient between the organic solvent and the acceptor solution, volume of the sample and acceptor solution, and the immobile boundary layer thickness between the sample solution and the SLM. The extraction process is comparatively slow, even if the above parameters are optimized. A new extraction procedure was proposed by Pedersen-Bjergaard and Rasmussen in the year 2006 to overcome this problem, which has been termed electromembrane extraction (EME). Initially, it was developed as a hybrid technique of HF-LPME and LLE, facilitated by application of an electric field. In this procedure, voltage was applied across the SLM that acts as the driving force to move charged analytes from the sample solution, through the SLM, into the acceptor solution.

10.2 Principle, EME Setup, and Procedure

A typical setup used for EME is shown in Figure 10.1. It contains a glass vial with a screw cap used to fill the sample solution (donor solution), and the pH of the sample solution is adjusted to charge the required analytes. The required length of the HF (usually made up of polypropylene or other porous hydrophobic material) is taken, and the lower end of the HF is sealed by applying mechanical pressure, whereas the upper end is coupled to a pipette tip as a guiding tube. The HF is dipped for a few seconds into a water-immiscible organic solvent to immobilize and form an SLM of the solvent in the wall of the HF. Excess of solvent in the SLM is gently removed by air-blowing with a medical syringe or with a

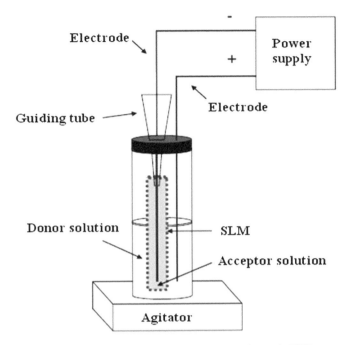

FIGURE 10.1 Schematic illustration of the setup for EME (Krishna Marothu et al. 2013).

medical wipe. Aqueous acceptor solution is filled into the lumen of the HF with a microsyringe through the guiding tube. The filled HF is placed into the sample solution through the vial cap. One platinum electrode is located in the sample solution, and another platinum electrode is located in the acceptor solution. Upon application of voltage over the electrodes, the charged analytes from the sample solution migrate through the SLM into the acceptor solution containing oppositely charged electrode. For cations, the cathode is placed in the acceptor solution and the anode is placed in the sample solution, whereas for anions, the anode is placed in the acceptor solution and the cathode is placed in the sample solution. Voltage is turned off to end the extraction, and acceptor solution is collected with the help of a microsyringe and transferred to a vial for analysis by HPLC, GC, CE, or any other suitable technique.

10.3 Parameters Influencing Flux of Analytes across SLM

10.3.1 Composition (Organic Solvent) of SLM, Viscosity, and Thickness

Success of EME mainly depends on the chemical nature of the SLM. The flux of analytes through the SLM is influenced by the difference in concentration of analytes across the SLM, which is determined partially by the distribution ratio of analyte from sample solution to the SLM; this, in turn, depends on the type of organic solvent used as the SLM. Selectivity, diffusion coefficient, and good clean-up during extraction are influenced by type of solvent used as an SLM. The organic phase used as an SLM in EME should support a relatively low current flow in the system by possessing a certain dipole moment or electrical conductivity. The organic solvent should facilitate electrokinetic migration and phase transfer of the analytes. In addition, the organic solvent should be immiscible with water to avoid the loss of solvent from SLM and dissolution in the sample and acceptor solution during stirring.

Nonpolar ($\log P > 2$) basic drugs are extracted by using nitro-aromatic, solvents such as 2-nitrophenyl octylether (NPOE) and nitrophenyl pentyl ether (NPPE), as SLMs. Phase transfer and electrokinetic migration of basic analytes was improved by the addition of hydrophobic alkylated phosphate reagents to the SLM.

Polar ($\log P < 1$) basic drugs are unable to migrate through the interface between the sample solution and the SLM formed by NPOE because the high polarity of these drugs counteracts the influence of the electric field. Transport of these drugs through SLM is facilitated by forming ion pairs with the analytes by adding ion pair reagents such as di-(2-ethylhexyl) phosphate (DEHP) or tris-(2-ethylhexyl) phosphate (TEHP) to the organic solvent. Basic drugs with a large log P window are extracted using an SLM comprising 10% DEHP and 10% TEHP in NPOE.

Nitro-aromatic solvents are not efficient for the extraction of acidic drugs. Long-chain alcohols such as 1-octanol and 1-heptanol have been used for the extraction of acidic drugs. They easily impregnate the membrane and are immiscible with water. These alcohols also offer suitable electrical resistance to the applied voltage, thereby avoiding excessive electrolysis, bubble formation, and stability issues of the analytes at the electrodes due to redox reactions. The type of organic solvent used as an SLM and voltage should be tuned and optimized for selective extraction of analytes and high recovery.

The viscosity of the organic solvent is another important parameter that influences the distribution of analytes through SLM: if the viscosity of the organic solvent is low, more diffusion of analytes is observed. Flux of analytes is also influenced by the thickness of the membrane: if the membrane thickness is high, a more diffuse path is observed, and this theoretically decreases the extraction recovery.

10.3.2 Extraction Voltage and Time

Electrokinetic migration of analytes through SLM into the acceptor solution is very much dependent on the applied voltage. The flux of analytes (J_i) is greatly influenced by the magnitude of the voltage applied. How different parameters influence the flux of analytes through the SLM is described in Equation 10.1.

$$J_i = \frac{-D_i}{h}\left(1 + \frac{v}{\ln X}\right)\left[\frac{X - 1}{X - exp(-v)}\right]\{c_i - c_{i0} exp(-v)\} \quad (10.1)$$

where
  D_i is the diffusion coefficient for the analyte,
  h is the thickness of the membrane,
  c_i is the analyte concentration at the SLM/sample interface
  c_{i0} is the concentration of the analyte at the SLM/acceptor interface
  v is a function of electrical potential
  and X is the ion balance (the ratio of the total ionic concentration in the sample solution to that in the acceptor solution).

Commonly, the EME voltage applied is in the range of 5–600 V. Recovery values will be decreased at higher voltages due to bubble formation, electrolysis, and degradation of analytes due to redox reactions at the electrodes. Time also influences the flux of analytes through SLM. An increase in extraction time or increase in voltage directly increases the flux of ions, and hence extraction recovery is increased, but if voltage and time are considered simultaneously, an antagonist effect is observed (decrease in the extraction recovery). Therefore, an increase in extraction time limits the applied voltage, and an increase in voltage limits the extraction time. Typically, at extraction times above 15 min, analyte recovery is declined due to the unsteadiness of the electrical current in the system, back migration of the analytes toward the sample solution (donor phase) due to alteration of pH from electrolysis, and a small loss of artificial liquid membrane. Voltage can be used for the selective extraction of analytes in EME. At higher voltages, all the analytes will be extracted, and at low voltages, some analytes will be extracted. This concept can be used for the selective extraction of analytes.

10.3.3 pH of Donor and Acceptor Phases

In the EME, analytes should be in the ionized state to be influenced by the electric field. For the fast extraction of basic analytes, the pH of the sample and acceptor solution is acidic to ionize the basic analytes; generally, hydrochloric acid, acetic acid, or formic acid is used. Extraction will be carried out by placing an anode in the sample solution and a cathode in the acceptor solution. The low pH in the sample solution ensures efficient flux of analytes through the SLM into the acceptor solution, while the low pH in the acceptor solution prevents the back diffusion of analytes into the sample solution. For the extraction of acidic analytes, alkaline pH is maintained in the sample and acceptor solution to ionize the analytes. A cathode is placed in the sample solution, and an anode is placed in the acceptor solution.

10.3.4 Volume of Sample and Acceptor Solution

High enrichment factors are observed by taking a large volume of sample solution and a small volume of acceptor solution. Although enrichment factors are high in EME, efficiency is decreased due to an increase in the sample volume. Higher sample volume causes an increase in distance between the electrodes, and hence a feeble electric field between the electrodes. Common sample volumes are in the range of 70 µL to 10 mL, and the acceptor solution volume is in the range of 10 to 100 µL.

10.3.5 Agitation/Stirring Speed

Agitation of the EME system leads to the increase in extraction efficiency by improving the kinetics or mass transfer of analytes from the sample solution to the SLM and reducing the thickness of the boundary layer around the SLM by means of convection. Commonly, a stirring speed of 0–1,250 rpm was used. An increase in extraction recovery was observed with stirring compared with no stirring, whereas at higher stirring rates, a decrease in extraction recovery was observed due to the formation of bubbles in the sample and acceptor solution and seepage of organic solvent from the SLM.

10.3.6 Presence of Salt/Salt Effect

Presence of salt or ionic substance in the sample solution leads to upsurge in the ionic balance of the system, and it causes a decrease in the flux of analytes through SLM. Hence, extraction efficiency is less in the presence of salt. However, few studies indicated that the addition of salt to the sample solution increased the efficiency of extraction. EME of acidic (non-steroidal anti-inflammatory drugs) and basic (β-blockers) drugs was achieved optimally by the addition of 30% (w/v) NaCl to the sample solution. The effect of salt on the EME of haloacetic acids was also investigated. NaCl in the concentration range of 3–15% was added to the sample solution and found that the addition of NaCl up to 5% increased the extraction recovery, and an above 5% decrease in the extraction recovery was observed due to variation in the conductivity of the sample solution and rise in the viscosity of the sample solution.

10.3.7 Temperature

The effect of temperature on the EME is also reported in the literature. The extraction process is rapid by increasing the temperature up to 40°C, and beyond 40°C partial degradation of SLM is observed.

10.4 Technical Developments in EME

In recent years, different EME setups were developed for improving recovery, sample enrichment, and throughput. EME devices are classified into two types depending on the configuration of the organic layer. One category of devices uses SLMs, where polymeric membranes are used for the impregnation of organic solvent, and the other category uses free liquid membranes (FLMs), where physical support is not used for the organic layer. Gel-electromembrane extraction (G-EME) is another recent technical development in EME, where gel membrane is used in place of SLM to carry out the extraction.

10.4.1 On-Chip EME

On-chip EME is an interesting development of EME. Advantages of on-chip EME are the requirement of very low sample volumes, continuous delivery of fresh samples to the SLM, low consumption of

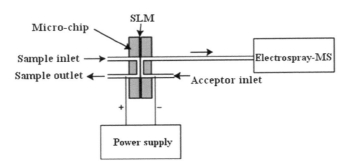

FIGURE 10.2 Schematic illustration of on-chip EME coupled to MS (Petersen et al. 2011).

reagents and chemicals, fast extractions due to very short diffusion path length, high extraction efficiency, and the possibility of online coupling to analytical instruments.

The on-chip EME (Figure 10.2) system consists of a porous polypropylene membrane impregnated with the organic solvent bonded between two poly(methyl methacrylate) (PMMA) substrates, each having channel structure toward the membrane. The sample solution is pumped through the sample channel of the chip with the help of a microsyringe pump. Analytes are diffused through SLM into the acceptor solution by the action of electrical potential. In one type of configuration, acceptor solution is stagnant and is removed by the pipette manually and analyzed offline. In another type of configuration, flow is introduced in the acceptor channel, and the acceptor solution is continuously pumped into an analytical instrument.

10.4.2 Low-Voltage EME

In low-voltage EME, 0–15 V is commonly used as a driving force to conduct the EME. Low-voltage EME is of interest because there is no chance of analyte degradation, interelectrode distance is less (few millimetres), extractions can be performed by common batteries, and there is the possibility for the development of portable extraction devices.

EME of 29 different basic model drug substances was performed at low voltage. The drug substances that had log P values of below 2.3 were not extracted at low voltages of less than 15 V. The drug substances that showed log P values of ≥2.3 and had two basic groups were also not extracted at low voltages of less than 15 V. Drug substances that had one basic group and log $P \geq 2.3$ were extracted at low voltages with strong selectivity.

10.4.3 Drop-to-Drop EME

Drop-to-drop EME is a miniaturized technique, performed under stagnant conditions by utilizing flat membranes. The setup of drop-to-drop EME is shown in Figure 10.3. It consists of aluminium foil with a well pressed into it, which acts as a sample compartment. The foil is coupled with the positive outlet of the power supply and acts as an anode. A platinum wire is placed in the acceptor droplet and connected to the negative outlet of the DC power supply, which acts as another electrode. Sample solution is placed in the well of the aluminium foil, and the foil is connected to the power supply. Organic solvent is immobilized in the membrane, which acts as an SLM and is placed on the top of the sample solution. The sample solution is sandwiched between the membrane and the aluminium foil. A droplet of acceptor solution is placed on the top of the membrane, and an electrode is inserted. Sample solution and acceptor solution are in contact with the SLM. Extraction is accomplished by applying a voltage for a certain period of time. After extraction, the acceptor solution is transferred to a vial with the help of a pipette for analysis. Advantages of drop-to-drop EME are direct use of samples without pre-concentration and selective extraction of analytes from a small volume of the samples. There is no chance of electrochemical degradation of the analytes because the extraction process is usually carried out at low voltages. The setup is simple and economical, and the carryover effect is not observed because the aluminium foil is used only for single extraction. The setup is stagnant and has no need for agitation.

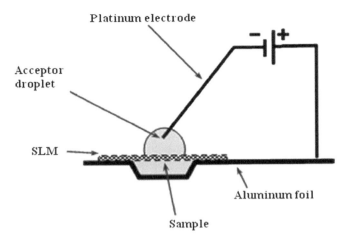

FIGURE 10.3 Schematic illustration of the setup of drop-to-drop EME (Petersen et al. 2009).

10.4.4 Pulsed EME

In the pulsed EME technique, pulsed voltage is used in the place of continuous DC voltage. The purpose of this is to provide a stable system when extraction is performed at higher voltages and to reduce the thickness of the ion double layer on both sides of the SLM. It improves the extractability by removing the mass transfer barrier. The principle involved in this is shown in Figure 10.4. Pulsed EME is performed by applying voltage in pulses for a short period of time (typically 15 s) with short breaks (2–10 s) in between. When voltage is applied, the duration of the pulse is enough to extract the analytes, and a double layer is formed at the SLM. Next, the voltage is turned off, whereas the sample solution is under stirring. Hence, ions accumulated at the SLM are dispersed again in the solution due to stirring, and the double layer will disappear. After the outage period, voltage will be applied again in a similar manner. Pre-concentration factors obtained with pulsed EME are higher than the values obtained with conventional EME.

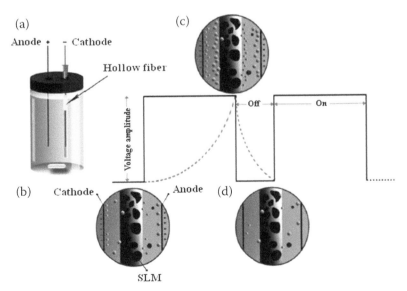

FIGURE 10.4 Principle involved in the pulsed EME. (A) EME setup; (B) beginning of the pulse duration; (C) end of the pulse duration; and (D) end of the outage period (Rezazadeh et al. 2012).

10.4.5 EME Followed by Low-Density Solvent-Based Ultrasound-Assisted Emulsification Micro-Extraction (EME-LDS-USAEME)

EME-LDS-USAEME is a two-step process. In the first step, extraction of analytes is carried out by EME, and the acceptor solution obtained in the EME is used in the second step (LDS-USAEME) as a sample solution. In the second technique, a soft Pasteur plastic pipette is used as an extraction device. The acceptor solution obtained in the first step (EME) is transferred to Pasteur pipette. If required, a suitable quantity of water is added to increase the volume of the sample solution, and a solvent with density lower than the water is injected into the pipette. The pipette is kept in an ultrasound water bath immediately to produce an emulsion for facilitating the analyte extraction. After extraction, the emulsion is centrifuged to separate into two phases. The bulb of the pipette is squeezed gently to raise the organic extract (upper layer) into the narrow stem of the pipette. Organic extract is collected with the help of microsyringe and transferred to a vial for introduction into the analytical system for qualitative or quantitative analysis. Extraction efficiency is high in this approach due to the combination of two techniques.

10.4.6 Parallel EME (Pa-EME)

Parallel EME is a high throughput sample preparation technique. It is performed with flat membranes in a 96 well-plate format. It consists of two well plates. The first plate has a conductive bottom, and the second plate has a polymeric membrane bottom. The first plate is used as a sample compartment, and the second plate is used to fill the acceptor solution. Donor solution is added to each sample well, polymeric membrane is impregnated with organic solvent, and acceptor solution is added to the wells in the acceptor plate. The second plate is inserted in the first plate, and voltage is applied between the plates. Recovery obtained with Pa-EME is comparable with that of the HF-EME.

10.4.7 EME Using Free Liquid Membranes (FLMs)

In this category of EME, physical support is not used for the organic layer, so they are called free liquid membranes (FLMs). Micro-EME (μ-EME) is a technique under this FLM category. It is a sandwich technique and is performed in horizontal configuration. It consists of a transparent polymeric tubing into which an organic layer (FLM) is sandwiched between the donor and acceptor solution. This forms a three-phase extraction system, which is stable and requires μL to sub-μL volumes of solutions. Among the different formats of EME, μ-EME is the only technique capable of handling very low sample volumes (≤1 μL). In the μ-EME, dilution of the sample is not required before extraction and can be used for extraction of low-volume biological samples directly without dilution. The disadvantage of μ-EME is FLM thickness is more and surface area is less. This leads to poor mass transfer of analytes through the FLM (organic layer). Compared with the SLM systems, FLM systems require longer extraction times to get the same recovery because agitation is also not possible in FLM systems.

10.4.8 Gel-Electromembrane Extraction (G-EME)

G-EME is a green extraction technique, and in place of SLM, gel membrane is used to carry out the extraction. Agarose gel is generally used to prepare the membrane without using any organic solvent. The membrane is prepared by dispersing agarose powder in deionized water and heating the solution at 90°C for 1 min using a microwave oven. The hot solution is immediately dropped into an Eppendorf tube using a micropipette and allowed to harden at 4°C for 30 min. After hardening, the end of the Eppendorf tube is cut carefully to make a compartment for the acceptor phase with a membrane sheet. The conical shape of the bottom of the Eppendorf tube helps to grip the gel and also close the compartment. Sample solution is taken into the glass vial, and an Eppendorf tube containing membrane is inserted into the sample. One electrode is introduced into the sample solution, and the other electrode is placed into the Eppendorf tube. Electrodes with ring-shaped ends are used to create a large electric field near the membrane. Voltage is applied across the membrane to initiate the extraction, and acceptor solution is withdrawn after extraction and introduced into an analytical technique. A schematic illustration of the G-EME setup is shown in Figure 10.5.

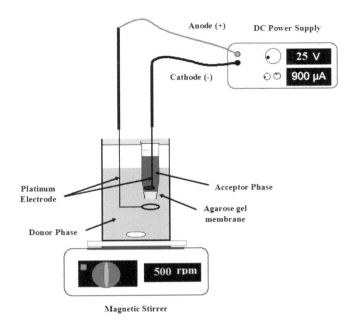

FIGURE 10.5 Schematic illustration of the G-EME setup (Tabani et al. 2017).

10.5 Applications

10.5.1 Extraction of Thebaine

EME of thebaine was performed on water samples, urine samples, street heroine, poppy capsules, and codeine tablets. 1 mM HCl was used as a sample solution, and 100 mM HCl was used as an acceptor solution. 2-Nitrophenyl octylether (NPOE) was used as an SLM, and the driving force used was 300 V. The extraction system was placed on an agitator and agitated at a speed of 1,250 rpm. Extraction was done for 15 min, and the acceptor solution was analyzed by HPLC with UV detection. Pre-concentration factors attained were in the range of 90–110.

10.5.2 Extraction of Six Basic Drugs

Exhaustive EME of six basic drugs (citalopram, loperamide, methadone, paroxetine, pethidine, and sertraline) from human plasma was performed by using three HFs in the same sample. Use of three HFs in the same sample compartment increased the surface area of the SLM and also the volume of the acceptor solution. The SLM used was NPOE, and the driving force used was a voltage of 200 V. The extraction time used was 10 min, and the system was agitated at 1,200 rpm. The acceptor solution used was 10 mM formic acid, and it was analyzed by LC-MS. Extraction recovery values obtained from 1,000 µL undiluted human plasma were in the range of 55–93% and from 50 µL undiluted human plasma were in the range of 56–102%.

10.5.3 Extraction of Mebendazole

EME of mebendazole from human plasma and urine samples was performed by using NPOE as an SLM. Sample and acceptor solution used was 100 mM HCl. Extraction was carried out with a voltage of 150 V as a driving force, and the time of extraction was 15 min. The extraction system was stirred at a speed of 700 rpm. The acceptor solution was analyzed by HPLC with UV detection, and pre-concentration factor obtained for plasma was 144 and for urine was 156.

10.5.4 Extraction of Nalmefene and Naltrexone

EME of nalmefene and naltrexone from untreated human urine and plasma samples was performed by using 85% NPOE and 15% di-(2-ethylhexyl) phosphate (DEHP) as an SLM. Donor and acceptor solution used was 10 and 100 mM HCl, respectively. Voltage of 100 V was used as a driving force for extraction, and time of extraction was 20 min. The system was agitated at 1,250 rpm, and the acceptor solution was analyzed by HPLC with UV detection. The observed pre-concentration factors were in the range of 109–149.

10.5.5 Extraction of Citalopram, Loperamide, Methadone, and Sertraline from Dried Blood Spots

10 μL of the whole blood was spiked with model drug substances and spotted on alginate and chitosan foams (sampling media). After drying at room temperature, the dried blood spot was punched out and dissolved in 1 mM HCl (sample solution). NPOE was used as an SLM, and 10 mM formic acid was used as an acceptor solution. Three HFs were used as acceptor compartments. Anode was positioned in the sample compartment, and three cathodes were positioned in the lumens of three HFs. The driving force for the extraction was voltage of 100 V, and extraction time was 10 min. The system was agitated at a speed of 3,000 rpm. Acceptor solution was collected, diluted with mobile phase, and injected into the LC-MS system.

10.5.6 Extraction of Six Basic Drugs of Abuse

EME of six basic drugs of abuse (cathinone, methamphetamine, 3,4-methylenedioxy-amphetamine, 3,4-methylenedioxy-methamphet-amine, ketamine, and 2,5-dimethoxy-4-iodoamphetamine) from undiluted whole blood and postmortem blood was performed by using 1-ethyl-2-nitrobenzene as an SLM. Acetic acid (10 mM) solution was used as an acceptor solution, and a voltage of 15 V was used as an extraction driving force. Extraction was performed under stagnant conditions, and the time of extraction was 5 min. Acceptor solution was analyzed by LC-MS, and recovery values obtained were in the range of 10–30%.

10.5.7 Extraction of Amphetamine-Type Stimulants from Human Urine

EME of amphetamine-type stimulants (amphetamine, methamphetamine, 3,4-methylenedioxymethamphetamine, 3,4-methylenedioxyethamphetamine, and methylbenzodioxolylbutanamine) from human urine samples was performed by using NPOE containing 15% tris-(2-ethylhexyl) phosphate (TEHP) as an SLM. 1 and 100 mM HCl were used as donor and acceptor phase solutions, respectively. Extraction was performed at 250 V, and the time of extraction was 7 min. The system was placed on an agitator with agitating speed of 1,000 rpm. Acceptor solution was analyzed by HPLC, and pre-concentration factors obtained were in the range of 108–140.

10.5.8 Extraction of Lithium from Human Body Fluids

EME of lithium from untreated human body fluids was performed by using 1-octanol as an SLM. The sample solution was prepared by diluting body fluids 100 times with 0.5 mM Tris solution, and 100 mM acetic acid was used as an acceptor solution. The driving force for the extraction was a potential of 75 V, and the sample solution was stirred at 750 rpm. The time of extraction was 10 min, and after extraction, the acceptor solution was analyzed by capillary electrophoresis with capacitively coupled contactless conductivity detection (CE-C^{4}D). Lithium recovery values obtained for whole blood, plasma, blood serum, and urine were 90, 107, 98, and 92%, respectively.

10.5.9 Extraction of Heavy Metal Cations

EME of heavy metal cations (Mn^{2+}, Cd^{2+}, Zn^{2+}, Co^{2+}, Pb^{2+}, Cu^{2+}, and Ni^{2+}) from aqueous samples was performed by using 1-octanol and 0.5% v/v bis(2-ethyl hexyl) phosphonic acid as an SLM. Water and 100 mM acetic acid solution were used as donor and acceptor solution, respectively. The driving force for the extraction was 75 V, and the sample solution was stirred at 750 rpm. Time for the extraction was 5 min, and the acceptor solution was analyzed by CE-C^4D. Extraction recovery values obtained were in the range of 15–42%. The method was extended to extraction of heavy metal cations from tap water and powdered milk samples.

10.5.10 Extraction of Nerve Agent Degradation Products

Electromembrane isolation of four nerve agent degradation products (methylphosphonic acid, ethyl methylphosphonic acid, isopropyl methylphosphonic acid, and cyclohexyl methylphosphonic acid) from spiked river water samples was performed by using 1-octanol as an SLM. Water was used as acceptor solution, and the pH of the donor (sample) and acceptor solution was 6.8. The driving force for the extraction was 300 V, and the time of extraction was 30 min. The system was agitated at 800 rpm, and the acceptor solution was analyzed by CE-C^4D. The limit of detection (LOD) values obtained for the analytes were in the range of 0.022 to 0.11 ng/mL.

10.5.11 Extraction of Chlorophenols from Sea Water

EME of chlorophenols (4-chlorophenol, 2,4-dichlorophenol, 2,4,6-trichlorophenol, and pentachlorophenol), which are major environmental pollutants, from sea water was performed by using 1-octanol as an SLM. The pH of donor (sample) and acceptor solution was 12, and the extraction was performed at a potential of 10 V. The sample solution was stirred at 1,250 rpm, and the time of extraction was 10 min. Acceptor solution was analyzed by HPLC-UV system and found that the proposed EME technique was highly selective toward pentachlorophenol.

10.6 Conclusion

EME is a rapid, selective, efficient, and cost-effective sample preparation technique. Various EME setups were developed in the last 15 years and employed for different applications. The development of on-chip EME permits rapid extractions and the possibility of online coupling with analytical instruments. Low-voltage EME allows the development of portable devices, and there is no chance of sample degradation. Drop-to-drop EME requires small sample volumes, and the carryover effect is avoided in these systems. Pulsed EME is suitable for the fast and efficient extraction of analytes from complex matrices. Extraction efficiency is high in EME-LDS-USAEME due to the combination of two extraction techniques. High-throughput sample preparation is possible with pa-EME systems. μ-EME is the only technique among the different formats of EME capable of handling very low sample volumes (≤1 μL). G-EME is a green extraction technique where organic solvent-free gel is used as a membrane to carry out the extraction. Surely, in the near future EME systems will be available commercially for routine use in analytical toxicology.

REFERENCES

Drouin, Nicolas, Pavel Kubáň, Serge Rudaz, Stig Pedersen-Bjergaard, and J. Schappler. "Electromembrane Extraction: Overview of the Last Decade." *TrAC - Trends in Analytical Chemistry* 113 (2019): 357–63.

Krishna Marothu, Vamsi, Madhavi Gorrepati, and Ramanasri Vusa. "Electromembrane Extraction—A Novel Extraction Technique for Pharmaceutical, Chemical, Clinical and Environmental Analysis." *Journal of Chromatographic Science* 51, no. 7 (2013): 619–31.

Petersen, Nickolaj Jacob, Sunniva Taule Foss, Henrik Jensen, Steen Honoré Hansen, Christian Skonberg, Detlef Snakenborg, Jörg P. Kutter, and Stig Pedersen-Bjergaard. "On-Chip Electro Membrane Extraction with Online Ultraviolet and Mass Spectrometric Detection." *Analytical Chemistry* 83, no. 1 (2011): 44–51.

Petersen, Nickolaj Jacob, Henrik Jensen, Steen Honoré Hansen, Knut Einar Rasmussen, and Stig Pedersen-Bjergaard. "Drop-to-Drop Microextraction Across a Supported Liquid Membrane by an Electrical Field Under Stagnant Conditions." *Journal of Chromatography A* 1216 (2009): 1496–502.

Rezazadeh, Maryam, Yadollah Yamini, Shahram Seidi, and Ali Esrafili. "Pulsed Electromembrane Extraction: A New Concept of Electrically Enhanced Extraction." *Journal of Chromatography A* 1262 (2012): 214–18.

Tabani, Hadi, Sakine Asadi, Saeed Nojavan, and Mitra Parsa. "Introduction of Agarose Gel as a Green Membrane in Electromembrane Extraction: An Efficient Procedure for the Extraction of Basic Drugs with a Wide Range of Polarities." *Journal of Chromatography A* 1497 (2017): 47–55.

11

Fabric Phase Sorptive Extraction in Analytical Toxicology

Natalia Manousi[1], Abuzar Kabir[2], and George A. Zachariadis[1]
[1]*Laboratory of Analytical Chemistry, Department of Chemistry, Aristotle University of Thessaloniki, GR-54124 Thessaloniki, Greece*
[2]*Department of Chemistry and Biochemistry, Florida International University, Miami, FL, USA*

CONTENTS

11.1	Introduction	183
11.2	Applications of FPSE in Analytical Toxicology	186
	11.2.1 Extraction of Benzodiazepines	186
	11.2.2 Extraction of Azole Antimicrobial Drug Residues	186
	11.2.3 Extraction of Aromatase Inhibitors	188
	11.2.4 Extraction of Inflammatory Bowel Disease Treatment Drugs	188
	11.2.5 Extraction of Antidepressant Drugs	189
	11.2.6 Extraction of Penicillin Antibiotics	190
11.3	Concluding Remarks and Future Perspectives of FPSE in Analytical Toxicology	190
References		191

11.1 Introduction

In forensic and clinical toxicology, sample preparation of biofluids is an important and demanding step in the overall analytical workflow. Sample preparation aims to extract the target analytes from complex biological matrices, as well as to exclude the response of interfering matrix constituents in the subsequent chromatographic determination (Pragst 2007). Usually, the target analytes, i.e., toxic chemical compounds and/or their metabolites, are extracted from whole blood, blood serum, and blood plasma.

The conventional extraction techniques for the analysis of samples of biological origin are solid-phase extraction (SPE) and liquid-liquid extraction (LLE); however, direct analysis and analysis after precipitation of proteins are also used. These techniques exhibit a plethora of drawbacks, including complicated and time-consuming steps that are prone to errors and the need for a high volume of sample and hazardous organic solvents, while they also present difficulties in automation (Kataoka 2003; Manousi and Zachariadis 2020; Samanidou et al. 2005). For simplification of sample preparation and eliminatation/ reduction of organic solvent use, solid-phase microextraction (SPME) was introduced in 1990 by Professor Janusz Pawliszyn (Arthur and Pawliszyn 1990). Subsequently, multiple microextraction techniques, based on solvent or solid sorbent, have emerged. Apart from SPME, notable microextraction techniques include liquid-phase microextraction (LPME) (Liu and Dasgupta 1996), pipette tip solid-phase extraction (PT-SPE) (Hasegawa et al. 2011), stir bar sorptive extraction (SBSE) (Nazyropoulou and Samanidou 2015), dispersive solid-phase extraction (d-SPE) (Manousi et al. 2020), magnetic solid-phase extraction (MSPE) (Filippou et al. 2017), and fabric phase sorptive extraction (FPSE).

The novel extraction techniques comply with the principles of green chemistry that are currently a trend in analytical toxicology and in other areas of analytical chemistry. The need for green chemical processes arose from increasing concern for human health and environmental protection, as well as for sustainability. The principles a chemical process should address to be characterized as environmentally friendly were defined by the researchers Anastas and Warner in 1998 (Turner 2013). Furthermore, green analytical chemistry (GAC) emerged from those principles, two years later. GAC deals with the contribution of analytical chemists in developing cheap and effective analytical methods that are friendly to the environment (Namieśnik 2000; Filippou et al. 2017; Armenta et al. 2015).

FPSE is an environmentally friendly sample preparation technique that was developed in 2014 by Kabir and Furton (Kabir and Furton 2014). In FPSE, extraction of the target compounds occurs onto the FPSE membrane that is directly introduced into the sample matrix. The inherent porous surface of the fabric substrate and the superior material properties of sol-gel derived sorbents that are uniformly dispersed as an ultra-thin film within the substrate, have made FPSE membranes very powerful and convenient sample preparation devices. This technique successfully incorporates the majority of the beneficial characteristics of SPME, which is an equilibrium-based extraction, and SPE, which is an exhaustive extraction technique (Samanidou et al. 2016). Moreover, FPSE exhibits a significant geometrical advantage of high primary contact surface area, and compared to SPME fibres, FPSE membrane contains approximately 400 times larger sorbent loading (Kabir et al. 2018; Kumar et al. 2014). Due to the variety of sol-gel derived hybrid sorbents and fabric substrates, various different FPSE membrane have been constructed with different selectivity toward the target analytes, different analyte retention capacity, and different extraction equilibrium points (Kazantzi and Anthemidis 2017). The steps of the FPSE procedure are illustrated in Figure 11.1.

As a result, FPSE can combine the benefits of sol-gel derived hybrid sorbents and the unique surface properties of (hydrophilic/hydrophobic/neutral) fabric substrates. FPSE is an environmentally friendly technique that has proved to exhibit performance superiority compared to other sample preparation techniques, while it reduces the consumption of hazardous solvents. Moreover, with FPSE both low and high sample volumes can be used, and it can be employed for the analysis of different samples, including environmental, biological, toxicological, and food samples (Kabir et al. 2017). In addition, the FPSE membranes are characterized by tunable selectivity and adjustable porosity, and they exhibit high flexibility and permeability, as well as high chemical and thermal stability (Kabir et al. 2018; Kumar et al. 2014). A comparison of the required steps in a conventional SPE method and a novel FPSE method, is shown in Figure 11.2.

FPSE has gained the attention of many analytical chemists working in the field of analytical toxicology due to its superior characteristics. Until now, FPSE has been used for the extraction of a plethora of analytes from different matrices, including the extraction of amphenicol residues from raw milk

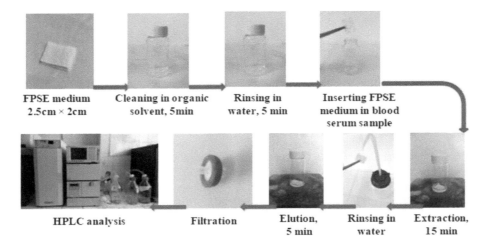

FIGURE 11.1 Typical steps involved in the FPSE procedure. Reproduced with permission from Zilfidou et al. (2019). Elsevier. Copyright Elsevier, 2019.

FPSE in Analytical Toxicology

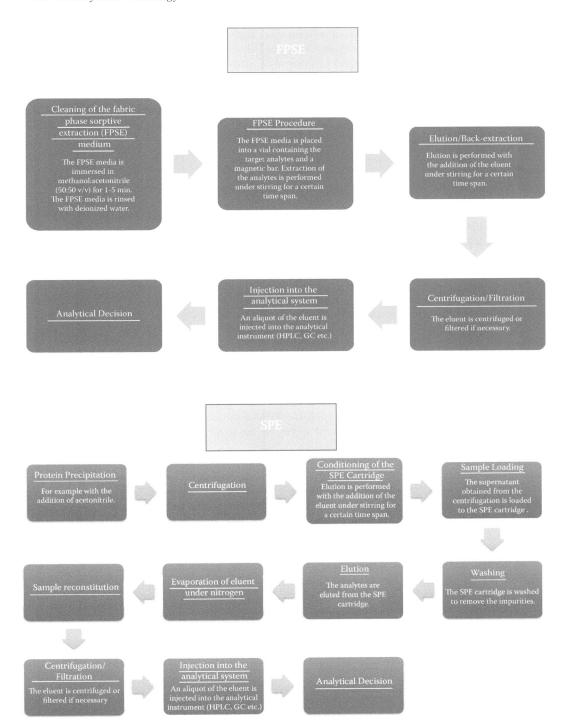

FIGURE 11.2 Comparison of the required steps in an FPSE method (top) and an SPE method (bottom).

(Samanidou et al. 2015), the extraction of sulfonamides from raw milk (Karageorgou et al. 2016), the extraction of triazine herbicides from water samples (Roldán-Pijuán et al. 2015), the extraction of estrogens from various kinds of samples (Kumar et al. 2014), the extraction of alkyl phenols from environmental samples (Kumar et al. 2015), and the extraction of inflammatory bowel disease treatment drugs from biofluids (Kabir et al. 2018). In analytical toxicology, FPSE is an important sample preparation technique that undoubtedly enriches the toolbox of analytical chemists who struggle to find a genuine solution for the analysis of complex sample matrices of bioanalytical interest. In this chapter, we aim to discuss the applications of FPSE in analytical toxicology.

11.2 Applications of FPSE in Analytical Toxicology

The application of FPSE in analytical toxicology was reported soon after its introduction for sample preparation in analytical chemistry (Samanidou et al. 2015). Table 11.1 presents the application of FPSE in analytical toxicology.

11.2.1 Extraction of Benzodiazepines

FPSE has been applied in the determination of benzodiazepines (i.e., bromazepam, lorazepam, diazepam, and alprazolam) in blood serum samples. Benzodiazepines are widely used drugs that exhibit antidepressive and tranquilizing properties, among others. Therefore, the determination of these drugs in biofluids is of high importance in toxicological studies.

In order to optimize the extraction process, the authors evaluated three different FPSE membranes, including sol-gel poly(ethylene glycol) (sol-gel PEG), coated on cellulose fabric substrate, sol-gel poly(tetrahydrofuran) (sol-gel PTHF) coated on hydrophilic cellulose fabric substrate, and sol-gel poly(dimethyldiphenylsiloxane) (sol-gel PDMDPS), coated on hydrophobic polyester substrate. In order to avoid contamination during sample preparation, the FPSE membrane was handled using tweezers. Sol-gel PEG-coated FPSE membrane was found to be the optimum FPSE medium, and extraction was mainly performed through hydrogen-bonding interactions.

First, the membranes were conditioned with a mixture of acetonitrile (ACN) and methanol (MeOH) (50:50, v/v) for 2 min and rinsed with water to remove residual organic solvents. Subsequently, the sample (50 μL) was mixed with water (500 μL) and transferred into a vial together with a magnetic stirrer. For the FPSE process, extraction of the analytes was performed for 20 min, while back extraction was performed with 500 μL of an ACN:MeOH (50:50, v/v) mixture for 10 min. The solution was collected, dried under nitrogen atmosphere, and reconstituted in the back-extraction solution mixture, prior to its analysis by high-performance liquid chromatography with diode-array detection (HPLC-DAD). After each application, the FPSE membrane was washed with ACN:MeOH mixture for 5 min. No carryover effect was observed, and the membranes were found to be reusable for approximately 30 times. Under optimum conditions, the absolute recoveries ranged between 27 and 63% for the target analytes.

The developed method could simplify the overall sample preparation workflow of blood serum samples, whereas it reduced the consumption of organic solvents. The proposed method was successfully used in the determination of benzodiazepines in serum samples, and the results indicated that it can be employed in routine analysis (Samanidou et al. 2016).

11.2.2 Extraction of Azole Antimicrobial Drug Residues

In 2017, Professor Marcello Locatelli and his research group used an FPSE technique in the determination of 12 azole antimicrobial drug residues in human plasma and urine samples. These drugs were organic compounds that are usually incorporated in pharmaceutical formulations, such as creams and shampoos for the treatment of fungal infections, and a clean-up procedure is generally required for their determination in complex biofluids.

The authors evaluated three different FPSE extraction membranes, i.e., sol-gel silica Carbowax® 20 M (sol-gel CW 20M), sol-gel poly(dimethylsiloxane) (sol-gel PDMS), and sol-gel polycaprolactone-

TABLE 11.1
Applications of FPSE in analytical toxicology

Analyte	Matrix	Sol-Gel Sorbent	Analytical Technique[2]	LOD (ng mL^{-1})	Extraction Recovery (%)	Reusability	Ref.
Benzodiazepines	Blood serum	Sol-gel poly(ethylene glycol)	HPLC-DAD	10	27–63	Approx. 30 times	(Samanidou, Kaltzi et al. 2016)
Azole antimicrobial drug	Blood serum, urine	Sol-gel silica Carbowax® 20 M	HPLC-DAD	30	NA	NA	(Locatelli et al. 2017)
Anticancer drugs	Whole blood, plasma, urine	Sol-gel PEG-PPG-PEG	HPLC-DAD	20–100	NA	NA	(Locatelli et al. 2018)
Inflammatory bowel disease treatment drugs	Whole blood, plasma, urine	Sol-gel silica Carbowax® 20 M	HPLC-DAD	20–100	NA	Approx. 30 times	(Kabir et al. 2018)
Antidepressant drugs	Urine	Sol-gel graphene	HPLC-DAD	150	25.5–67.0	Up to 30 times	(Lioupi et al. 2019)
	Blood serum	Sol-gel PCL-PDMS-PCL	HPLC-DAD	150	9.4–88.1	At least 30 times	(Zilfidou et al. 2019)
Penicillin antibiotics	Blood serum	Sol-gel poly (tetrahydrofuran)	HPLC-DAD	150	10.8–65.5	At least 35 times	(Alampanos et al. 2019)

polydimethylsiloxane-polycaprolactone (sol-gel PCL-PDMS-PCL), with the aim of finding the optimum conditions for the extraction of the drugs that exhibit a wide range of log K_{ow}. Higher extraction efficiencies were reported with the sol-gel PCL-PDMS-PCL coated FPSE media; however, unwanted peaks were present in the chromatograms. Therefore, the authors finally chose the sol-gel CW 20M extraction membrane. After the selection of the optimum FPSE material, different dimensions: 2.5 × 2 cm blocks, as well as circular discs with a diameter of 0.6 cm and 1 cm were investigated. Authors found that reducing the dimension of the FPSE membrane enabled them to handle a smaller sample volume, and the optimum results were obtained with the use of circular disk FPSE membrane with an diameter of 1 cm.

For the sample preparation, the urine and blood plasma samples were mixed with the analytes and internal standard (IS) solution and vortexed. Initially, the FPSE membrane was conditioned with a mixture of ACN and MeOH and with Milli-Q water. Afterward, the FPSE media were placed in a vial containing the sample solution, and the drugs were extracted from plasma (500 μL) or urine (1 mL) samples within 30 min and eluted with MeOH (150 μL) for 10 min. After centrifugation, the eluates were analyzed by HPLC-DAD.

With the developed protocol, the target analytes were extracted from the biofluids after simple immersion of the FPSE membrane without any requirement for previous treatment, e.g., protein precipitation. Moreover, the FPSE methodology was found to be a simple, fast, and green procedure that complies with the GAC principles (Locatelli et al. 2017).

11.2.3 Extraction of Aromatase Inhibitors

The application of FPSE for the extraction of aromatase inhibitors from human whole blood, plasma, and urine samples prior to HPLC analysis, has been also reported. These drugs are employed in the treatment of breast cancer, and their determination in biological matrices is important in analytical toxicology (Locatelli et al. 2018).

The authors evaluated the performance of six FPSE membranes, i.e., sol-gel Octadecyl (sol-gel C8), sol-gel Sucrose (sol-gel SUC), sol-gel PCL-PDMS-PCL, sol-gel poly(caprolactone) (sol-gel PCL), sol-gel poly(ethylene glycol)-block-poly(propylene glycol)-block-poly(ethylene glycol) (sol-gel PEG-PPG-PEG), and sol-gel CW 20M. Among the examined membranes, the former three exhibited higher extraction efficiency, and different dimensions were evaluated. The optimum extraction membrane was found to be sol-gel PEG-PPG-PEG circular discs with a diameter of 1 cm.

Prior to the FPSE protocol, the biofluids were mixed with an analyte working solution and a solution of the IS, followed by dilution with deionized water. The FPSE medium was activated with a mixture of ACN:MeOH (50:50, v/v) and washed with Milli-Q water. Subsequently, the FPSE membrane was employed for the extraction of the drugs from the sample at a rotator within 30 min. Elution of the compounds was achieved by the addition of 150 μL of MeOH into 10 min, followed by analysis of the extract by HPLC-DAD.

The developed method was used for the analysis of biological samples obtained from patients through normal medical treatment practice. The overall methodology was found to be proficient, simple, rugged, and green, while it enabled the extraction of small organic molecules directly from whole blood without interferences. Whole blood is an important sample matrix in analytical toxicology since it is rich in information. However, because of the complex nature of this matrix, an analytical toxicology analysis of blood serum or blood plasma is usually preferred. In this case, partial loss of analyte may take place. Therefore, sample preparation techniques that enable the analysis of whole blood without any need for protein precipitation before the extraction are of high importance in analytical toxicology and other bioanalytical applications (Kabir et al. 2018).

11.2.4 Extraction of Inflammatory Bowel Disease Treatment Drugs

FPSE has also been used for the extraction of inflammatory bowel disease treatment drugs from whole blood, plasma, and urine samples before their analysis by HPLC-DAD. Because of the low concentration levels of the residual drugs in the biological sample in combination with the small available sample quantity, a step is required to reduce the interferences and pre-concentrate the compounds in order to obtain satisfactory method sensitivity.

In order to select the most appropriate FPSE sorbent for the simultaneous extraction of the three drugs, five different FPSE membranes were evaluated, which included sol-gel SUC, sol-gel PCL-PDMS-PCL, sol-gel PCL, sol-gel PEG-PPG-PEG, and sol-gel CW 20M. Circular discs of two different diameters (i.e., 1 cm and 0.6 cm) were evaluated. Among the initially examined membranes, the sol-gel CW 20M and the sol-gel PCAP-PDMS-PCAP media showed better enrichment factors and were selected for further optimization. Both coatings exhibited biocompatibility that referred to the tendency toward adsorption of protein and adhesion of platelets during the exposure of the FPSE membrane to the physiological fluid. Therefore, no previous protein precipitation was required to prevent clogging or irreversible adhesion of macromolecules and platelets to the surface of the FPSE membrane.

Regarding sample pre-treatment, whole blood (180 μL) was mixed with 10 μL of the standard solution containing the analytes and 10 μL of the IS solution, followed by 5-fold dilution with Milli-Q water, and vortex mixing. For plasma samples, a 450 μL aliquot was mixed with 25 μL of the standard solution containing the analytes and 25 μL of IS solution, while for urine samples, a 900 μL aliquot of sample was mixed with 50 μL of the standard solution containing the analytes and 50 μL of the IS solution.

For the extraction, the sol-gel Carbowax® 20 M media circular disk membranes were initially cleaned with a mixture of ACN:MeOH (50:50, v/v) and then rinsed with Milli-Q water. Extraction of the drugs was performed within 30 min in a rotator for 30 min, while back extraction was performed in 10 min with the addition of 150 μL of methanol. Afterward, the eluent was centrifuged and injected into the HPLC-PDA system.

The developed analytical protocol exhibited good performance characteristics, and it was able to eliminate the required samples (i.e., precipitation of proteins, evaporation of solvent, and reconstitution of sample) that are usually applied in conventional sample preparation workflow (Kabir et al. 2018).

11.2.5 Extraction of Antidepressant Drugs

Extraction of antidepressant drugs from urine samples before their determination by HPLC-DAD has also been suggested. In this work, the authors reported the simultaneous extraction of five widely used antidepressant drugs. The determination of antidepressants in biofluids is important for multiple research areas, including analytical toxicology (Lioupi et al. 2019).

For the development of the FPSE method, various extraction solvent systems and nine different media were investigated, including sol-gel PEG, sol-gel PTHF, sol-gel octadecyl (sol-gel C18), sol-gel C8, sol-gel PEG-PPG-PEG, sol-gel PDMS, sol-gel graphene (sol-gel GRP), etc. The best performance was seen with the sol-gel graphene FPSE membrane and a mixture of ACN and MeOH (50:50 v/v) as eluent.

Under optimum conditions, the selected FPSE membrane was treated with ACN:MeOH (50:50 v/v) for 2 min for activation and rinsed with Milli-Q water to dispose of remaining organic solvents. Subsequently, the FPSE media were put into a mixture of 500 μL urine sample and 500 μL of Milli-Q water. No previous treatment of urine samples prior to the FPSE procedure was required. Extraction of the analytes was achieved in 20 min, whereas elution was performed within 10 min with the addition of the elution solvent system. Direct injection of the eluent into the HPLC-DAD system was carried out or filtration with syringe filters was done if necessary.

As a result, an efficient, user-friendly, and time-efficient method was developed and successfully applied for the determination of antidepressants in human urine. Moreover, the FPSE membranes were found reusable for up to 30 times, when a washing step with a mixture of ACN:MeOH was employed after each extraction cycle.

In 2019, the same research group reported an improved FPSE protocol for the extraction of the same antidepressants from blood serum before their determination by HPLC-DAD. The development of efficient analytical protocols for the rapid determination of antidepressants in blood serum samples is of high importance in toxicological evaluations and therapeutic drug monitoring and other pharmacodynamic and pharmacokinetic applications. In this work, sol-gel PCL-PDMS-PCL FPSE membranes coated onto a polyester substrate were used.

Prior to the FPSE process, blood serum (50 μL) was placed in a glass vial and mixed with a standard solution containing the target analytes (500 μL) and MQ water (450 μL). Moreover, the FPSE membrane was placed in a mixture of ACN:MeOH (50:50 v/v) to avoid potential impurities and rinsed with Milli-Q water to avoid organic solvent residues. For the FPSE protocol, the FPSE membranes were placed into the sample, and the drugs were extracted in 15 min under stirring. After the extraction, the FPSE device was removed and rinsed with Milli-Q water, while back extraction of the target analytes was performed with the addition of 500 μL of a mixture composed of MeOH and ACN within 5 min. Subsequently, the eluent was filtered and analyzed by HPLC-DAD.

The FPSE membranes were found reusable for at least 30 times. For this purpose, after each extraction cycle, the FPSE membrane was washed with MeOH:ACN for 5 min, left to dry, and kept in an air-tight vial in order to avoid potential carryover effects. The developed FPSE protocol could efficiently extract the target analytes from blood serum samples without any need for a protein precipitation step prior to the sample preparation process, while it also avoided the need for evaporation of an organic solvent and sample reconstitution that are error-prone steps. Moreover, the developed method was rapid, was simple in operation, and reduced the consumption of organic solvents. As a result, it could be a useful analytical tool for analytical toxicology applications (Zilfidou et al. 2019).

11.2.6 Extraction of Penicillin Antibiotics

Penicillin antibiotics have been determined in human blood serum by FSPE followed by HPLC-DAD analysis (Alampanos et al. 2019). Penicillins are β-lactam antibiotics that are widely used against bacterial infections, and as a result, they are considered important in veterinary and human medicine. The determination of penicillin drugs in biological fluids is a complex procedure due to their low concentration in combination with the complexity of the biofluid.

The authors tested 14 different sol-gel FPSE membranes, including sol-gel PTHF, sol-gel octadecyl, etc. Among the examined FPSE membranes, sol-gel PTHF FPSE media coated on a substrate made of cellulose was finally chosen as the optimal extraction membrane. That sorbent is normally recommended for the extraction of target analytes with medium or high polarity.

For the FPSE protocol, the FPSE membranes were washed with a mixture of ACN:MeOH (50:50 v/v) for 5 min in order to remove unwanted residues, followed by immersion in deionized water for another 5 min to remove residues of the solvents. Subsequently, the FPSE membrane was immersed in a solution containing blood serum (50 μL), deionized water (450 μL), and standard solution with the target analytes (or deionized water for blanks) (500 μL) that were placed in a 5 mL vial. Extraction of the penicillin drugs was achieved in 25 min while back extraction was performed with the addition of a mixture of 90:10 v/v ACN:0.05 M ammonium acetate under stirring.

The FPSE membrane was found to be reusable for at least 35 times without observing carryover effects or loss of extraction efficiency after washing with ACN:MeOH (50:50 v/v) for 5 min and drying in an airtight glass container. The developed method was environmentally friendly and of low cost, and it can be easily applied in laboratories for determining penicillins in blood for various purposes, including analytical toxicology applications.

11.3 Concluding Remarks and Future Perspectives of FPSE in Analytical Toxicology

FPSE is a simple and rapid, recently introduced sample preparation technique that serves as a useful tool that enriches the toolbox of analytical scientists working in the scientific field of analytical toxicology. FPSE complies with GAC principles. Therefore, FPSE can successfully be applied for the analysis of complex sample matrices of bioanalytical interest. Various FPSE membranes have been developed and successfully used for sample preparation in analytical toxicology.

Among the advantages of FPSE are ease of operation, reduced consumption of organic solvents, and overall performance superiority. Moreover, FPSE membranes are characterized by high chemical resistance, stability, and reusability. A wide variety of novel sol-gel coatings can be used as the sorbent, and a wide variety of organic solvents can be used for the desorption of target analytes, so FPSE can successfully extract a plethora of organic compounds from complex matrices. Additionally, FPSE opens up a new direction toward whole blood analysis.

Future perspectives in the field of analytical toxicology should focus on expanding applications of the existing FPSE membranes and developing new coatings for the determination of any type of compounds of interest in complex biofluids. Furthermore, the use of FPSE for the analysis of alternative matrices (e.g., hair, nails, saliva, cerebrospinal fluids) should also be evaluated. Other future challenges are automation of the whole FPSE procedure and the application of FPSE for in situ sampling in toxicological research.

REFERENCES

Alampanos, Vasileios, Abuzar Kabir, Kenneth G. Furton, Victoria Samanidou, and Ioannis Papadoyannis. "Fabric Phase Sorptive Extraction for Simultaneous Observation of Four Penicillin Antibiotics from Human Blood Serum Prior to High Performance Liquid Chromatography and Photo-Diode Array Detection." *Microchemical Journal* 149 (2019): 103964. 10.1016/j.microc.2019.103964.

Armenta, Sergio, Salvador Garrigues, and Miguel de la Guardia. "The Role of Green Extraction Techniques in Green Analytical Chemistry." *TrAC - Trends in Analytical Chemistry* 71 (2015): 2–8. 10.1016/j.trac.2014.12.011.

Arthur, Catherine L., and Janusz Pawliszyn. "Solid Phase Microextraction with Thermal Desorption Using Fused Silica Optical Fibers." *Analytical Chemistry* 62, no. 19 (1990): 1145–2148. 10.1021/ac00218a019.

Filippou, Olga, Dimitrios Bitas, and Victoria Samanidou. "Green Approaches in Sample Preparation of Bioanalytical Samples Prior to Chromatographic Analysis." *Journal of Chromatography B: Analytical Technologies in the Biomedical and Life Sciences* 1043 (2017): 44–62. 10.1016/j.jchromb.2016.08.040.

Filippou, Olga, Eleni A. Deliyanni, and Victoria F. Samanidou. "Fabrication and Evaluation of Magnetic Activated Carbon as Adsorbent for Ultrasonic Assisted Magnetic Solid Phase Dispersive Extraction of Bisphenol A from Milk Prior to High Performance Liquid Chromatographic Analysis with Ultraviolet Detection." *Journal of Chromatography A* 1479 (2017): 20–31. 10.1016/j.chroma.2016.12.002.

Hasegawa, Chika, Takeshi Kumazawa, Seisaku Uchigasaki, Xiao Pen Lee, Keizo Sato, Masaru Terada, and Kunihiko Kurosaki. "Determination of Dextromethorphan in Human Plasma Using Pipette Tip Solid-Phase Extraction and Gas Chromatography-Mass Spectrometry." *Analytical and Bioanalytical Chemistry* 401, no. 7 (2011): 2215–23. 10.1007/s00216-011-5324-5.

Kabir, Abuzar, and Kenneth G. Furton. "Fabric Phase Sorptive Extractor (FPSE)." *U.S. Patent and Trademark Office* 14 (2014): 116–21.

Kabir, Abuzar, Kenneth G. Furton, Nicola Tinari, Laurino Grossi, Denise Innosa, Daniela Macerola, Angela Tartaglia, Valentina Di Donato, Cristian D'Ovidio, and Marcello Locatelli. "Fabric Phase Sorptive Extraction-High Performance Liquid Chromatography-Photo Diode Array Detection Method for Simultaneous Monitoring of Three Inflammatory Bowel Disease Treatment Drugs in Whole Blood, Plasma and Urine." *Journal of Chromatography B: Analytical Technologies in the Biomedical and Life Sciences* 1084 (2018): 53–63. 10.1016/j.jchromb.2018.03.028.

Kabir, Abuzar, Marcello Locatelli, and Halil Ibrahim Ulusoy. "Recent Trends in Microextraction Techniques Employed in Analytical and Bioanalytical Sample Preparation." *Separations* 4 (2017): 36. 10.3390/separations4040036.

Karageorgou, Eftychia, Natalia Manousi, Victoria Samanidou, Abuzar Kabir, and Kenneth G. Furton. "Fabric Phase Sorptive Extraction for the Fast Isolation of Sulfonamides Residues from Raw Milk Followed by High Performance Liquid Chromatography with Ultraviolet Detection." *Food Chemistry* 196 (2016): 428–36. 10.1016/j.foodchem.2015.09.060.

Kataoka, Hiroyuki. "New Trends in Sample Preparation for Clinical and Pharmaceutical Analysis." *TrAC - Trends in Analytical Chemistry* 22, no. 4 (2003): 232–44. 10.1016/S0165-9936(03)00402-3.

Kazantzi, Viktoria, and Aristidis Anthemidis. "Fabric Sol–Gel Phase Sorptive Extraction Technique: A Review." *Separations* 4, no. 2 (2017): 20. 10.3390/separations4020020.

Kumar, Rajesh, Gaurav, Abuzar Kabir, Kenneth G. Furton, and Ashok Kumar Malik. "Development of a Fabric Phase Sorptive Extraction with High-Performance Liquid Chromatography and Ultraviolet Detection Method for the Analysis of Alkyl Phenols in Environmental Samples." *Journal of Separation Science* 38, no. 18 (2015): 3228–38. 10.1002/jssc.201500464.

Kumar, Rajesh, Gaurav, Heena, Ashok Kumar Malik, Abuzar Kabir, and Kenneth G. Furton. "Efficient Analysis of Selected Estrogens Using Fabric Phase Sorptive Extraction and High Performance Liquid Chromatography-Fluorescence Detection." *Journal of Chromatography A* 1359 (2014): 16–25. 10.1016/j.chroma.2014.07.013.

Lioupi, Artemis, Abuzar Kabir, Kenneth G. Furton, and Victoria Samanidou. "Fabric Phase Sorptive Extraction for the Isolation of Five Common Antidepressants from Human Urine Prior to HPLC-DAD Analysis." *Journal of Chromatography B: Analytical Technologies in the Biomedical and Life Sciences* 1118–1119 (2019): 171–9. 10.1016/j.jchromb.2019.04.045.

Liu, Hanghui, and Pumendu K. Dasgupta. "Analytical Chemistry in a Drop. Solvent Extraction in a Microdrop." *Analytical Chemistry* 68, no. 11 (1996): 1817–21. 10.1021/ac960145h.

Locatelli, Marcello, Abuzar Kabir, Denise Innosa, Teresa Lopatriello, and Kenneth G. Furton. "A Fabric Phase Sorptive Extraction-High Performance Liquid Chromatography-Photo Diode Array Detection Method for the Determination of Twelve Azole Antimicrobial Drug Residues in Human Plasma and Urine." *Journal of Chromatography B: Analytical Technologies in the Biomedical and Life Sciences* 1040 (2017): 192–8. 10.1016/j.jchromb.2016.10.045.

Locatelli, Marcello, Nicola Tinari, Antonino Grassadonia, Angela Tartaglia, Daniela Macerola, Silvia Piccolantonio, Elena Sperandio, et al. "FPSE-HPLC-DAD Method for the Quantification of Anticancer Drugs in Human Whole Blood, Plasma, and Urine." *Journal of Chromatography B: Analytical Technologies in the Biomedical and Life Sciences* 1095 (2018): 204–13. 10.1016/j.jchromb.2018.07.042.

Manousi, Natalia, Beatrice Gomez-Gomez, Yolanda Madrid, E.A. Deliyanni, and George A. Zachariadis. "Determination of Rare Earth Elements by Inductively Coupled Plasma-Mass Spectrometry After Dispersive Solid Phase Extraction with Novel Oxidized Graphene Oxide and Optimization with Response Surface Methodology and Central Composite Design." *Microchemical Journal* 152 (2020): 104428. 10.1016/j.microc.2019.104428.

Manousi, Natalia, and George A. Zachariadis. "Recent Advances in the Extraction of Polycyclic Aromatic Hydrocarbons from Environmental Samples." *Molecules (Basel, Switzerland)* 25, no. 9 (2020): 1–29. 10.3390/molecules25092182.

Namieśnik, Jacek. "Trends in Environmental Analytics and Monitoring." *Critical Reviews in Analytical Chemistry* 30, no. 2 (2000): 221–69. 10.1080/10408340091164243.

Nazyropoulou, Chrysoula, and Victoria Samanidou. "Stir Bar Sorptive Extraction Applied to the Analysis of Biological Fluids." *Bioanalysis* 7, no. 17 (2015): 2241–50. 10.4155/bio.15.129.

Pragst, Fritz. "Application of Solid-Phase Microextraction in Analytical Toxicology." *Analytical and Bioanalytical Chemistry* 388, no. 7 (2007): 1393–414. 10.1007/s00216-007-1289-9.

Roldán-Pijuán, Mercedes, Rafael Lucena, Soledad Cárdenas, Miguel Valcárcel, Abuzar Kabir, and Kenneth G. Furton. "Stir Fabric Phase Sorptive Extraction for the Determination of Triazine Herbicides in Environmental Waters by Liquid Chromatography." *Journal of Chromatography A* 1376 (2015): 35–45. 10.1016/j.chroma.2014.12.027.

Samanidou, Victoria F., Eleni A. Christodoulou, and Ioannis N. Papadoyannis. "Determination of Fluoroquinolones in Edible Animal Tissue Samples by High Performance Liquid Chromatography After Solid Phase Extraction." *Journal of Separation Science* 28, no. 6 (2005): 555–65. 10.1002/jssc.200401910.

Samanidou, Victoria, Ioanna Kaltzi, Abuzar Kabir, and Kenneth G. Furton. "Simplifying Sample Preparation Using Fabric Phase Sorptive Extraction Technique for the Determination of Benzodiazepines in Blood Serum by High-Performance Liquid Chromatography." *Biomedical Chromatography* 30, no. 6 (2016): 829–36. 10.1002/bmc.3615.

Samanidou, Victoria, Lavrentis Demetrios Galanopoulos, Abuzar Kabir, and Kenneth G. Furton. "Fast Extraction of Amphenicols Residues from Raw Milk Using Novel Fabric Phase Sorptive Extraction Followed by High-Performance Liquid Chromatography-Diode Array Detection." *Analytica Chimica Acta* 855 (2015): 41–50. 10.1016/j.aca.2014.11.036.

Turner, Charlotta. "Sustainable Analytical Chemistry-More Than Just Being Green." *Pure and Applied Chemistry* 85, no. 12 (2013): 2217–29. 10.1351/PAC-CON-13-02-05.

Zilfidou, Eirini, Abuzar Kabir, Kenneth G. Furton, and Victoria Samanidou. "An Improved Fabric Phase Sorptive Extraction Method for the Determination of Five Selected Antidepressant Drug Residues in Human Blood Serum Prior to High Performance Liquid Chromatography with Diode Array Detection." *Journal of Chromatography B: Analytical Technologies in the Biomedical and Life Sciences* 1125 (2019): 121720. 10.1016/j.jchromb.2019.121720.

12
Sorbent-Based Microextraction Using Molecularly Imprinted Polymers

Cecilia Ortega-Zamora[1], Gabriel Jiménez-Skrzypek[1], Javier González-Sálamo[1,2], and Javier Hernández-Borges[1,2,3]
[1]*Departamento de Química, Unidad Departamental de Química Analítica, Facultad de Ciencias, Universidad de La Laguna (ULL), Avda. Astrofísico Fco. Sánchez, San Cristóbal de La Laguna, España*
[2]*Instituto Universitario de Enfermedades Tropicales y Salud Pública de Canarias, Universidad de La Laguna (ULL), Avda. Astrofísico Fco. Sánchez, San Cristóbal de La Laguna, España*
[3]*Universidad de La Laguna, España*

CONTENTS

12.1	Introduction	193
12.2	Molecularly Imprinted Polymer Synthesis	194
	12.2.1 Covalent Imprinting Method	195
	12.2.2 Non-Covalent Imprinting Method	195
	12.2.3 Semi-Covalent Imprinting Method	195
12.3	Application in Analytical Toxicology	195
	12.3.1 Miniaturized Solid-Phase Extraction	195
	12.3.2 Solid-Phase Microextraction	198
	12.3.3 Stir Bar Sorptive Extraction	198
	12.3.4 Miscellaneous	199
12.4	Conclusions	200
Acknowledgements		200
References		200

12.1 Introduction

Miniaturization is currently one of the most important trends in sample preparation, which is still, despite efforts to eliminate it, a key factor of any analytical method. The downscaling of sample treatment has yielded to the appearance of solvent- and sorbent-based extraction techniques. In both cases, the final objective is to reduce the amount of samples, reagents, and solvents, with the aim of also simplifying the procedures and making them more simple and straightforward. Sample throughput is also a key aspect to be considered.

Regarding sorbent-based microextraction techniques, the introduction of new sorbents or coatings with high extraction capacity, high surface-to-volume ratios, and high porosity is an important research field that is daily contributing to its consolidation and wide application. However, apart from such relevant characteristics, sorbents/coatings should also possess another inherent and important feature: selectivity. Extraction should be selective enough to separate the target analytes from the sample matrix, or at least from as many components as possible.

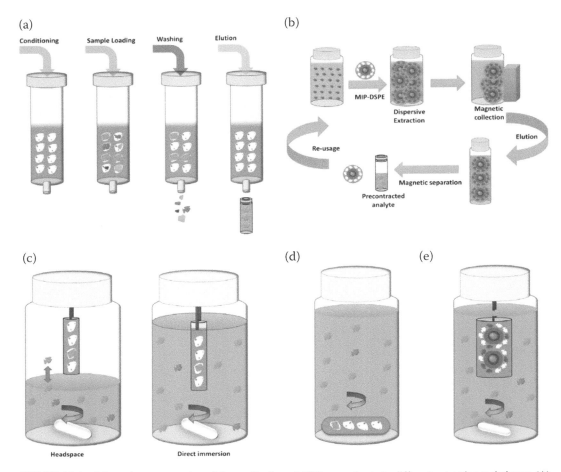

FIGURE 12.1 Schematic representation of the application of MIPs as sorbents in different extraction techniques: (A) MIP-SPE; (B) MIP-dSPE; (C) MIP-SPME; (D) MIP-SBSE; (E) supported liquid membrane (SLM)-MIPs. Reprinted from Azizi and Bottaro (2020) with permission of Elsevier.

In the search for highly selective sorbents, chemists have looked for specific materials based on one of the most selective existing interaction mechanisms: molecular recognition. As a result, molecularly imprinted polymers (MIPs) appeared, since previous (non-imprinted) polymeric materials demonstrated a high sorption capacity that could be enhanced or changed by trying to include such specific recognition.

Nowadays, there exists a wide variety of MIP materials that have been applied with success in different areas, including toxicology, but specially in miniaturized extraction techniques, to which they add more value as selectivity is also incorporated (Figure 12.1). This chapter provides a general overview of the toxicological applications of MIPs in sorbent-based extraction techniques. In particular, their use in miniaturized solid-phase extraction (SPE), solid-phase microextraction (SPME), as well as stir bar sorptive extraction (SBSE) is reviewed in more detailed, since they are the miniaturized techniques most commonly applied in this field.

12.2 Molecularly Imprinted Polymer Synthesis

Molecular imprinting is a complex process that can be achieved through three different pathways: covalent, non-covalent, or semi-covalent imprinting methods. Despite the existence of these different approaches (which will be discussed later), the general procedure is fairly similar among them and can be summarized as follows: first, before any polymerization takes place, a monomer-template complex is generated; then, polymerization is initiated – usually triggered by a radical initiator – in the presence of suitable crosslinkers (which grant stability to the polymer matrix, control polymer morphology, and stabilize the imprinted binding sites); and finally, the template is removed from the binding sites, delivering the final imprinted polymer.

12.2.1 Covalent Imprinting Method

Covalent imprinting, as its name suggests, involves the formation of covalent bonds between monomers and templates before polymerization takes place. In order to obtain a usable MIP, this covalent bond must be reversible so that the template can be removed from the binding site. The method's main advantage comes from the homogeneity of the binding sites and cavities generated, resulting from the fixed stoichiometric ratios and well-defined bonding. Nevertheless, the homogeneity comes at a cost, and that is the difficulty in the design of such monomer-template complexes, since they require reversibility in the formation and cleavage of the covalent bonds under mild conditions while assuring, simultaneously, specific chemical and geometric characteristics for target molecule retention (Speltini et al. 2017; Azizi and Bottaro 2020).

12.2.2 Non-Covalent Imprinting Method

Non-covalent imprinting opts for a different approach in the formation of the binding sites, implementing secondary bonds (e.g., ionic interactions, hydrogen bonding, among others) between monomers and templates, prior to polymerization. The main advantages of the method include simplification of experimental procedures, easier template removal, and greater functional diversity in the MIPs' binding sites. The main drawback comes from equilibrium processes in the monomer-template interactions. To obtain the desired product, excess amounts of monomers are used to shift the equilibrium, which frequently remains in the imprinted polymer matrix (randomly incorporated), leading to the formation of non-specific binding sites, reducing the selectivity of the MIPs (Speltini et al. 2017; Azizi and Bottaro 2020).

12.2.3 Semi-Covalent Imprinting Method

Semi-covalent imprinting is an intermediate approach for MIP synthesis. In this method, as in covalent imprinting, the monomer-template complex is covalently bonded; however, once the template is removed, rebinding is driven by non-covalent interactions. The semi-covalent pathway combines the advantages of both covalent and non-covalent imprinting methods. First, since the monomer-template is covalently bonded, the binding sites show greater homogeneity (increasing selectivity). Second, once the template is removed, rebinding takes place through secondary bonds, which facilitate the extraction of the analyte and subsequent elution from the imprinted polymer, reducing the long equilibrium times of covalently imprinted polymers (Speltini et al. 2017; Azizi and Bottaro 2020).

12.3 Application in Analytical Toxicology

12.3.1 Miniaturized Solid-Phase Extraction

As previously indicated, one of the current trends in sample preparation is miniaturization in order to comply with Green Analytical Chemistry principles. MIPs have also been applied with success to miniaturized sorbent-based techniques. Some examples are compiled in Table 12.1. In particular, some applications can be found in μ-SPE using different formats, such as classic cartridges and discs, but with reduced amounts of sorbent (lower than 100 mg), or more recent miniaturized devices, such as pipette tips, spin columns, well filter plates, or so-called membrane envelopes (Turiel and Martín-Esteban 2019). As examples, Jing and coworkers (Jing et al. 2014) developed a straightforward and selective spin-column

TABLE 12.1

Some applications of MIPs in sample treatment

Analytes	Samples (Amount)	Extraction Technique (Sorbent Amount)	Determination Technique	Recovery (RSD)	LODs	Comments	Reference
20 synthetic cannabinoids	Urine (1 mL)	μ-SPE (50 mg of sorbent)	HPLC-MS/MS	91–102% (1.0–7.0%)	0.032–0.748 μg/L	The MIP was introduced into a cone-shape PP membrane. JWH015 was used as the template. MIP-μ-SPE devices can be reused for 30–35 cycles.	(Sánchez-González et al. 2018)
5 aflatoxins	Cultured fish (1.5 mL of fish extracts)	μ-dSPE (40 mg of sorbent)	HPLC-MS/MS	83–102% (<19.0%)	0.029–0.060 μg/kg	5,7-dimethoxycoumarin was used as the template.	(Jayasinghe et al. 2020)
16 PAHs	River water and produced water (20 mL)	m-μ-dSPE (10 mg of sorbent)	AP-GC-MS/MS	72–135% (1.2–28.0%)	0.000001–0.0001 μg/L	The MMIP was prepared via RAFT polymerization.	(Azizi et al. 2020)
5 estrogens	Milk (15 mL)	DI-SPME	UHPLC-MS/MS	84–105% (2.4–7.8%)	0.00008–0.00026 μg/kg	Estradiol was used as template.	(Wang et al. 2020)
5 organophosphorus pesticides	Fruits and vegetables (1–2 g)	HS-SPME	GC-NPD	75–123% (1.1–11.8%)	0.0052–0.23 μg/kg	Diazinon, parathion-methyl, and isocarbophos were used as templates.	(Xiang et al. 2020)
11 phenolic compounds	Water samples (30 mL)	TFME	UHPLC-PDA	81–107% (0.1–13.9%)	0.1–2.0 μg/L	1,2-dihydroxybenzene was used as a pseudo-template.	(Abu-Alsoud and Bottaro 2021)
5 oestrogenic compounds	Water and packing samples (15 mL samples/extract)	SBSE	HPLC-DAD	67–102% (1.4–11.0%)	1.0–5.0 μg/L	Plastic packing samples were first extracted with acetone and methanol.	(Xu et al. 2014)
9 fluoroquinolones	Meat and fish samples (10 mL extract)	SBSE	HPLC-DAD	67–100% (<9.8%)	0.1–0.3 μg/L	Samples were first extracted with a mixture of ACN/trichloroacetic acid (7:3, v/v).	(Yang et al. 2017)
3 chlorophenols	Sea water (10 mL)	SBSE	HPLC-UV	84–99% (1.9–4.0%)	0.17–0.33 μg/L	–	(Hashemi and Najari 2019)

ACN: acetonitrile; AP: atmospheric pressure chemical ionization; DAD: diode array detector; DI: direct immersion; μ-dSPE: micro-dispersive solid-phase extraction; GC: gas chromatography; HPLC: high-performance liquid chromatography; HS: headspace; LOD: limit of detection; m-μ-dSPE: magnetic micro-dispersive solid-phase extraction; MIP: molecularly imprinted polymer; MMIP: magnetic molecularly imprinted polymer; MS/MS: tandem mass spectrometry; NPD: nitrogen phosphorus detector; PAH: polycyclic aromatic hydrocarbon; PDA: photodiode array detector; PP: polypropylene; RAFT: reversible addition fragmentation chain transfer; RSD: relative standard deviation; μ-SPE: micro solid-phase extraction; SBSE: stir bar sorptive extraction; SPME: solid-phase microextraction; TFME: thin-film microextraction; UHPLC: ultra high-performance liquid chromatography; UV: ultraviolet.

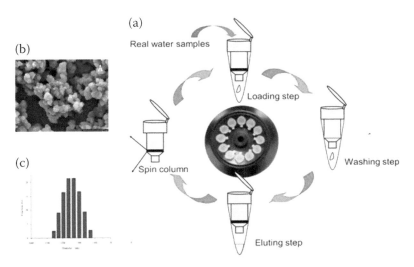

FIGURE 12.2 (A) The extraction procedure of hydrophilic MIPs packed spin column, (B) scanning electron micrographs, and (C) MIPs particle size distribution. Reprinted from Jing et al. (2014) with the permission of Wiley Online Library.

technique using MIPs as the sorbent for quantifying nitrophenol pollutants in wastewater, lake water, and river water samples by centrifugation of the spin column between the different loading, washing, and elution steps (Figure 12.2); Teixeira and coworkers (Teixeira et al. 2018) developed a pipette tip SPE method (part of the pipette tips are filled with the MIP sorbent) for the analysis of two macrocyclic lactones in mineral water and grape juice samples; and Feng and coworkers (Feng et al. 2009) determined phenolic compounds in tap, river, and raw sewage waters using MIPs enclosed within a porous polypropylene membrane sheet. On the other hand, another miniaturized format of SPE is the microextraction by packed sorbents (MEPSs) technique, in which the packing material is not located in a separate column, but directly incorporated into the syringe barrel as a plug or between the needle and the barrel as a cartridge (Moein et al. 2015).

MIPs have also been used in other miniaturized sample pre-treatment methods based on SPE. One of them is dispersive μ-SPE (μ-dSPE), a technique that presents greater simplicity and time saving compared to conventional μ-SPE, since the stages of conditioning the sorbent and loading the sample are not necessary. This is due to the high porosity and surface area, as well as the good dispersibility and chemical stability that they present under the conditions given during extraction (Chisvert et al. 2019). In addition, it is the recommended technique for the analysis of samples containing microparticles or microorganisms in order to avoid clogging the cartridges used in conventional μ-SPE. An example is the work of Ostovan and co-workers (Ostovan et al. 2017), in which they prepared hollow porous MIPs (HPMIPs) for the determination of glibenclamide in human urine samples. In the synthesis of HPMIPs, glibenclamide was used as a template, methacrylic acid (MAA) as a functional monomer, ethylene glycol dimethacrylate (EGDMA) as a crosslinker, and mesoporous MCM-48 nanospheres as a support. It was demonstrated that HPMIPs had a higher adsorption capacity and a lower equilibrium time of adsorption than core-shell MIPs due to greater accessibility to the HPMIP-specific cavities. The developed method (μ-dSPE high-performance liquid chromatography (HPLC)-ultraviolet (UV)) provided recovery values between 87.7 and 104.3% with high precision (relative standard deviations (RSDs) in the range 2.3–4.4%).

In addition, magnetic MIPs (MMIPs) have also been used as sorbents in the magnetic μ-dSPE (m-μ-dSPE) procedure. Magnetic nanoparticles (mNPs) frequently have a Fe_3O_4 core, which is later coated with the MIP. In m-μ-dSPE, after the first extraction step, which is the same as in the non-magnetic version, the sorbent is retained and isolated from the sample matrix with ease using an external magnetic field, thus avoiding any centrifugation step or retention of the sorbent (Płotka-Wasylka et al. 2015). As examples, MMIPs have been widely used as sorbent materials in this extraction technique to preconcentrate and determine polycyclic aromatic hydrocarbons (PAHs) in different types of water samples

(Azizi et al. 2020), kaempferol from apple samples (Cheng et al. 2020), patulin from juice samples (Zhao et al. 2020), or phenoxy carboxylic acid herbicides from cereals (Yuan et al. 2020), among others.

12.3.2 Solid-Phase Microextraction

SPME was first introduced by Arthur and Pawliszyn in 1990 (Arthur and Pawliszyn 1990) as an alternative to traditional exhaustive extraction techniques. SPME offers multiple advantages (greenness, simplicity, rapidity, etc.), which have yielded to its extensive application in both sampling and sample preparation (Li and Row 2018; Azizi and Bottaro 2020). Furthermore, SPME provides high accuracy in trace analysis, and it is compatible with different separation techniques (gas chromatography (GC), liquid chromatography (LC), and capillary electrophoresis (CE)) (Li and Row 2018). The use of MIP-coated fibres in SPME was first applied by Koster et al. (Koster et al. 2001) in 2001 for biological samples. Ever since, multiple publications have made use of MIP coatings in SPME, both in the modalities of direct immersion (DI) and headspace (HS), although in recent years, thin-film microextraction (TFME), as a variant of SPME, has also been applied.

DI-SPME involves introducing the coated fibre into the sample matrix to extract the target analytes. Analytes from different families of compounds, such as oestrogens (Wang et al. 2020), opioids (El-Beqqali and Abdel-Rehim 2016), antibiotics (Zhao et al. 2015), or polyphenolic flavonoids (Rahimi et al. 2019), have been analyzed through this technique (using various types of MIPs) in matrices of different natures and complexities (beverages, biological samples, etc.), showing selectivity toward the studied analytes and acceptable recovery values. Concerning HS-SPME, it can be considered a solvent-free extraction technique when thermal desorption is carried out. In this case, the coated fibre is suspended over the sample matrix in order to retain volatile analytes. Organophosphorus pesticides (OPPs) (Xiang et al. 2020), phenolic compounds (Abolghasemi and Yousefi 2014), acetaldehyde (Rajabi Khorrami and Narouenezhad 2011), and phthalate esters (PAEs) (He et al. 2010), among others, are some of the analytes that have been analyzed using this technique in numerous samples (water, beverages, fruits, vegetables, etc.). HS-SPME shows important advantages, including reduced effect of interferences and improved efficiency (Azizi and Bottaro 2020).

TFME is another variation of traditional SPME, where better extraction efficiencies are obtained without dramatically affecting the overall extraction time, as a result of the larger surface area to extraction-phase volume ratio (Olcer et al. 2019). Although the total number of publications employing TFME is not large, there are some examples where it has been applied for the analysis of compounds such as phenols (Abu-Alsoud and Bottaro 2021), PAHs (Shahhoseini et al. 2020), or polycyclic aromatic sulphur heterocycles (Hijazi and Bottaro 2020) in complex environmental samples (seawater, produced water, etc.).

Table 12.1 compiles some examples of publications where one of the SPME modalities (DI, HS, or TFME) has been employed. As can be seen, these techniques have been successfully applied in the analysis of some of the previously mentioned compounds (oestrogens (Wang et al. 2020), OPPs (Xiang et al. 2020), and phenolic compounds (Abu-Alsoud and Bottaro 2021)) in complex matrices (milk (Wang et al. 2020), fruits and vegetables (Xiang et al. 2020), and water samples (Abu-Alsoud and Bottaro 2021)). Overall, results showed acceptable recovery values (75.1–123.2%) with low RSDs (0.1–13.9%) and limits of detection (LODs) in the ppb range.

12.3.3 Stir Bar Sorptive Extraction

SBSE was introduced for the first time by Baltussen et al. in 1999 (Baltussen et al. 1999) as an alternative to SPME, trying to solve sorption competence problems that frequently take place in SPME between the extraction vessel walls, the stir bar used, and the fibre coating. In SBSE, the fibre is eliminated, and the stir bar is directly coated with the sorbent, which can be applied by immersing it directly into the sample or its HS, similarly to SPME (Soares Da Silva Burato et al. 2020). In general terms, SBSE is simpler, more robust, and shows an improved extraction efficiency than SPME due to a higher amount of sorbent being used. However, this higher extraction capacity results in longer equilibration times, which limits its use in certain applications (Hasan et al. 2020; Trujillo-Rodríguez et al. 2020).

Despite SBSE being employed for the analysis of samples of a very different nature, only three coatings are currently commercially available, including polyacrylate (PA), polyethylene glycol (PEG), and polydimethylsiloxane (PDMS). As a consequence of the low commercial availability of coatings, research effort has been invested in the development of new coatings (also those based on MIPs) in order to extend the applicability and versatility of the technique. In this sense, different methods, such as adhesion methods, sol-gel based approaches, or solvent exchange processes, are some of the most commonly used (Hasan et al. 2020). Among them, molecular imprinting technology has gained great interest. The use of MIPs as SBSE coatings have several remarkable advantages, such as great selectivity, high chemical and mechanical stability, and fast adsorption kinetics, as well as good reproducibility and simplicity, and cost effective preparation (Hasan et al. 2020). However, some limitations are commonly found when MIPs are used as coatings: multiple polymerization processes are usually required to maximize the adsorption capacity, which could affect the extraction efficiency; tough conditions are often used to remove templates, which could result in reduction of desorption kinetics or even bleeding; and the high cost of some templates forces the use of dummy templates, which is detrimental to the selectivity of the material (Hasan et al. 2020). In this sense, it is important to highlight that, despite the thermal stability of these polymeric coatings not being investigated in the last years, their chemical stability has been studied in different aqueous (acidic and basic media) and organic solvents (acetonitrile, dichloromethane, methanol, benzene, or acetone, among others) under different stirring speed and time, and no flacking or cracking was shown in any case (Hasan et al. 2020).

Regarding extraction devices, two main strategies are followed, which include the coating of glass capillaries or stir bars directly with the polymer (Gomez-Caballero et al. 2016) or composites composed by mNPs coated with MIPs and magnetically retained or embedded in the polymeric monoliths (Díaz-Álvarez et al. 2016). Independently of the device, selectivity of MIPs have allowed them to be applied to the extraction of a wide variety of analytes (see Table 12.1), including oestrogenic compounds (Xu et al. 2014), phenols (Hashemi and Najari 2019), herbicides (Gomez-Caballero et al. 2016), or pharmaceuticals (Yang et al. 2017) from biological (Fan et al. 2016), environmental (Gomez-Caballero et al. 2016; Xu et al. 2014), and food samples (Yang et al. 2017). As an example of the applicability of MIPs as coatings in SBSE, Xu and co-workers (Xu et al. 2014) developed a dual-template MIP using bisphenol A (BPA) and estradiol as templates in order to generate two different specific cavities. This MIP was used to coat a silylated glass capillary, in which a magnetic core was introduced, sealing both ends with a flame. This device allowed the extraction of five oestrogenic compounds from lake water, river water, a disposable lunch box cover, a biscuit box, and a yoghurt bottle, with recovery values in the range 67–102%, which prove the good extraction capacity of MIPs when applied as sorbents in SBSE.

12.3.4 Miscellaneous

In addition to the microextraction techniques previously described in which MIPs are used as sorbents, other variants designated as stir cake or rotating disk extraction and matrix solid-phase dispersion (MSPD), among others, have also been employed in numerous occasions.

In order to overcome the inherent limitations of SPME and SBSE procedures, stir cake sorptive extraction (SCSE) and rotating disk sorptive extraction (RDSE) modes were developed. Both techniques are similar from an operational point of view and present an easy design of the extraction medium-monolithic cake as well as have high cost-efficiency, straightforward operation, high extraction capacity, and high environmental friendliness. Concerning RDSE (so-called when using a Teflon disk with a miniature magnetic stirring bar embedded (Manzo et al. 2015)), the fact that the sorptive phase is only in contact with the liquid sample and not with the extraction vessel, allows higher stirring speeds to be used than in SBSE without causing damage in the extraction phase, thus facilitating the transfer of the analyte to the surface of the sorbent (Jachero et al. 2014). However, in spite of these good features, to our knowledge, only one study related to the use of MIPs in SCSE (Sorribes-Soriano et al. 2019) and another work where this sorbent is utilized in RDSE (Manzo et al. 2015) have been described.

Another miniaturized technique used as an alternative sample pre-treatment is MSPD. It is characterized by being simple and cheap, and because it involves the disruption and extraction of different liquid, viscous, semi-solid, and solid samples by a sorbent. In this process, the sample and the sorbent are blended and homogenized together, and subsequently, the adsorbed analytes are eluted with a suitable solvent (Turiel and Martín-Esteban 2019). Like SPE, MSPD also has different variants characterized by the way in which the second stage is performed. For example, in a case similar to conventional SPE, the homogeneous mixture was transferred to an SPE cartridge and subjected to elution, whereas in the magnetic version, it was added without the need for column packing, as Gholami and co-workers (Gholami et al. 2019) did for the analysis of melamine in various milk samples.

On the other hand, interest in the applications of membrane-based liquid-phase microextraction (M-LPME) for sample preparation has increased, especially in reinforced hollow-fibre (HF)-LPME (Chimuka et al. 2011). Compared with conventional SPME fibres, HFs have a greater surface area, since they can be covered by the sorbent on the internal and external walls, which provides a higher extraction efficiency, and HFs also are more robust than the single drop microextraction (SDME) technique (Kokosa 2019). MIPs-HF-LPME has been used in the determination of a wide variety of analytes, such as PAEs (Mirzajani et al. 2020), antibiotics (Barahona et al. 2019), and triazines (Barahona et al. 2016), among others, in aqueous and biological samples.

12.4 Conclusions

New trends in Analytical Chemistry are focused on minimizing the negative effects derived from the application of previous methodologies. In this sense, efforts have been made to reduce the amount of solvents and reagents used, especially during the application of extraction techniques.

The introduction of miniaturized versions of the sorbent-based extraction techniques classically used has posed a great advance to achieve such an objective. These techniques brought with them the implementation of new materials with improved properties that allowed better extraction efficiency and selectivity. In this sense, MIPs have generated great interest for their excellent performance in these two aspects, which together with their great versatility, have made possible their use as sorbents in different extraction techniques with remarkable results, such as μ-SPE, SPME, or SBSE, among others.

Since their introduction, MIPs have grabbed the attention of the Analytical Chemistry field for their outstanding properties and versatility to be applied not only as extraction sorbents, but also in sensors or even stationary phases in separation techniques. However, these materials have shown great potential, so their applicability in different areas will continue to be explored in the coming years.

Acknowledgements

J.G.S. would like to thank "Cabildo de Tenerife" for the Agustín de Betancourt contract at the Universidad de La Laguna.

REFERENCES

Abolghasemi, Mir Mahdi, and Vahid Yousefi. "Three Dimensionally Honeycomb Layered Double Hydroxides Framework as a Novel Fiber Coating for Headspace Solid-Phase Microextraction of Phenolic Compounds." *Journal of Chromatography A* 1345 (2014): 9–16. 10.1016/j.chroma.2014.04.018.

Abu-Alsoud, Ghadeer F., and Christina S. Bottaro. "Porous Thin-Film Molecularly Imprinted Polymer Device for Simultaneous Determination of Phenol, Alkylphenol and Chlorophenol Compounds in Water." *Talanta* 223 (2021): 121727. 10.1016/j.talanta.2020.121727.

Arthur, Catherine L., and Janusz Pawliszyn. "Solid Phase Microextraction with Thermal Desorption Using Fused Silica Optical Fibers." *Analytical Chemistry* 62, no. 19 (1990): 2145–8. 10.1021/ac00218a019.

Azizi, Ali, and Christina S. Bottaro. "A Critical Review of Molecularly Imprinted Polymers for the Analysis of Organic Pollutants in Environmental Water Samples." *Journal of Chromatography A* 1614 (2020): 460603. 10.1016/j.chroma.2019.460603.

Azizi, Ali, Fereshteh Shahhoseini, and Christina S. Bottaro. "Magnetic Molecularly Imprinted Polymers Prepared by Reversible Addition Fragmentation Chain Transfer Polymerization for Dispersive Solid Phase Extraction of Polycyclic Aromatic Hydrocarbons in Water." *Journal of Chromatography A* 1610 (2020): 460534. 10.1016/j.chroma.2019.460534.

Baltussen, Erik, Pat Sandra, Frank David, and Carel Cramers. "Stir Bar Sorptive Extraction (SBSE), a Novel Extraction Technique for Aqueous Samples: Theory and Principles." Edited by Department of Chemical Engineering and Chemistry and Instrumental Analysis. *Journal of Microcolumn Separations* 11, no. 10 (1999): 737–47 10.1002/(SICI)1520-667X(1999)11:10<737::AID-MCS7>3.0.CO;2-4.

Barahona, Francisco, Beatriz Albero, José Luis Tadeo, and Antonio Martín-Esteban. "Molecularly Imprinted Polymer-Hollow Fiber Microextraction of Hydrophilic Fluoroquinolone Antibiotics in Environmental Waters and Urine Samples." *Journal of Chromatography A* 1587 (2019): 42–9. 10.1016/j.chroma.2018.12.015.

Barahona, Francisco, Myriam Díaz-Álvarez, Esther Turiel, and Antonio Martín-Esteban. "Molecularly Imprinted Polymer-Coated Hollow Fiber Membrane for the Microextraction of Triazines Directly from Environmental Waters." *Journal of Chromatography A* 1442 (2016): 12–18. 10.1016/j.chroma.2016.03.004.

Cheng, Yang, Jiyun Nie, Hongdi Liu, Lixue Kuang, and Guofeng Xu. "Synthesis and Characterization of Magnetic Molecularly Imprinted Polymers for Effective Extraction and Determination of Kaempferol from Apple Samples." *Journal of Chromatography A* 1630 (2020): 461531. 10.1016/j.chroma.2020.461531.

Chimuka, Luke, Ewa Cukrowska, Monika Michel, and Boguslaw Buszewski. "Advances in Sample Preparation Using Membrane-Based Liquid-Phase Microextraction Techniques." *TrAC, Trends in Analytical Chemistry (Regular Ed.)* 30, no. 11 (2011): 1781–92. 10.1016/j.trac.2011.05.008.

Chisvert, Alberto, Soledad Cárdenas, and Rafael Lucena. "Dispersive Micro-Solid Phase Extraction." *TrAC, Trends in Analytical Chemistry (Regular Ed.)* 112 (2019): 226–33. 10.1016/j.trac.2018.12.005.

Díaz-Álvarez, Myriam, Esther Turiel, and Antonio Martín-Esteban. "Molecularly Imprinted Polymer Monolith Containing Magnetic Nanoparticles for the Stir-Bar Sorptive Extraction of Triazines from Environmental Soil Samples." *Journal of Chromatography A* 1469 (2016): 1–7. 10.1016/j.chroma.2016.09.051.

El-Beqqali, Aziza, and Mohamed Abdel-Rehim. "Molecularly Imprinted Polymer-Sol-Gel Tablet Toward Micro-Solid Phase Extraction: I. Determination of Methadone in Human Plasma Utilizing Liquid Chromatography–Tandem Mass Spectrometry." *Analytica Chimica Acta* 936 (2016): 116–22. 10.1016/j.aca.2016.07.001.

Fan, Wenying, Man He, Linna You, Xuewei Zhu, Beibei Chen, and Bin Hu. "Water-Compatible Graphene Oxide/Molecularly Imprinted Polymer Coated Stir Bar Sorptive Extraction of Propranolol from Urine Samples Followed by High Performance Liquid Chromatography-Ultraviolet Detection." *Journal of Chromatography A* 1443 (2016): 1–9. 10.1016/j.chroma.2016.03.017.

Feng, Qinzhong, Lixia Zhao, and Jin-Ming Lin. "Molecularly Imprinted Polymer as Micro-Solid Phase Extraction Combined with High Performance Liquid Chromatography to Determine Phenolic Compounds in Environmental Water Samples." *Analytica Chimica Acta* 650, no. 1 (2009): 70–6. 10.1016/j.aca.2009.04.016.

Gholami, Habibeh, Maryam Arabi, Mehrorang Ghaedi, Abbas Ostovan, and Ahmad Reza Bagheri. "Column Packing Elimination in Matrix Solid Phase Dispersion by Using Water Compatible Magnetic Molecularly Imprinted Polymer for Recognition of Melamine from Milk Samples." *Journal of Chromatography A* 1594 (2019): 13–22. 10.1016/j.chroma.2019.02.015.

Gomez-Caballero, Alberto, Goretti Diaz-Diaz, Olatz Bengoetxea, Amaia Quintela, Nora Unceta, María Aranzazu Goicolea, and Ramón José Barrio. "Water Compatible Stir-Bar Devices Imprinted with Underivatised Glyphosate for Selective Sample Clean-Up." *Journal of Chromatography A* 1451 (2016): 23–32. 10.1016/j.chroma.2016.05.017.

Hasan, Chowdhury Kamrul, Alireza Ghiasvand, Trevor W. Lewis, Pavel N. Nesterenko, and Brett Paull. "Recent Advances in Stir-Bar Sorptive Extraction: Coatings, Technical Improvements, and Applications." *Analytica Chimica Acta* 1139 (2020): 222–40. 10.1016/j.aca.2020.08.021.

Hashemi, Seyed Hossein, and Fahimeh Najari. "Response Surface Methodology of Pre-Concentration of Chorophenols from Seawater Samples by Molecularly Imprinted Stir Bar Sorptive Extraction

Combined with HPLC: Box–Behnken Design." *Journal of Chromatographic Science* 57, no. 3 (2019): 279–89. 10.1093/chromsci/bmy107.

He, Juan, Ruihe Lv, Haijun Zhan, Huizhi Wang, Jie Cheng, Kui Lu, and Fengcheng Wang. "Preparation and Evaluation of Molecularly Imprinted Solid-Phase Micro-Extraction Fibers for Selective Extraction of Phthalates in an Aqueous Sample." *Analytica Chimica Acta* 674, no. 1 (2010): 53–8. 10.1016/j.aca.2010.06.018.

Hijazi, Hassan Y., and Christina S. Bottaro. "Molecularly Imprinted Polymer Thin-Film as a Micro-Extraction Adsorbent for Selective Determination of Trace Concentrations of Polycyclic Aromatic Sulfur Heterocycles in Seawater." *Journal of Chromatography A* 1617 (2020): 460824. 10.1016/j.chroma.2019.460824.

Jachero, Lourdes, Inés Ahumada, and Pablo Richter. "Rotating-Disk Sorptive Extraction: Effect of the Rotation Mode of the Extraction Device on Mass Transfer Efficiency." *Analytical and Bioanalytical Chemistry* 406, no. 12 (2014): 2987–92. 10.1007/s00216-014-7693-z.

Jayasinghe, G.D. Thilini Madurangika, Raquel Domínguez-González, Pilar Bermejo-Barrera, and Antonio Moreda-Pieiro. "Miniaturized Vortex Assisted-Dispersive Molecularly Imprinted Polymer Micro-Solid Phase Extraction and HPLC-MS/MS for Assessing Trace Aflatoxins in Cultured Fish." *Analytical Methods* 12, no. 35 (2020): 4351–62. 10.1039/d0ay01259a.

Jing, Tao, Yusun Zhou, Wei Wu, Min Liu, Yikai Zhou, and Surong Mei. "Molecularly Imprinted Spin Column Extraction Coupled with High-performance Liquid Chromatography for the Selective and Simple Determination of Trace Nitrophenols in Water Samples." *Journal of Separation Science* 37, no. 20 (2014): 2940–6. 10.1002/jssc.201400625.

Kokosa, John M. "Selecting an Extraction Solvent for a Greener Liquid Phase Microextraction (LPME) Mode-Based Analytical Method." *TrAC, Trends in Analytical Chemistry (Regular Ed.)* 118 (2019): 238–47. 10.1016/j.trac.2019.05.012.

Koster, Emile H.M., Carlo Crescenzi, Widia den Hoedt, Kees Ensing, and Gerhardus J. de Jong. "Fibers Coated with Molecularly Imprinted Polymers for Solid-Phase Microextraction." Edited by Faculty of Science and Engineering. *Analytical Chemistry* 73, no. 13 (2001): 3140–5. 10.1021/ac001331x.

Li, Guizhen, and Kyung Ho Row. "Recent Applications of Molecularly Imprinted Polymers (MIPs) on Micro-Extraction Techniques." *Separation and Purification Reviews* 47 (2018): 1–18. 10.1080/15422119.2017.1315823.

Manzo, Valentina, Karla Ulisse, Inés Rodríguez, Eduardo Pereira, and Pablo Richter. "A Molecularly Imprinted Polymer as the Sorptive Phase Immobilized in a Rotating Disk Extraction Device for the Determination of Diclofenac and Mefenamic Acid in Wastewater." *Analytica Chimica Acta* 889, no. C (2015): 130–7. 10.1016/j.aca.2015.07.038.

Mirzajani, Roya, Fatemeh Kardani, and Zahra Ramezani. "Fabrication of UMCM-1 Based Monolithic and Hollow Fiber – Metal-Organic Framework Deep Eutectic Solvents/Molecularly Imprinted Polymers and Their Use in Solid Phase Microextraction of Phthalate Esters in Yogurt, Water and Edible Oil by GC-FID." *Food Chemistry* 314 (2020): 126179. 10.1016/j.foodchem.2020.126179.

Moein, Mohammad Mahdi, Abbi Abdel-Rehim, and Mohamed Abdel-Rehim. "Microextraction by Packed Sorbent (MEPS)." *TrAC, Trends in Analytical Chemistry (Regular Ed.)* 67 (2015): 34–44. 10.1016/j.trac.2014.12.003.

Olcer, Yekta Arya, Marcos Tascon, Ahmet E. Eroglu, and Ezel Boyacı. "Thin Film Microextraction: Towards Faster and More Sensitive Microextraction." *TrAC, Trends in Analytical Chemistry* 113 (2019): 93–101. 10.1016/j.trac.2019.01.022.

Ostovan, Abbas, Mehrorang Ghaedi, Maryam Arabi, and Arash Asfaram. "Hollow Porous Molecularly Imprinted Polymer for Highly Selective Clean-up Followed by Influential Preconcentration of Ultra-Trace Glibenclamide from Bio-Fluid." *Journal of Chromatography A* 1520 (2017): 65–74. 10.1016/j.chroma.2017.09.026.

Płotka-Wasylka, Justyna, Natalia Szczepańska, Miguel de la Guardia, and Jacek Namieśnik. "Miniaturized Solid-Phase Extraction Techniques." *TrAC, Trends in Analytical Chemistry* 73 (2015): 19–38. 10.1016/j.trac.2015.04.026.

Rahimi, Marzieh, Soleiman Bahar, Rouhollah Heydari, and Seyed Mojtaba Amininasab. "Determination of Quercetin Using a Molecularly Imprinted Polymer as Solid-Phase Microextraction Sorbent and High-

Performance Liquid Chromatography." *Microchemical Journal* 148 (2019): 433–41. 10.1016/j.microc.2019.05.032.

Rajabi Khorrami, Afshin, and E. Narouenezhad. "Synthesis of Molecularly Imprinted Monolithic Fibers for Solid-Phase Microextraction of Acetaldehyde from Head-Space of Beverages Stored in PET Bottles." *Talanta* 86 (2011): 58–63. 10.1016/j.talanta.2011.08.002.

Sánchez-González, Juan, Sara Odoardi, Ana María Bermejo, Pilar Bermejo-Barrera, Francesco Saverio Romolo, Antonio Moreda-Piñeiro, and Sabina Strano-Rossi. "Development of a Micro-Solid-Phase Extraction Molecularly Imprinted Polymer Technique for Synthetic Cannabinoids Assessment in Urine Followed by Liquid Chromatography–Tandem Mass Spectrometry." *Journal of Chromatography A* 1550 (2018): 8–20. 10.1016/j.chroma.2018.03.049.

Shahhoseini, Fereshteh, Ali Azizi, Stefana N. Egli, and Christina S. Bottaro. "Single-Use Porous Thin Film Extraction with Gas Chromatography Atmospheric Pressure Chemical Ionization Tandem Mass Spectrometry for High-Throughput Analysis of 16 PAHs." *Talanta* 207 (2020): 120320. 10.1016/j.talanta.2019.120320.

Soares Da Silva Burato, Juliana, Deyber Arley Vargas Medina, Ana Lúcia Toffoli, Edvaldo Vasconcelos Soares Maciel, and Fernando Mauro Lanças. "Recent Advances and Trends in Miniaturized Sample Preparation Techniques." *Journal of Separation Science* 43, no. 1 (2020): 202–25. 10.1002/jssc.201900776.

Sorribes-Soriano, Aitor, R. Arráez-González, Francesc Albert Esteve-Turrillas, Sergio Armenta, and José Manuel Herrero-Martínez. "Development of a Molecularly Imprinted Monolithic Polymer Disk for Agitation-Extraction of Ecgonine Methyl Ester from Environmental Water." *Talanta (Oxford)* 199 (2019): 388–95. 10.1016/j.talanta.2019.02.077.

Speltini, Andrea, Andrea Scalabrini, Federica Maraschi, Michela Sturini, and Antonella Profumo. "Newest Applications of Molecularly Imprinted Polymers for Extraction of Contaminants from Environmental and Food Matrices: A Review." *Analytica Chimica Acta* 974 (2017): 1–26. 10.1016/j.aca.2017.04.042.

Teixeira, Roseane Andrade, Diego Hernando Ângulo Flores, Ricky Cássio Santos Da Silva, Flávia Viana Avelar Dutra, and Keyller Bastos Borges. "Pipette-Tip Solid-Phase Extraction Using Poly(1-Vinylimidazole-Co-Trimethylolpropane Trimethacrylate) as a New Molecularly Imprinted Polymer in the Determination of Avermectins and Milbemycins in Fruit Juice and Water Samples." *Food Chemistry* 262 (2018): 86–93. 10.1016/j.foodchem.2018.04.076.

Trujillo-Rodríguez, María José, Idaira Pacheco-Fernández, Iván Taima-Mancera, Juan Heliodoro Ayala Díaz, and Verónica Pino. "Evolution and Current Advances in Sorbent-Based Microextraction Configurations." *Journal of Chromatography A* 1634 (2020): 461670. 10.1016/j.chroma.2020.461670.

Turiel, Esther, and Antonio Martín-Esteban. "Molecularly Imprinted Polymers-Based Microextraction Techniques." *TrAC, Trends in Analytical Chemistry (Regular Ed.)* 118 (2019): 574–86. 10.1016/j.trac.2019.06.016.

Wang, Shenling, Yanling Geng, Xiaowei Sun, Rongyu Wang, Zhenjia Zheng, Shenghuai Hou, Xiao Wang, and Wenhua Ji. "Molecularly Imprinted Polymers Prepared from a Single Cross-Linking Functional Monomer for Solid-Phase Microextraction of Estrogens from Milk." *Journal of Chromatography A* 1627 (2020): 461400. 10.1016/j.chroma.2020.461400.

Xiang, Xiaozhe, Yulong Wang, Xiaowei Zhang, Mingquan Huang, Xiujuan Li, and Siyi Pan. "Multifiber Solid-Phase Microextraction Using Different Molecularly Imprinted Coatings for Simultaneous Selective Extraction and Sensitive Determination of Organophosphorus Pesticides." *Journal of Separation Science* 43, no. 4 (2020): 756–65. 10.1002/jssc.201900994.

Xu, Zhigang, Zailei Yang, and Zhimin Liu. "Development of Dual-Templates Molecularly Imprinted Stir Bar Sorptive Extraction and Its Application for the Analysis of Environmental Estrogens in Water and Plastic Samples." *Journal of Chromatography A* 1358 (2014): 52–9. 10.1016/j.chroma.2014.06.093.

Yang, Kun, Geng Nan Wang, Hui Zhi Liu, Jing Liu, and Jian Ping Wang. "Preparation of Dual-Template Molecularly Imprinted Polymer Coated Stir Bar Based on Computational Simulation for Detection of Fluoroquinolones in Meat." *Journal of Chromatography. B, Analytical Technologies in the Biomedical and Life Sciences* 1046 (2017): 65–72. 10.1016/j.jchromb.2017.01.033.

Yuan, Xucan, Yunxia Yuan, Xun Gao, Zhili Xiong, and Longshan Zhao. "Magnetic Dummy-Template Molecularly Imprinted Polymers Based on Multi-Walled Carbon Nanotubes for Simultaneous Selective Extraction and Analysis of Phenoxy Carboxylic Acid Herbicides in Cereals." *Food Chemistry* 333 (2020): 127540. 10.1016/j.foodchem.2020.127540.

Zhao, Minjuan, Hua Shao, Jun Ma, Hui Li, Yahui He, Miao Wang, Fen Jin, et al. "Preparation of Core-Shell Magnetic Molecularly Imprinted Polymers for Extraction of Patulin from Juice Samples." *Journal of Chromatography A* 1615 (2020): 460751. 10.1016/j.chroma.2019.460751.

Zhao, Tong, Xiujuan Guan, Wanjin Tang, Ying Ma, and Haixia Zhang. "Preparation of Temperature Sensitive Molecularly Imprinted Polymer for Solid-Phase Microextraction Coatings on Stainless Steel Fiber to Measure Ofloxacin." *Analytica Chimica Acta* 853, no. 1 (2015): 668–75. 10.1016/j.aca.2014.10.019.

13 Applications of Ionic Liquids in Microextraction

Devendra Kumar Patel[1,2], Neha Gupta[1,2], Sandeep Kumar[1,2], and Juhi Verma[1]
[1]*Analytical Chemistry Laboratory, CSIR-Indian Institute of Toxicology Research, Vishvigyan Bhawan, UP, India*
[2]*Academy of Scientific & Innovative Research (AcSIR), Ghaziabad, UP, India*

CONTENTS

13.1 Introduction 205
13.2 IL-Based Microextraction Techniques 209
    13.2.1 IL-DLLME (Ionic Liquid-Based Dispersive Liquid-Liquid Microextraction) 209
    13.2.2 IL-SDME (Ionic Liquid Single Drop Microextraction) 209
    13.2.3 IL-SPME (Ionic Liquid Solid-Phase Microextraction) 210
    13.2.4 IL-SBSE (Ionic Liquid Stir Bar Sorptive Extraction) 210
    13.2.5 IL-SCSE (Ionic Liquid Stir Cake Sorptive Extraction) 210
13.3 Application of IL-Based Microextraction Techniques 211
13.4 Conclusion 212
References 216

13.1 Introduction

Microextraction is defined as an analytical extraction technique that is non-exhaustive and utilizes a very small volume of the extracting phases in relation to the volume of the sample. Microextraction techniques improve sample preparation by miniaturizing the steps, provide onsite analysis, and are automatic and economical for time. There are three main reasons that supported the evolution of microextraction techniques to provide miniaturization and generate new methods for extraction and determination. First, interested analytes are present in trace quantities in the real sample and are not enough to determine through the macroscopic method. Second, there was a need to save time and meet the requirement of rapid determination on a small sample volume to increase the number of samples processed and their analysis. Third, techniques needed to achieve green analytical chemistry (GAC) of reducing toxic chemicals and lowering waste generation from the laboratory.

GAC defines the miniaturization of the analytical protocol and use of chemicals to reduce the negative impact on the environment and the health of analytical chemists performing laboratory work. The green chemistry concept was introduced by Anastas in 1998 (Anastas and Beach 2007). The basic principles are to make the processes greener by increasing the safety of operators, using less toxic reagents and auxiliaries; to decrease energy consumption by using mild reaction conditions; to improve waste management; to limit or eradicate the use of hazardous chemicals; and to substitute them with benign ones wherever possible, like avoiding derivatization and using renewable resources as a substrate (Marcinkowska, Namieśnik, and Tobiszewski 2019; Tobiszewski et al. 2015). GAC is known as an analytical wing of sustainable development whose main principle is providing a framework for proper chemical processes that are environmentally friendly. It is steadily gaining popularity in today's scenario because its implementation takes analytical chemistry toward sustainable growth by facilitating the inimical effects of its techniques and methodologies on human health and environment. Researchers follow the principles as stated under the GAC to develop straightforward, uncomplicated, and efficacious techniques to extract substances of interest from composite matrices. The designed methods are generally proclaimed as environmentally harmless by reducing the use of potential toxic compounds and lowering energy consumption. There are many ways to reduce pollution by utilizing GAC in order to generate 'clean waste' rather than 'hazardous waste' by unification of analytical processes and steps to miniaturize it.

Ionic liquids (ILs) are organic salt, a combination of organic cations that are bonded with organic or inorganic anions, having a melting point equal or less than 100°C, with tuneable physiochemical properties by changing the structure of cation and anion molecules.

ILs mostly consists of (Yavir et al. 2019):

a. Bulky organic nitrogen-containing cations: imidazolium, pyrrolidinium, pyridinium, tetra alkylphosphonium, and tetraalkylammonium.
b. Halogen-based organic or inorganic anions: bis (trifluoromethylsulfonyl) imide, tetrafluoroborate, bromide, acetate, chloride, trifluoromethylsulfonate, trifluoroethanoate, and hexafluorophosphate.

Figure 13.1 shows the structure of the most common cations and anions used in ILs.

IL has been considered a 'designer solvent' due to its involvement of diverse anions and cations with distinct alkyl substituent to cations. As a substitute for conventional organic solvents, an eco-friendly solvent IL is used because of its less toxic nature and because it does not release harmful, poisonous

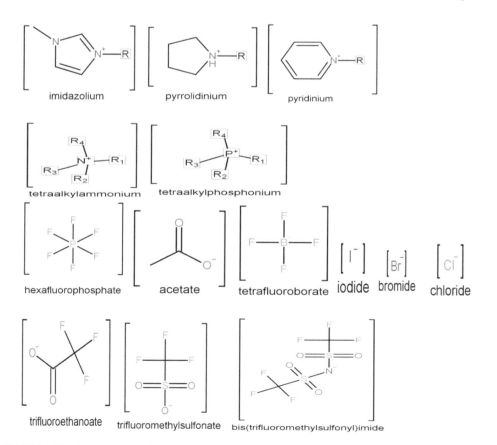

FIGURE 13.1 Chemical structures of commonly used cations and anions of ionic liquids (ILs).

vapours to the surroundings (Marcinkowska et al. 2019). ILs have been extensively used in the field of analytical chemistry due to their following unique characteristics (Weingärtner 2008; Roth 2009; Ruiz-Aceituno et al. 2013; Han et al. 2012; Freire et al. 2012):

1. Very low vapour pressure
2. High viscosity
3. High thermal stability (around $300°C$)
4. Negligible flammability
5. Capability of dissolving a wide spectra of organic and inorganic compounds (i.e., strong salvation power)
6. Specific electrochemical characteristics
7. High ionic conductivity

These unique properties are due to the interaction that exists between the cations and anions forming the ILs. Usually an asymmetrical arrangement of cations and anions forms the ILs by the ionic interactions along with the convectional interactions, like hydrogen bonding, Van deer Waal's, and dipole-dipole interactions. Their solubility in polar solvents is determined due to the ionic interactions, and an alkyl chain on cations determines their solubility in non-polar solvents. The changes in the length and branching of an alkyl group make it possible to fine-tune the properties of the ILs. Hydrogen bonding is a readily observed interaction in IL because oxygen or a halide group of anion can easily interact with the hydrogen atom present on the imidazolium, pyrrolidinium, or pyridinium ring of the cations. IL thus shows great results where trace impurities can change the results for catalysis, separation, and extraction. Besides providing a combination of structures that become a 'green' solvent, there are some combinations of anions and cations that are toxic to both abiotic and biotic components and are non-biodegradable.

Some of the important reasons for using IL in an extraction technique are as follows:

1. The densities of ILs are higher than water and organic solvents, which provides easy removal of the extraction phase after centrifugation.
2. ILs have negligible vapour pressure so the extraction phase does not evaporate during ultrasound, temperature, time, and microwave-assisted extraction techniques.
3. ILs have high thermal stability, thus preventing them from degradation during the thermal desorption of the extracted analytes and preventing any type of contamination during the analysis.
4. ILs have tuneable solubility that makes them disperse well in aqueous solution, enhancing the mass transfer of the targeted analytes to the IL phase with its easy retrieval in dispersive liquid-liquid microextraction (DLLME).
5. Hydrophilic magnetic ionic liquids (MILs) are successful in the extraction of hydrophobic samples and are easily retrieved due to their strong attraction toward magnets.

ILs are used as both a liquid extractant and sorption material in stationary phases, expanding their range of utilization in the field of extraction. In recent years, microextraction techniques have grown rapidly because of numerous advantages: cost-effectiveness, simplicity, miniaturization, low expenditure of sample, easy automation, and being ecologically sound. Many ILs were designed according to the needs of the extraction with the required properties. The principles as proposed under GAC and eco-friendly properties of ILs are replacement of organic solvents and designing the process of liquid-liquid microextraction (LLE). In the past decade, numerous papers have emphasized applications of ILs in sample preparation techniques. They can provide multiple sites for interactions of the target analyte and are therefore regarded as 'ideal extraction media' (Tables 13.1 and 13.2).

TABLE 13.1

Physical Properties of Commonly Used Ionic Liquids (ILs)

Ionic Liquid	Molecular Weight	Melting Point (°C)	Density (g/mL^{-1}) 25°C	Viscosity (cP) 25°C
[C$_2$MIM][BF$_4$]	197.8	15	1.248	66
[C$_2$MIM][PF$_6$]	256.13	58–60	1.373	450
[C$_2$MIM][NTf$_2$]	391.3	4	1.425	323
[C$_4$MIM][BF$_4$]	225.80	−81	1.208	233
[C$_4$MIM][PF$_6$]	284.18	10	1.373	400
[C$_4$MIM][Br]	218.9	73	1.134	Solid
[C$_4$MIM][Cl]	146.50	41	1.120	Solid
[C$_4$MIM][NTf$_2$]	487.9	−25	1.420	52
[C$_6$MIM][BF$_4$]	254.08	−71	1.075	211
[C$_6$MIM][PF$_6$]	312.00	−73.5	1.304	800
[C$_6$MIM][NTf$_2$]	534.9	–	1.423	674
[C$_8$MIM][BF$_4$]	281.8	−88	1.11	440
[C$_8$MIM][Cl]	230.50	−55	1.000	16000
[C$_8$MIM][PF$_6$]	445.0	14	1.212	4232
[C$_8$MIM][NTf$_2$]	645.7	–	1.242	5234

TABLE 13.2

Milestones in the Discovery of IL-Based Microextraction

1887: 'Red oil', first IL

1914: Synthesis of protic ethylammonium nitrate (mp 12.5°C)

1934: First application of IL (1-ethylpyridinium chloride) in dissolving cellulose

1948: Molten mixtures of ethylpyridinium halides and ammonium chloride with the electrodeposition of aluminium

1972: Use of ammonium ILs in homogeneous catalysis

1981: The use of phosphonium ILs to make ethylene glycol

1982: New type of IL made of imidazolium cation and aluminium chloride anion

1982: Use of ethylammonium nitrate as a stationary phase in Gas Liquid Chromatography (GLC)

1990: Solid-phase microextraction (SPME) introduced in analytical chemistry

1992: First air- and water-stable ILs, [C$_2$C$_1$IM][BF$_6$] and [C$_2$C$_1$IM][PF$_6$]

1996: Development of single drop microextraction (SDME)

1998: ILs as novel media for 'clean' LLE

2003: ILs in the extraction solvent for SDME

2004: Discovery of MILs [C$_4$C$_1$IM][FeC$_{l4}$]

2005: First approach for using ILs as a sorbent coating in SPME

2006: Introduction of DLLME for organic and inorganic analytes

2008: Application of ILs in DLLME

2008: Introduction of polymeric ionic liquids (PILs)

2009: First approach using ILs as solid-phase extraction (SPE) sorbent

2014: MIL-based DLLME

2017: MILs used in vacuum and magnetic headspace SDME (HS-SDME)

2018: MIL based in situ and in situ stir bar DLLME

13.2 IL-Based Microextraction Techniques

13.2.1 IL-DLLME (Ionic Liquid-Based Dispersive Liquid-Liquid Microextraction)

DLLME as an easy, effective, and novel extraction technique for the extraction of organic compounds from water samples was first proposed by Rezaee et al. in 2006 (Passos et al. 2012). IL utilization in DLLME for quantification of organophosphorus (OP) pesticides was proposed by Zhou et al. (S. Li et Hal. 2009). Baghdadi and Shemirani reported the extraction of mercury in different environmental samples (Yao et al. 2011). In DLLME, the aqueous sample having the desired analytes is mixed with a few micro litres of the extracting solvents that are immiscible into the aqueous phase, followed by the addition of dispersive solvent that has solubility in both aqueous and extracting solvents. These three solvents all together are mixed using a syringe or micropipette followed by gentle shaking, which leads to the formation of various micro droplets to form a homogeneous cloudy solution. The samples are then subjected to centrifugation; the sediment phase is collected, and analysis is done by using any sophisticated analytical techniques (Abujaber et al. 2018a). In DLLME, the dispersive solvent is basically used to increase the extraction efficiency by increasing the contact surface between the analytes and the extraction solvents. The choice of extraction solvent is important because for the formation of the cloudy solution in the presence of the dispersive solvent; the extraction solvent must have a density greater than water's density (Almeida et al. 2017a; Vichapong et al. 2016a).

Figure 13.2 shows the schematic representation of IL-DLLME.

13.2.2 IL-SDME (Ionic Liquid Single Drop Microextraction)

IL-SDME, or ionic liquid single drop microextraction, replaced the conventional method of analyte preconcentration and extraction from discrete samples, i.e., LLE and SPE, that required excessive use of solvents for extraction, which evaporated at last to concentrate the analytes into a known amount of solvent – a tedious and tiresome process. In order to overcome the problem of solvent evaporation, an alternative method was proposed by Liu and Dasgupta in 1995 called the single drop microextraction (SDME) technique (Liu and Dasgupta 1996). The basic principle of the technique is the utilization of extraction solvent in droplet form suspended from the tip of a microsyringe needle. These suspended microdroplets extract the target analyte from the aqueous solution, thereby reducing the chances of interference due to sample mixing. The stability of the suspended drop is also influenced by the shape of the needle tip, or the drop holder (if a microsyringe is not used). The major factor in SDME extraction is the stability of the microdrop by modifications of the solvent holder so that the method becomes faster, potent, and solvent free. The type of extraction solvent used influences the choice of the final determinative technique. Therefore, the extractant solvent used should have comparatively low solubility in water, low toxicity for both environment and human health, and good stability of drop. Besides these required qualities, the extraction solvent should also be able to extract analytes efficiently, and peak chromatograms should be clear enough for proper differentiation.

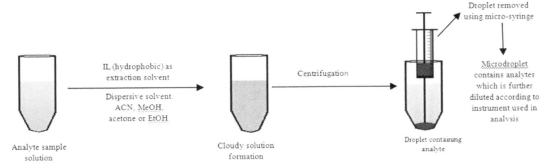

FIGURE 13.2 Schematic diagrammatic representation of IL-DLLME (IL-based dispersive liquid-liquid microextraction).

13.2.3 IL-SPME (Ionic Liquid Solid-Phase Microextraction)

Solid-phase microextraction (SPME) comes under the green analytical method due to the fact that it does not use chemicals or solvents for the extraction of analytes from the diverse sample. In 1989, Belardi and Pawliszyn established the concept of SPME, which opened the area for its association with a variety of commercially available sorbent coatings depending on the analyte, such as polyacrylate, carboxen, divinylbenzene (DVB) called PDMS carboxen, and polydimethylsiloxane (PDMS) (Ho et al. 2011; Souza Silva et al. 2013). This works on the principle of absorption or adsorption of the target analyte on the fibre that is coated with a thin layer of any polymer designated for the target analyte. After equilibrium is attained between all the phases, these fibres are directly analysed on any sophisticated analytical instrument. IL can be employed as an effective SPME fibre coating due to its regulate able physical and chemical properties, such as variable viscosity, tuneable salvation interactions, high thermal stability, and negligible vapour pressure. In broad spectra, SPME is an expeditious, straightforward, and solvent-free method with high sensitive when coupled with a suitable technique like gas chromatography. IL-coated sorbent can be combined with either HS-SPME or direct immersion SPME (DI-SPME), depending on the requirements. Merdivan et al. developed the first polymeric ionic liquid (PIL) by use of monomer (VBHDIM-NTf$_2$ IL) and ((DVBIM)$_2$C$_{12}$-2NTf$_2$), forming benzyl functionalized cross-linked PILs for extraction and quantification of seven volatile polycyclic aromatic hydrocarbons (PAHs) in environmental water samples using gas chromatography with flame ionization detector (GC-FID) (Merdivan et al. 2017).

13.2.4 IL-SBSE (Ionic Liquid Stir Bar Sorptive Extraction)

Baltussen and co-workers (Baltussen et al. n.d.) introduced the concept of stir bar sorptive extraction (SBSE) in 1999, which was very similar to SPME as it also works on the phenomenon of absorption and adsorption of the target analyte molecule on the sorptive material, usually consisting of polydimethylsiloxane (PDMS) or C-18 that is placed on glass that covers a magnet. The magnetic bar is stirred continuously unless equilibrium is obtained between the target analyte on sorbent material and sample matrix. After the extraction, the magnetic bar is removed and transferred to a vial for analysis of the target analyte using sophisticated analytical tools (Camino-Sánchez et al. 2014).

Fan and his co-workers synthesized an IL using (methacryloxypropyl) trimethoxysilane (KH-570) instead of PDMS or C-18 as a bridging agent due to its unique properties.The IL synthesized was 1-allylimidazolium tetrafluoroborate ([AIM][BF$_4$]), used for the extraction and quantification of nonsteroidal anti-inflammatory drugs (NSAIDs) using HPLC-UV (Fan et al. 2014). With the fast-moving research for improvement and development, two or more techniques can be clubbed together for significant enhancement of the extraction protocol, thereby reducing the extraction time, cost, and sensitivity. In this perspective, SBSE and DLLME are conflated, introducing stir bar dispersive liquid microextraction (SBDME) with the advent of MIL and a neodymium-core magnetic stirrer as the extraction phase by Chisvert et al. (Chisvert et al. 2017). At a higher stirring rate, the MIL is dispersed into the solution in accordance with DLLME principles, and at a lower stirring rate, it acts according to SBSE principles. After the extraction is done, MILs are easily retrieved from the solution using magnets. This was applied for the extraction and determination lipophilic organic UV filters from the environmental water samples.

Another similar combination of two techniques was introduced by Benede et al. (Benedé et al. 2018) for the determination of PAHs in water samples. The technique appeared to be more successful than the previously used extraction techniques as it required less sample processing time and manipulation in samples. Being an emerging technique of microextraction, it is still in its evolving phase, so much application is not yet reported in literature.

13.2.5 IL-SCSE (Ionic Liquid Stir Cake Sorptive Extraction)

In 2011, SBSE was improved by placing a stationary phase in a holder contained of iron and rest steps similar to stir bar sorptive microextraction. It was termed stir cake sorptive extraction (SCSE), which are monolithic cakes designed and prepared properly according to the requirement of the target analyte (He et al. 2012). These designed monolithic cakes are added to the solution and stirred properly after the extraction is over; they are retrieved and run directly to any sophisticated analytical instrument. A PIL monolith formed for the analysis and determination of trace benzimidazoles residues in water, milk, and honey samples in the presence of N,N-dimethylformamide of IL 1-allyl-3-methylimidazoliumbis [(trifluoro methyl)sulfonyl] imide (AMII) and divinylbenzene (DVB) by in-situ copolymerization as a new approach in SCSE for determining trace benzimidazoles (Bas) residues in water, milk, and honey samples. There are few applications available in literature regarding SCSE that provide better extraction for analytes with satisfactory results (Wang et al. 2014).

The application of an IL-based SBSE method is yet to be discovered. The literature reports suggested some of its uses in determining inorganic elements, heavy metals, preservatives used in fruit juices and tea drinks, and oestrogen-level analysis in water samples. For the measurement of antimony in environmental matrices, monolith of 3-(1-ethyl imidazolium-3-yl) propyl-methacrylamido bromide and ethylene dimethacrylate by in-situ polymerization is preferred because the cross linker provides stability to its three-dimensional structure and thus has a good life span. Antimony is an analyte of concern due to its biological toxicity, and a PIL-based SBSE method proved to be a good extraction method, with limits of detection (LODs) as low as 0.048 μg/L (Zhang et al. 2016). A monolith synthesised by the copolymerization reaction between 1-ally-3-vinylimidazolium chloride (AV) and divinylbenzene (DVB) with the help of porogen solvent containing 1-propanol and 1,4-butanediol as crosslinker is used for the analysis of preservatives (Chen and Huang 2016). The above proposed method for the preservative showed better analytical characteristics compared to other analytical methods with increased sensitivity, better reproducibility, good cost-effectiveness, and environmental friendliness.

A monolith cake consisting of PIL-based poly (1-ally-3-vinylimidazolium chloride-co-ethylene dimethacrylate)-AVED for determining oestrogen in water samples showed good values for analytical parameters, providing a wide linear range, low values of LODs, acceptable reproducibility, and better recoveries for real water samples. The method gives LODs in the range of 0.024-0.057 mg/L and limits of quantification (LOQs) with a range 0.08-0.19 mg/L (Chen et al. 2016). The novel SBSE solvent is synthesised with IL 1-ally-3-methylimidazolium chloride as a monomer with in-situ copolymerization with ethylene dimethacrylate in the presence of 1-propanol and dimethylformamide as porogen for inorganic element determination (Huang et al. 2012).

13.3 Application of IL-Based Microextraction Techniques

IL-based microextraction has been successfully utilized for the extraction and determination of various organic and inorganic analytes from a wide spectra of matrices, including water, food, environmental samples, cosmetics, biological samples, etc., in analytical chemistry due to its tremendous use increasing day by day. It moves toward GAC by replacing traditional solvents, which were drained in litres to the environment, and by dealing with a wide range of analytes from various samples. Analysts are keenly interested in this technique due to its harmless nature toward the environment and human beings. DLLME combined with IL provides a low limit of detection, high selectivity for the desired analyte, high recovery percentage, and high enrichment factor, besides providing various other advantages like reduction in the amount of organic solvents, reduction of cost, reduction in the number of steps for sample preparation, and increased efficiency and environmental friendliness. IL-based microextraction reduces both air and water pollution because several studies have been conducted on the toxic effect of IL on aquatic environments, and the results showed that IL is less toxic than traditional solvents to both the environment and human health. Being low volatile by nature, these ILs do not contribute to air pollution.

The drawbacks associated with the use of organic solvents were overcome by non-volatile IL use, as it provides advantages to the already-established method by being accurate, reproducible, time sovereign, and linear over a wide concentration range. For better results of experiment extraction efficiency, the variables were optimized using a multifaceted strategy. Several recent studies have reported using experimental designs for screening diverse parameters – using Plackett-Burman design (PBD) and central composite design (CCD) – for optimization of major factors that show direct relation with extraction efficacy. Thus, it is clear that IL-DLLME has been widely used in the study of diverse organic and inorganic analytes in variable research areas of environmental, food, and biomedical studies. DLLME-based research papers use mainly hydrophobic IL for most analysis, but in some cases of the DLLME method, hydrophilic IL is used. It is clearly visible from the number of research articles published in this field that the scientific community is highly interested in different IL-DLLME analytical applications and novel IL-DLLME technical solutions.

Since the discovery of SDME, it has been the most popular solvent-based microextraction technique for sample preparation in chemical, pharmaceutical, clinical, biological, food, and forensic analysis. Unending strong interest stimulated new development as an alternative to organic solvent extractants. Hydrophilic and hydrophobic ILs have found numerous applications for all types of analytes in varied samples with different modes of SDME using ILs as an extracting solvent. In the extraction process, distribution of the featured analyte under the optimized condition occurred between the donor aqueous phase and the acceptor organic solvent drop. In view of the fact that a large portion of the target analyte is unmoving from the donor phase, inspection is carried for migration of matrix substances. The debarring of matrix substances provides better sample clean-up in a non-equilibrium state of reaction between the analyte and extractant. SDME appears to have special value in sample clean-up due to rapid extraction and a large enrichment factor, both attained delinquent of the great surface area of the extraction solvent.

IL-SDME was successfully applied for the extraction and determination of pesticides in the agricultural sector and drugs from household drainage and others, as they are discharged to the environment accidentally or deliberately, leading to serious threat to human health and animals. Even various research papers reported about the unbeaten utilization of IL-SDME in PAHs and phthalate esters due to their substantial applications in both industrial and domestic products. This class of contamination is highly toxic for the environment and human health if released intentionally or unintentionally; therefore, its monitoring is of utmost importance. Several human health-related problems, such as cancer, asthma, and cardiovascular diseases, have been reported as side effects of long-term exposure. Different sources of the emissions in the environment lead to different types of health issues; for example, sources can be combustion processes like volcanic eruptions, forest fires, biomass burning, waste incineration, and various industrial and anthropogenic activities.

The inherent inconveniences due to LLE were overcome by SPME to accomplish extraction and preconcentration of analytes through the use of less volume of toxic and environmentally unfriendly solvents, reduction in waste generation, and speeding up of the slow and labour-intensive workup. IL-SPME adds up in the list of analysts as an extraction method and is widely used in scientific and technological fields. SPME is used as an extraction technique in determining various classes of pollutants, ranging from the pesticides used in agricultural farmland and their seepage and contamination in different water bodies, to analysis of PAHs in environmental samples. SPME is mainly utilized in biological samples as they are highly sensitive matrices and can be easily degraded in no time. SPME is mainly performed on solid matrices because the matrix interference in this technique is minimal. Table 13.3 shows the application of IL in microextraction.

13.4 Conclusion

Due to its unique and tunable physical properties, IL is becoming popular in sample preparation technology. IL-based microextraction is gaining attention due to the availability of designing and synthesizing target-specific ILs and expanding the research field of ILs to microextraction technology. In accordance with the GAC principle, ILs have gained popularity due to miniaturization in the amount used for extraction. ILs are costlier than the traditionally available extraction solvents, so an approach is needed in future perspectives for lowering the cost of ILs in order to replace the traditional solvents with

TABLE 13.3

Applications of Ionic Liquid-Based Microextraction

Analyte	Sample	Ionic Liquid	LODs	Recovery (%)	Instrument	Reference
Aromatic amines	Water	[C$_6$MIM][PF$_6$]	0.17–0.49 g/L	92.2–119.3	HPLC	(Zhou et al. 2009)
Bisphenol A	Human fluids	ABS	–	–	HPLC-MS	(Passos et al. 2012)
Cadmium	Water	[C$_6$MIM][PF$_6$]	7.4 ng/L	87.2–106	ETAAS	(S. Li et al. 2009)
Emerging contaminants	Water	[C$_4$MIM][PF$_6$]+LiNTf$_2$	0.1–55.8 µg/L	91–110	HPLC-UV/Vis	(Yao et al. 2011)
Emerging pollutants	Water	[C$_4$MIM][PF$_6$]	3.2–7.2 µg/L	85–116	HPLC-UV	(Abujaber et al. 2018b)
Fluoroquinolones and NSAIDs	Water	ABS	–	–	HPLC-DAAD	(Almeida et al. 2017b)
Neonicotinoids	Honey	[C$_4$MIM][PF$_6$]	0.01 mg/L	86–100	HPLC-DAAD	(Vichapong et al. 2016b)
Organophosphate esters	Water	[HMIM][FAP]	0.13–7.40 ng/L	84.0–108	GC-MS	(Shi et al. 2016)
Organophosphorus pesticides	Water	[C$_4$MIM][Cl]+LiNTf$_2$	5–16 ng/L	97–113	GC-MS	(Cacho et al. 2018)
Organophosphorus pesticides and aromatic compounds	Tap, rain, and river water	[C$_4$MIM][NTf$_2$]	0.01–1.0 µg/L	82.7–118.3	HPLC-UV	(Wang et al. 2016)
Pollutants	Water	ABS	–	–	HPLC-UV	(Dinis et al. 2018)
Polycyclic aromatic hydrocarbons	Water	[C$_8$MIM][PF$_6$]	0.0005–0.88 mg/L	83.5–118	HPLC	(Q. Zhou and Gao 2014)
Sulfonylurea herbicides	Wine samples	[C$_6$MIM][PF$_6$]	3.2–6.6 µg/kg	–	HPLC	(Gure et al. 2015)
Triazine herbicides	Water	[C$_8$MIM][PF$_6$]	0.05–0.06 mg/L	–	HPLC	(Q. X. Zhou and Gao 2014)
Lipids	Serum	[C$_4$MIM][PF$_6$]	0.012–0.034 ng/mL	90.9–114	LC-MS/MS	(Panchal et al. 2016)
Bisphenols	Thermal papers	[C$_4$MIM][PF$_6$]	0.93–1.25 µg/kg	99–101	LC-MS/MS	(Asati et al. 2017)
Neurotransmitters	Rat plasma, brain, and cell	[C$_4$MIM][PF$_6$]	0.021–0.978 µg/L	88–128	LC-MS/MS	(Asati et al. 2017)
Zinc	Water and milk	[HPy][PF$_6$]	0.22 µg/L	97.5–102.5	FAAS	(Jha, Singh, Pant, et al. 2018)
Pesticides	Alcoholic and non-alcoholic drinks	[C$_4$MIM][PF$_6$]	0.001–0.348 µg/L	92–138	LC-MS/MS	(Jha, Singh, Kumari, et al. 2018)

(Continued)

TABLE 13.3 (Continued)
Applications of Ionic Liquid-Based Microextraction

Analyte	Sample	Ionic Liquid	LODs	Recovery (%)	Instrument	Reference
Antidepressant drugs	Human plasma	[C$_8$MIM][PF$_6$]	6-10 μg/L	90-92	HPLC-PDA	(Vaghar-Lahijani et al. n.d.)
3-Hydroxybenzo[a]pyrene	Urine	[C$_8$MIM][PF$_6$] Disperser solvent:Acetone	0.2 pg/mL	92.0	HPLC-HRMS/MS	(Hu et al. 2016)
Alkaloids	Macleayacordata	[C$_6$MIM][PF$_6$]+[K][PF$_6$]	0.080 ng/mL	86.42-112.48	UPLC-MS/MS	(L. Li et al. 2017)
Antihypertensive drugs	Urine	[C$_8$MIM][PF$_6$] Disperser solvent:Acetone	–	83.5-89.6	HPLC-DAD	(Z. Li et al. 2013)
Celastrol	Urine	[C$_6$MIM][PF$_6$] Disperser solvent:MeOH	1.6 μg/L	93.2-109.3	HPLC	(Sun et al. 2011)
Curcuminoids	Curcuma longa	[C$_6$MIM][PF$_6$]	3000 ng/L	93	HPLC	(Shu et al. 2016)
Danazol	Serum	[C$_6$MIM][PF$_6$] Disperser solvent: NaCl	0.054 μg/mL	90.5-103.4	UV spectroscopy	(Gong and Zhu 2016)
DNA	Aqueous solution of DNA	[C$_{16}$C$_3$(OH)$_2$Im][Br] +[Li][NTf$_2$]	–	97	HPLC-UV	(T. Li et al. 2013)
Drugs	Urine	[C$_4$MIM][PF$_6$] Disperser solvent:MeOH	8.3-32 ng/L	99.6-107	LC-UV	(Cruz-Vera et al. 2009)
Emodin and its metabolites	Urine	[C$_6$MIM][PF$_6$] Disperser solvent:ACN	0.1-1 ng/mL	94.8-103	HPLC-UV	(Tian et al. 2012)
Ofloxacin	Urine and plasma	[C$_6$MIM][PF$_6$]+ [Na][PF$_6$]	0.029 μg/L	91.5-106.9 (urine) 89.5-97.1 (plasma)	Spectrofluorometric	(Zeeb et al. n.d.)
DNA	Biological samples	PIL-SPME	–	–	qPCR	(Nacham, Clark, and Anderson 2016)
DNA	Biological samples containing mRNA	[(C$_8$)$_3$BnN][FeCl$_3$], [(C$_{16}$BnIM)$_2$C$_{12}$] [TFSI], [P$_{6,6,6,14}$] [FeCl$_4$]	–	–	RT-qPCR	(Nacham et al. 2017)

Analyte	Sample	IL	LOD	Recovery	Technique	Reference
DNA	Mycobacterium in artificial sputum samples	PIL-IMSA	–	–	IMSA	(Varona et al. 2018)
Methylcyclopentadienyl-manganese tricarbonyl (MMT)	Water and gasoline	[C₄MIM][PF₆]	10.0 ng/L	81.2–101.0	ETAAS	(Rahmani and Kaykhaii 2011)
Copper	Water and food	[C₄MIM][PF₆]	0.15 μg/L	98.3–105	Spectrophoto meter	(Wen et al. 2011)
Trimethylamine	Fish	CdSe/ZnS QDs +[C₆MIM][PF₆]	0.014 mg/L	92–105.6	Fluorimetric	(Carrillo-Carrión et al. 2012)
2,4,6-Trichloroanisole	Wine and water	[C₆MIM][NTf₂]	0.2 ng/L	–	IMS	(Márquez-Sillero et al. 2011)
2,4,6-Trichloroanisole	Wine	[C₆MIM][NTf₂]	0.01 ng/L	–	MCC-IMS	(Márquez-Sillero, Cárdenas, and Valcárcel 2011)
Benzophenone-3	Human urine	[C₆MIM][PF₆]	1.3 ng/mL	–	HPLC-DAD	(Vidal et al. 2007)
UV filters	Surface water samples	[C₆MIM][PF₆]	0.06–3.0 μg/L	92–115	HPLC-UV	(Vidal et al. 2010)
Essential oil	Forsythia suspensa	[C₈MIM][PF₆]	–	–	GC-FID	(Yang et al. 2014)

ILs. In this chapter, different extraction techniques employing ILs were discussed with necessary analytical parameters. It is clear that within no time ILs found a significant place in the extraction of a variety of analytes from diverse sample types with generation of less waste in the laboratory, thereby making the processes environmentally friendly.

REFERENCES

Abujaber, Feras, Mohammed Zougagh, Shehdeh Jodeh, Ángel Ríos, Francisco Javier Guzmán Bernardo, and Rosa C. Rodríguez Martín-Doimeadios. "Magnetic Cellulose Nanoparticles Coated with Ionic Liquid as a New Material for the Simple and Fast Monitoring of Emerging Pollutants in Waters by Magnetic Solid Phase Extraction." *Microchemical Journal* 137 (March, 2018a): 490–5. doi: 10.1016/j.microc.2017.12.007

Abujaber, Feras, Mohammed Zougagh, Shehdeh Jodeh, Ángel Ríos, Francisco Javier Guzmán Bernardo, and Rosa C. Rodríguez Martín-Doimeadios. "Magnetic Cellulose Nanoparticles Coated with Ionic Liquid as a New Material for the Simple and Fast Monitoring of Emerging Pollutants in Waters by Magnetic Solid Phase Extraction." *Microchemical Journal* 137 (March, 2018b): 490–5. doi: 10.1016/j.microc.2017.12.007.

Almeida, Hugo F.D., Mara G. Freire, and Isabel M. Marrucho. "Improved Monitoring of Aqueous Samples by the Preconcentration of Active Pharmaceutical Ingredients Using Ionic-Liquid-Based Systems." *Green Chemistry* 19, no. 19 (2017a): 4651–9. doi: 10.1039/c7gc01954h.

Almeida, Hugo F.D., Mara G. Freire, and Isabel M. Marrucho. "Improved Monitoring of Aqueous Samples by the Preconcentration of Active Pharmaceutical Ingredients Using Ionic-Liquid-Based Systems." *Green Chemistry* 19, no. 19 (2017b): 4651–9. doi: 10.1039/c7gc01954h.

Anastas, Paul T., and Evan S. Beach. "Green Chemistry: The Emergence of a Transformative Framework." *Green Chemistry Letters and Reviews* 1, no. 1 (2007): 9–24. doi: 10.1080/17518250701882441.

Asati, Ankita, G.N.V. Satyanarayana, Smita Panchal, Ravindra Singh Thakur, Nasreen G. Ansari, and Devendra K. Patel. "Ionic Liquid Based Vortex Assisted Liquid–Liquid Microextraction Combined with Liquid Chromatography Mass Spectrometry for the Determination of Bisphenols in Thermal Papers with the Aid of Response Surface Methodology." *Journal of Chromatography A* 1509 (August, 2017): 35–42. doi: 10.1016/j.chroma.2017.06.040.

Baltussen, Erik, Pat Sandra, Frank David, and Carel Cramers. "Stir Bar Sorptive Extraction SBSE, a Novel Extraction Technique for Aqueous Samples: Theory and Principles." n.d.

Benedé, Juan L., Jared L. Anderson, and Alberto Chisvert. "Trace Determination of Volatile Polycyclic Aromatic Hydrocarbons in Natural Waters by Magnetic Ionic Liquid-Based Stir Bar Dispersive Liquid Microextraction." *Talanta* 176 (January, 2018): 253–61. doi: 10.1016/j.talanta.2017.07.091.

Cacho, Juan Ignacio, Natalia Campillo, Pilar Viñas, and Manuel Hernández-Córdoba. "In Situ Ionic Liquid Dispersive Liquid-Liquid Microextraction Coupled to Gas Chromatography-Mass Spectrometry for the Determination of Organophosphorus Pesticides." *Journal of Chromatography A* 1559 (July, 2018): 95–101. doi: 10.1016/j.chroma.2017.12.059.

Camino-Sánchez, F. J., Rocío Rodríguez-Gómez, A. Zafra-Gómez, A. Santos-Fandila, and José Luis Vílchez. "Stir Bar Sorptive Extraction: Recent Applications, Limitations and Future Trends." *Talanta* 130 (2014): 388–99. doi: 10.1016/j.talanta.2014.07.022.

Carrillo-Carrión, Carolina, Bartolomé M. Simonet, and Miguel Valcárcel. "(CdSe/ZnS QDs)-Ionic Liquid-Based Headspace Single Drop Microextraction for the Fluorimetric Determination of Trimethylamine in Fish." *Analyst* 137, no. 5 (2012): 1152–9. doi: 10.1039/c2an15914g.

Chen, Lei, and Xiaojia Huang. "Preparation of a Polymeric Ionic Liquid-Based Adsorbent for Stir Cake Sorptive Extraction of Preservatives in Orange Juices and Tea Drinks." *Analytica Chimica Acta* 916 (April, 2016): 33–41. doi: 10.1016/j.aca.2016.02.030.

Chen, Lei, Meng Mei, Xiaojia Huang, and Dongxing Yuan. "Sensitive Determination of Estrogens in Environmental Waters Treated with Polymeric Ionic Liquid-Based Stir Cake Sorptive Extraction and Liquid Chromatographic Analysis." *Talanta* 152 (May, 2016): 98–104. doi: 10.1016/j.talanta.2016.01.044.

Chisvert, Alberto, Juan L. Benedé, Jared L. Anderson, Stephen A. Pierson, and Amparo Salvador. "Introducing a New and Rapid Microextraction Approach Based on Magnetic Ionic Liquids: Stir Bar Dispersive Liquid Microextraction." *Analytica Chimica Acta* 983 (August, 2017): 130–40. doi: 10.1016/j.aca.2017.06.024.

Cruz-Vera, Marta, Rafael Lucena, Soledad Cárdenas, and Miguel Valcárcel. "One-Step in-Syringe Ionic Liquid-Based Dispersive Liquid-Liquid Microextraction." *Journal of Chromatography A* 1216, no. 37 (2009): 6459–65. doi:10.1016/j.chroma.2009.07.040.

Dinis, Teresa B.V., Helena Passos, Diana L.D. Lima, Ana C.A. Sousa, João A.P. Coutinho, Valdemar I. Esteves, and Mara G. Freire. "Simultaneous Extraction and Concentration of Water Pollution Tracers Using Ionic-Liquid-Based Systems." *Journal of Chromatography A* 1559 (July, 2018): 69–77. doi:10.1016/j.chroma.2017.07.084.

Fan, Wenying, Xiangju Mao, Man He, Beibei Chen, and Bin Hu. "Development of Novel Sol-Gel Coatings by Chemically Bonded Ionic Liquids for Stir Bar Sorptive Extraction - Application for the Determination of NSAIDS in Real Samples." *Analytical and Bioanalytical Chemistry* 406, no. 28 (2014): 7261–73. doi:10.1007/s00216-014-8141-9.

Freire, Mara G., Ana Filipa M. Cláudio, João M.M. Araújo, João A.P. Coutinho, Isabel M. Marrucho, José N. Canongia Lopes, and Luís Paulo N. Rebelo. "Aqueous Biphasic Systems: A Boost Brought about by Using Ionic Liquids." *Chemical Society Reviews* 41, no. 14 (2012): 4966–95. doi:10.1039/c2cs35151j.

Gong, Aiqin, and Xiashi Zhu. "Miniaturized Ionic Liquid Dispersive Liquid-Liquid Microextraction in a Coupled-Syringe System Combined with UV for Extraction and Determination of Danazol in Danazol Capsule and Mice Serum." *Spectrochimica Acta - Part A: Molecular and Biomolecular Spectroscopy* 159 (April, 2016): 163–8. doi:10.1016/j.saa.2016.01.053.

Gure, Abera, Francisco J. Lara, Ana M. García-Campaña, Negussie Megersa, and Monsaluddel Olmo-Iruela. "Vortex-Assisted Ionic Liquid Dispersive Liquid-Liquid Microextraction for the Determination of Sulfonylurea Herbicides in Wine Samples by Capillary High-Performance Liquid Chromatography." *Food Chemistry* 170 (March, 2015): 348–53. doi:10.1016/j.foodchem.2014.08.065.

Han, Dandan, Baokun Tang, Yu Ri Lee, and Kyung Ho Row. "Application of Ionic Liquid in Liquid Phase Microextraction Technology." *Journal of Separation Science* 00 (2012): 1–13. doi:10.1002/jssc.201200486.

He, Xiaowen, Fucheng Zhang, and Ye Jiang. "An Improved Ionic Liquid-Based Headspace Single-Drop Microextraction-Liquid Chromatography Method for the Analysis of Camphor and Trans-Anethole in Compound Liquorice Tablets." *Journal of Chromatographic Science* 50, no. 6 (2012): 457–63. doi:10.1093/chromsci/bms038.

Ho, Tien D., Anthony J. Canestraro, and Jared L. Anderson. "Ionic Liquids in Solid-Phase Microextraction: A Review." *Analytica Chimica Acta* 695 (2011): 18–43. doi: 10.1016/j.aca.2011.03.034.

Hu, Huan, Baizhan Liu, Jun Yang, Zuomin Lin, and Wuer Gan. "Sensitive Determination of Trace Urinary 3-Hydroxybenzo[a]Pyrene Using Ionic Liquids-Based Dispersive Liquid-Liquid Microextraction Followed by Chemical Derivatization and High Performance Liquid Chromatography-High Resolution Tandem Mass Spectrometry." *Journal of Chromatography B: Analytical Technologies in the Biomedical and Life Sciences* 1027 (August, 2016): 200–6. doi:10.1016/j.jchromb.2016.05.041.

Huang, Xiaojia, Linli Chen, Dongxing Yuan, and Shangshang Bi. "Preparation of a New Polymeric Ionic Liquid-Based Monolith for Stir Cake Sorptive Extraction and Its Application in the Extraction of Inorganic Anions." *Journal of Chromatography A* 1248 (July, 2012): 67–73. doi:10.1016/j.chroma.2012.06.004.

Jha, Rakesh Roshan, Chetna Singh, Aditya B. Pant, and Devendra K. Patel. "Ionic Liquid Based Ultrasound Assisted Dispersive Liquid-Liquid Micro-Extraction for Simultaneous Determination of 15 Neurotransmitters in Rat Brain, Plasma and Cell Samples." *Analytica Chimica Acta* 1005 (April, 2018): 43–53. doi:10.1016/j.aca.2017.12.015.

Jha, Rakesh Roshan, Nivedita Singh, Rupender Kumari, and Devendra Kumar Patel. "Dispersion-Assisted Quick and Simultaneous Extraction of 30 Pesticides from Alcoholic and Non-Alcoholic Drinks with the Aid of Experimental Design." *Journal of Separation Science* 41, no. 7 (2018): 1625–34. doi:10.1002/jssc.201701155.

Li, Linqiu, Mingyuan Huang, Junli Shao, Bokun Lin, and Qing Shen. "Rapid Determination of Alkaloids in MacleayaCordata Using Ionic Liquid Extraction Followed by Multiple Reaction Monitoring UPLC–MS/MS Analysis." *Journal of Pharmaceutical and Biomedical Analysis* 135 (February, 2017): 61–6. doi:10.1016/j.jpba.2016.12.016.

Li, Shengqing, Shun Cai, Wei Hu, Hao Chen, and Hanlan Liu. "Ionic Liquid-Based Ultrasound-Assisted Dispersive Liquid-Liquid Microextraction Combined with Electrothermal Atomic Absorption

Spectrometry for a Sensitive Determination of Cadmium in Water Samples." *Spectrochimica Acta - Part B Atomic Spectroscopy* 64, no. 7 (2009): 666–71. doi: 10.1016/j.sab.2009.05.023.

Li, Tianhao, Manishkumar D. Joshi, Donald R. Ronning, and Jared L. Anderson. "Ionic Liquids as Solvents for in Situ Dispersive Liquid-Liquid Microextraction of DNA." *Journal of Chromatography A* 1272 (January, 2013): 8–14. doi: 10.1016/j.chroma.2012.11.055.

Li, Zhiling, Fan Chen, Xuedong Wang, and Chengjun Wang. "Ionic Liquids Dispersive Liquid-Liquid Microextraction and High-Performance Liquid Chromatographic Determination of Irbesartan and Valsartan in Human Urine." *Biomedical Chromatography* 27, no. 2 (2013): 254–8. doi: 10.1002/bmc.2784.

Liu, Hanghui, and Purnendu K. Dasgupta. "Analytical Chemistry in a Drop. Solvent Extraction in a Microdrop." *Accelerated Articles* 68 (1996): 1817–21.

Marcinkowska, Renata, Karolina Konieczna, Łukasz Marcinkowski, J. Namieśnik, and Adam Kloskowski. "Application of Ionic Liquids in Microextraction Techniques: Current Trends and Future Perspectives." *TrAC-Trends in Analytical Chemistry* 119 (2019): 115614. doi: 10.1016/j.trac.2019.07.025.

Marcinkowska, Renata, Jacek Namieśnik, and Marek Tobiszewski. "Green and Equitable Analytical Chemistry." *Current Opinion in Green and Sustainable Chemistry* 19 (2019): 19–23. doi: 10.1016/j.cogsc.2019.04.003.

Márquez-Sillero, Isabel, Eva Aguilera-Herrador, Soledad Cárdenas, and Miguel Valcárcel. "Determination of 2,4,6-Tricholoroanisole in Water and Wine Samples by Ionic Liquid-Based Single-Drop Microextraction and Ion Mobility Spectrometry." *Analytica Chimica Acta* 702, no. 2 (2011): 199–204. doi: 10.1016/j.aca.2011.06.046.

Márquez-Sillero, Isabel, Soledad Cárdenas, and Miguel Valcárcel. "Direct Determination of 2,4,6-Tricholoroanisole in Wines by Single-Drop Ionic Liquid Microextraction Coupled with Multicapillary Column Separation and Ion Mobility Spectrometry Detection." *Journal of Chromatography A* 1218, no. 42 (2011): 7574–80. doi: 10.1016/j.chroma.2011.06.032.

Merdivan, Melek, Verónica Pino, and Jared L. Anderson. "Determination of Volatile Polycyclic Aromatic Hydrocarbons in Waters Using Headspace Solid-Phase Microextraction with a Benzyl-Functionalized Crosslinked Polymeric Ionic Liquid Coating." *Environmental Technology (United Kingdom)* 38, no. 15 (2017): 1897–1904. doi: 10.1080/09593330.2016.1240242.

Nacham, Omprakash, Kevin D. Clark, and Jared L. Anderson. "Extraction and Purification of DNA from Complex Biological Sample Matrices Using Solid-Phase Microextraction Coupled with Real-Time PCR." *Analytical Chemistry* 88, no. 15 (2016): 7813–20. doi: 10.1021/acs.analchem.6b01861.

Nacham, Omprakash, Kevin D. Clark, Marcelino Varona, and Jared L. Anderson. "Selective and Efficient RNA Analysis by Solid-Phase Microextraction." *Analytical Chemistry* 89, no. 20 (2017): 10661–6. doi: 10.1021/acs.analchem.7b02733.

Panchal, Smita, Ankita Asati, G.N.V. Satyanarayana, Alok Raghav, Jamal Ahmad, and Devendra K. Patel. "Ionic Liquid Based Microextraction of Targeted Lipids from Serum Using UPLC-MS/MS with a Chemometric Approach: A Pilot Study." *RSC Advances* 6, no. 94 (2016): 91629–40. doi: 10.1039/c6ra17408f.

Passos, Helena, Ana C.A. Sousa, M. Ramiro Pastorinho, António J.A. Nogueira, Luís Paulo N. Rebelo, João A.P. Coutinho, and Mara G. Freire. "Ionic-Liquid-Based Aqueous Biphasic Systems for Improved Detection of Bisphenol A in Human Fluids." *Analytical Methods* 4, no. 9 (2012): 2664–7. doi: 10.1039/c2ay25536g.

Rahmani, Mashaallah, and Massoud Kaykhaii. "Determination of Methylcyclopentadienyl-Manganese Tricarbonyl in Gasoline and Water via Ionic-Liquid Headspace Single Drop Microextraction and Electrothermal Atomic Absorption Spectrometry." *Microchimica Acta* 174, no. 3 (2011): 413–19. doi: 10.1007/s00604-011-0648-6.

Roth, Michal. "Partitioning Behaviour of Organic Compounds between Ionic Liquids and Supercritical Fluids." *Journal of Chromatography A* 1216 (2009): 1816–80. doi: 10.1016/j.chroma.2008.10.032.

Ruiz-Aceituno, Laura, María L. Sanz, and Lourdes Ramos. "Use of Ionic Liquids in Analytical Sample Preparation of Organic Compounds from Food and Environmental Samples." *TrAC - Trends in Analytical Chemistry* 43 (2013). doi: 10.1016/j.trac.2012.12.006.

Shi, Fengqiong, Jingfu Liu, Kang Liang, and Rui Liu. "Tris(Pentafluoroethyl)Trifluorophosphate-Basd Ionic Liquids as Advantageous Solid-Phase Micro-Extraction Coatings for the Extraction of Organophosphate Esters in Environmental Waters." *Journal of Chromatography A* 1447 (May, 2016): 9–16. doi: 10.1016/j.chroma.2016.04.021.

Shu, Yang, Mingcen Gao, Xueying Wang, Rusheng Song, Jun Lu, and Xuwei Chen. "Separation of Curcuminoids Using Ionic Liquid Based Aqueous Two-Phase System Coupled with in Situ Dispersive Liquid-Liquid Microextraction." *Talanta* 149 (March, 2016): 6–12. doi:10.1016/j.talanta.2015.11.009.

Souza Silva, Erica A., Sanja Risticevic, and Janusz Pawliszyn. "Recent Trends in SPME Concerning Sorbent Materials, Configurations and in Vivo Applications." *TrAC - Trends in Analytical Chemistry* 43 (2013): 24–36. doi:10.1016/j.trac.2012.10.006.

Sun, Jian Nan, Yan Ping Shi, and Juan Chen. "Ultrasound-Assisted Ionic Liquid Dispersive Liquid-Liquid Microextraction Coupled with High Performance Liquid Chromatography for Sensitive Determination of Trace Celastrol in Urine." *Journal of Chromatography B: Analytical Technologies in the Biomedical and Life Sciences* 879, no. 30 (2011): 3429–33. doi:10.1016/j.jchromb.2011.09.019.

Tian, Jie, Xuan Chen, and Xiaohong Bai. "Comparison of Dispersive Liquid-Liquid Microextraction Based on Organic Solvent and Ionic Liquid Combined with High-Performance Liquid Chromatography for the Analysis of Emodin and Its Metabolites in Urine Samples." *Journal of Separation Science* 35, no. 1 (2012): 145–52. doi:10.1002/jssc.201100729.

Tobiszewski, Marek, Mariusz Marć, Agnieszka Gałuszka, and Jacek Namieśnik. "Green Chemistry Metrics with Special Reference to Green Analytical Chemistry." *Molecules* 20 (2015): 10928–46. doi:10.3390/molecules200610928.

Vaghar-Lahijani, Gholamreza, Parviz Aberoomand-Azar, Mohammad Saber-Tehrani, and Mojtaba Soleimani. "Application of Ionic Liquid-Based Ultrasonic-Assisted Microextraction Coupled with HPLC for Determination of Citalopram and Nortriptyline in Human Plasma." n.d.

Varona, Marcelino, Xiong Ding, Kevin D. Clark, and Jared L. Anderson. "Solid-Phase Microextraction of DNA from Mycobacteria in Artificial Sputum Samples to Enable Visual Detection Using Isothermal Amplification." *Analytical Chemistry* 90, no. 11 (2018): 6922–8. doi:10.1021/acs.analchem.8b01160.

Vichapong, Jitlada, Rodjana Burakham, Yanawath Santaladchaiyakit, and Supalax Srijaranai. "A Preconcentration Method for Analysis of Neonicotinoids in Honey Samples by Ionic Liquid-Based Cold-Induced Aggregation Microextraction." *Talanta* 155 (August, 2016a): 216–21. doi:10.1016/j.talanta.2016.04.045.

Vichapong, Jitlada, Rodjana Burakham, Yanawath Santaladchaiyakit, and Supalax Srijaranai. "A Preconcentration Method for Analysis of Neonicotinoids in Honey Samples by Ionic Liquid-Based Cold-Induced Aggregation Microextraction." *Talanta* 155 (August, 2016b): 216–21. doi:10.1016/j.talanta.2016.04.045.

Vidal, Lorena, Alberto Chisvert, Antonio Canals, and Amparo Salvador. "Sensitive Determination of Free Benzophenone-3 in Human Urine Samples Based on an Ionic Liquid as Extractant Phase in Single-Drop Microextraction Prior to Liquid Chromatography Analysis." *Journal of Chromatography A* 1174, no. 1–2 (2007): 95–103. doi:10.1016/j.chroma.2007.07.077.

Vidal, Lorena, Alberto Chisvert, Antonio Canals, and Amparo Salvador. "Ionic Liquid-Based Single-Drop Microextraction Followed by Liquid Chromatography-Ultraviolet Spectrophotometry Detection to Determine Typical UV Filters in Surface Water Samples." *Talanta* 81, no. 1–2 (2010): 549–55. doi:10.1016/j.talanta.2009.12.042.

Wang, Ya Li, Li Qin You, Yu Wen Mei, Jian Ping Liu, and Li Jun He. "Benzyl Functionalized Ionic Liquid as New Extraction Solvent of Dispersive Liquid-Liquid Microextraction for Enrichment of Organophosphorus Pesticides and Aromatic Compounds." *Chinese Journal of Analytical Chemistry* 44, no. 6 (2016): 942–9. doi:10.1016/S1872-2040(16)60937-4.

Wang, Yulei, Jie Zhang, Xiaojia Huang, and Dongxing Yuan. "Preparation of Stir Cake Sorptive Extraction Based on Polymeric Ionic Liquid for the Enrichment of BenzimidazoleAnthelmintics in Water, Honey and Milk Samples." *Analytica Chimica Acta* 840 (August, 2014): 33–41. doi:10.1016/j.aca.2014.06.039.

Weingärtner, Hermann. "Understanding Ionic Liquids at the Molecular Level: Facts, Problems, and Controversies." *Angewandte Chemie - International Edition* 47 (2008): 654–70. doi:10.1002/anie.200604951.

Wen, Xiaodong, Qingwen Deng, and Jie Guo. "Ionic Liquid-Based Single Drop Microextraction of Ultra-Trace Copper in Food and Water Samples before Spectrophotometric Determination." *Spectrochimica Acta - Part A: Molecular and Biomolecular Spectroscopy* 79, no. 5 (2011): 1941–5. doi:10.1016/j.saa.2011.05.095.

Yang, Jinjuan, Hongmin Wei, Xiane Teng, Hanqi Zhang, and Yuhua Shi. "Dynamic Ultrasonic Nebulisation Extraction Coupled with Headspace Ionic Liquid-Based Single-Drop Microextraction for the Analysis of the Essential Oil in Forsythia Suspensa." *Phytochemical Analysis* 25, no. 2 (2014): 178–84. doi:10.1002/pca.2490.

Yao, Cong, Tianhao Li, Pamela Twu, William R. Pitner, and Jared L. Anderson. "Selective Extraction of Emerging Contaminants from Water Samples by Dispersive Liquid-Liquid Microextraction Using Functionalized Ionic Liquids." *Journal of Chromatography A* 1218, no. 12 (2011): 1556–66. doi:10.1016/j.chroma.2011.01.035.

Yavir, Kateryna, Łukasz Marcinkowski, Renata Marcinkowska, Jacek Namieśnik, and Adam Kloskowski. "Analytical Applications and Physicochemical Properties of Ionic Liquid-Based Hybrid Materials: A Review." *Analytica Chimica Acta* S0003-2670, no. 18 (2019): 31303–5. doi:10.1016/j.aca.2018.10.061.

Zeeb, Mohsen, Mohammad Reza Ganjali, and Parviz Norouzi "Modified Ionic Liquid Cold-Induced Aggregation Dispersive Liquid-Liquid Microextraction Combined with Spectrofluorimetry for Trace Determination of Ofloxacin in Pharmaceutical and Biological Samples." n.d. www.tums.ac.ir.

Zhang, Yong, Meng Mei, Tong Ouyang, and Xiaojia Huang. "Preparation of a New Polymeric Ionic Liquid-Based Sorbent for Stir Cake Sorptive Extraction of Trace Antimony in Environmental Water Samples." *Talanta* 161 (December, 2016): 377–83. doi:10.1016/j.talanta.2016.08.063.

Zhou, Qing Xiang, and Yuan Yuan Gao. "Combination of Ionic Liquid Dispersive Liquid-Phase Microextraction and High Performance Liquid Chromatography for the Determination of Triazine Herbicides in Water Samples." *Chinese Chemical Letters* 25, no. 5 (2014): 745–8. doi:10.1016/j.cclet.2014.01.026.

Zhou, Qingxiang, and Yuanyuan Gao. "Determination of Polycyclic Aromatic Hydrocarbons in Water Samples by Temperature-Controlled Ionic Liquid Dispersive Liquid-Liquid Microextraction Combined with High Performance Liquid Chromatography." *Analytical Methods* 6, no. 8 (2014): 2553–9. doi:10.1039/c3ay42254b.

Zhou, Qingxiang, Xiaoguo Zhang, and Junping Xiao. "Ultrasound-Assisted Ionic Liquid Dispersive Liquid-Phase Micro-Extraction: A Novel Approach for the Sensitive Determination of Aromatic Amines in Water Samples." *Journal of Chromatography A* 1216, no. 20 (2009): 4361–5. doi:10.1016/j.chroma.2009.03.046.

14

Deep Eutectic Solvent-Based Microextraction

Amir M. Ramezani[1], Yadollah Yamini[2], and Raheleh Ahmadi[1]
[1]*Healthy Ageing Research Centre, Neyshabur University of Medical Sciences, Neyshabur, Iran*
[2]*Department of Chemistry, Faculty of Sciences, Tarbiat Modares University, P.O. Box 14115-175, Tehran, Iran*

CONTENTS

14.1	Introduction	221
14.2	Deep Eutectic Solvents: A Concise Overview	222
	14.2.1 Structure of Deep Eutectic Solvents	222
	14.2.2 Preparation of Deep Eutectic Solvents	223
14.3	Utilization of Deep Eutectic Solvents in the Microextraction of Drugs and Poisons from Biological Matrices	223
	14.3.1 Utilization of Deep Eutectic Solvents in Liquid-Based Microextraction Methods	224
	14.3.2 Utilization of Deep Eutectic Solvents in Solid-Based Microextraction Methods	229
	14.3.3 Utilization of Deep Eutectic Solvents in Combined Microextraction Methods	231
14.4	Conclusion and Future Perspectives	233
References		233

14.1 Introduction

Sample pre-treatment is an essential step of any biological sample analysis because of the complexity of the biological matrices as well as the low concentration of purpose analytes (Li et al. 2018). Therefore, sample purification and enrichment of the targeted analytes in a complex matrix can be one of the main aims of sample analysis methods (Li et al. 2018). In this regard, different types of organic solvents, including esters, ethers, alcohols, aromatic and aliphatic hydrocarbons, halogenated hydrocarbons, aldehydes, and ketones, are mainly applied as extraction or adsorbent solvents (Płotka-Wasylka et al. 2017). It is well known that most of these solvents are highly toxic, flammable, volatile, and detrimental to the environment (Płotka-Wasylka et al. 2017).

By the development of green chemistry, the main trend of sample pre-treatment methods has been focused on limiting the use of hazardous organic solvents and replacing them with new eco-friendly green solvents (Ahmadi et al. 2019; Makoś et al. 2020). In this context, deep eutectic solvents (DESs), a new subgroup of ionic liquids (ILs), recently have attracted growing interest as an important class of green solvents due to their unique properties (Huang et al. 2019). They have the most significant features of ILs, including low volatility, high chemical and thermal stability, non-flammability, and reusability (Ramezani et al. 2020). Moreover, DESs are frequently proposed as a worthy and green alternative to traditional ILs. DESs not only share most of ILs' unique advantages, but they also overcome the ILs' limitations of toxicity, complex and time-consume preparation process, bio-incompatibility, and costliness (Huang et al. 2019).

14.2 Deep Eutectic Solvents: A Concise Overview

The term *eutectic* was first used in 1884 by the British physicist Frederick Guthrieto to describe metal alloys that have lower melting points than their constituent components (Guthrie 1884). This description has developed, and nowadays, the term *eutectic mixture* is described as a mixture of two or three compounds at a certain molar ratio, which illustrates a minimum melting temperature in the corresponding phase diagram (Gill and Vulfson 1994). This point in the phase diagrams is called the eutectic point (Gill and Vulfson 1994). The melting points of the eutectic mixtures are significantly lower than their pure constituents due to the presence of the strong intermolecular bonds between the components (Gill and Vulfson 1994).

In 2003, Abbott and co-workers demonstrated that the eutectic mixtures of amides with quaternary ammonium salts are liquid at ambient temperatures and have unusual solvent properties (Abbott et al. 2003). They coined the term *deep eutectic solvents* (DESs) to describe these eutectic mixtures. In a more complete definition, DES has been introduced as a eutectic mixture of a hydrogen bond acceptor (HBA) and a hydrogen bond donor (HBD) compound. Having a eutectic point temperature below that of an ideal liquid mixture distinguishes DESs from any other eutectic mixtures (Martins et al. 2019). In 2004, Abbott and co-workers introduced DESs as a versatile alternative for ILs (Abbott et al. 2004). After that, this group of solvents received special attention in various research fields. In subsequent years, a wide variety of DESs have been reported (Smith et al. 2014). Nowadays, there are many reviews available on the DESs and their applications in the literature (Kalhor and Ghandi 2019; Smith et al. 2014). With the expansion of the study on the DESs, new subgroups of these solvents have been introduced. In 2011, Choi and co-workers developed a new subclass of DESs named natural deep eutectic solvent (NADES), which is prepared from natural chemicals, namely, organic acids, sugars, sugar alcohols, polyalcohols, and amino acids (Choi et al. 2011). In 2015, a new subgroup of DESs with hydrophobic features was reported by the Marrucho (Ribeiro et al. 2015) and the Kroon groups (van Osch et al. 2015).

14.2.1 Structure of Deep Eutectic Solvents

The DES can be represented by the general formula, $C^+X^-.zY$, where C^+ is a sulfonium, phosphonium, or ammonium cation; X^- is a halide anion, in generala Lewis base; Y is a Lewis or Brønsted acid; and z is the number of Y molecules (Smith et al. 2014). The complex anionic species are formed between X and Y. The DESs can be classified into four main groups based on the nature of their complexing agents: type I is a combination of a quaternary ammonium salt and a non-hydrated metal chloride, type II is a combination of a quaternary ammonium salt and a hydrated metal chloride, type III is a combination of a quaternary ammonium salt as an HBA and an HBD, and type IV is a combination of HBD and metal chloride (Table 14.1) (Smith et al. 2014). All the DES types I and II are limited to hydrophilic solvents, but types III and IV can be hydrophilic or hydrophobic, depending on the nature of their components (Makoś et al. 2020). An attractive characteristic of DES type III is the possibility of a huge number of these solvents with different physical and chemical properties due to the availability of a wide range of

TABLE 14.1

Classification of DESs

Types	General Formula	Terms
Type I	$C^+X^- zMCl_x$	M = Zn, Sn, Fe, Al, Ga, In
Type II	$C^+X^- zMCl_x.yH_2O$	M = Cr, Co, Cu, Ni, Fe
Type III	$C^+X^- zRZ$	Z = OH, COOH, $CONH_2$
Type IV	$MCl_x + RZ = MCl_{x-1} + RZ + MCl_{x+1}$	M= Al, Zn and Z = OH, $CONH_2$

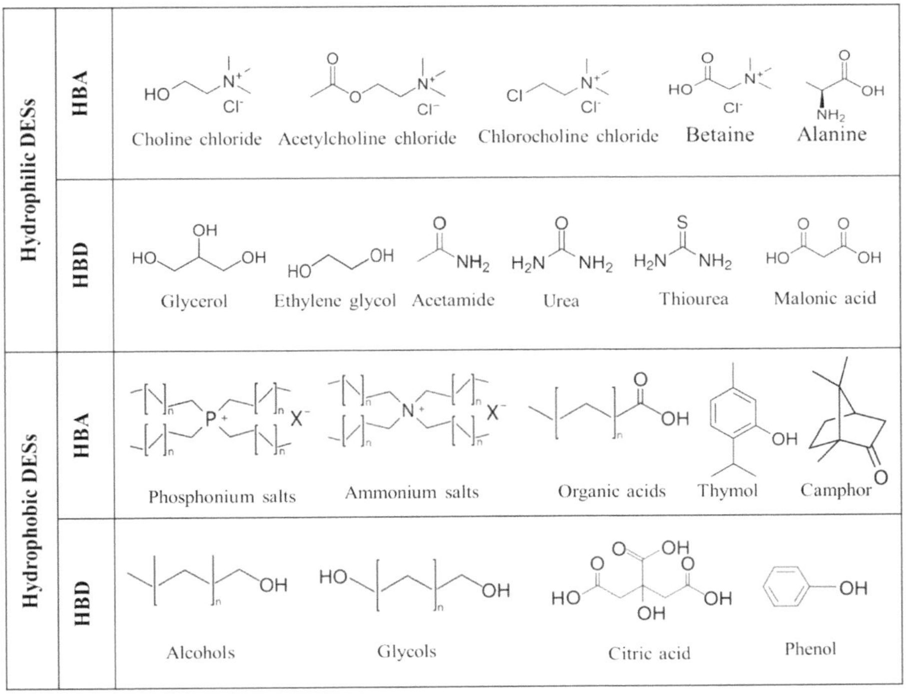

FIGURE 14.1 Structures of some HBD and HBA compounds that are utilized in the formation of hydrophilic and hydrophobic DESs. (Adapted from Makoś et al. 2020; Smith et al. 2014).

HBD and HBA compounds (Cunha and Fernandes 2018; Smith et al. 2014). Therefore, the properties of this class of DESs (type III) can be simply adjusted for special applications by changing one or both of their components (Smith et al. 2014). Figure 14.1 demonstrates a number of common HBD and HBA compounds that are used to produce hydrophilic and hydrophobic DESs.

14.2.2 Preparation of Deep Eutectic Solvents

The simplicity of DES preparation methods is one of their primary attractions. In their preparation procedures, no organic solvent is required, and there is no need for purification of the final product due to the lack of by-product formation. The DESs can be prepared by simply mixing components at an appropriate molar ratio via heating, freeze-drying, and grinding (Li and Row 2019; Li et al. 2020). Heating is the most common method of DES synthesis. In this method, the mixture of components is heated under stirring at 80–100 °C until a homogeneous and clear solution is produced (Ahmadi et al. 2018a). The freeze-drying approach includes two steps: (1) dissolution of the DES components in water with the help of heat, ultrasound, or vortex motion, and (2) removing water via freeze-drying (Gutiérrez et al. 2009). Also, water can be evaporated through a rotary evaporator or centrifugal vacuum. In the grinding approach, the mixture of DES components is ground in a mortar at room temperature until a uniform liquid is formed (Florindo et al. 2014).

14.3 Utilization of Deep Eutectic Solvents in the Microextraction of Drugs and Poisons from Biological Matrices

The DESs, a green alternative to volatile organic solvents and also ILs, have shown unique solvent properties, such as low vapour pressure, a broad range of polarities, recycling possibility, low toxicity, high biodegradability, high biocompatibility, air and moisture stability, high chemical and thermal stability, the ability to stay in the liquid phase even at a temperature well below 0 °C, and a high ability to dissolve many chemicals (Ahmadi et al. 2018b). These great features of DESs led to their widespread application in separation science (Li and Row 2016), such as designing the mobile phase compositions for reversed-phase liquid chromatography (Li et al. 2015; Ramezani et al. 2020, 2018; Ramezani and Absalan 2020; Tan et al. 2016). Their majority applications have been based on their use as the extraction phases in liquid-based extraction techniques and as the desorption media in solid-based extraction methods (Cunha and Fernandes 2018). In addition, DESs are successfully used for the surface modification of some materials (i.e., silica or polymers) (Tang et al. 2015). These DES-modified materials have been successfully applied as sorbents in the extraction methods due to their excellent interactions with the target analytes, including π–π interactions, hydrogen bonding, anion exchange, etc. (Tang et al. 2015). In this chapter, some remarkable examples of the DES applications in the microextraction of drugs and poisons from biological matrices are reviewed.

14.3.1 Utilization of Deep Eutectic Solvents in Liquid-Based Microextraction Methods

Liquid-phase microextraction (LPME) is one of the most popular pre-treatment methods, and it was introduced for the first time in 1996 (Liu and Dasgupta 1996). In this microextraction method, a few microliters of an extraction solvent (acceptor phase) is suspended in an aqueous solution of analytes (donor phase) (Liu and Dasgupta 1996). In the last few years, various LPME modes have been developed, including single drop microextraction (SDME), hollow-fibre LPME (HF-LPME), dispersive liquid-liquid microextraction (DLLME), and so forth. In SDME, a single drop of extractive solvent is hanging from a syringe tip that can be placed directly in the donor phase or located in the headspace of the sample solution (HS-SDME). In HF-LPME, usually the acceptor phase is injected into the lumen of a polypropylene hollow fibre and then soaked in the donor phase. The DLLME is based on a triple solvent system, including an aqueous sample solution, a water-immiscible organic extraction solvent, and a disperser solvent that is miscible in both aqueous and extraction phases. By injecting the mixture of extraction and disperser solvents into the sample solution, the cloudy solution is formed. Currently, in some DLLME procedures, supplementary energy or reagents are used to accelerate the formation of the cloudy solution, such as injection of air bubbles (air-assisted LLME, AA-DLLME) (Farajzadeh and AfsharMogaddam 2012), injection of argon gas (gas-assisted DLLME, GA-DLLME) (Akhond et al. 2016), ultrasonication (ultrasonic-assisted DLLME, UA-DLLME) (Malaei et al. 2018), vortex mixing (vortex-assisted DLLME, VA-DLLME) (Safavi et al. 2018), shaker mixing (shaker-assisted DLLME, SA-DLLME) (Ahmadi et al. 2019), bubbling as a result of an effervescent reaction (effervescence-assisted DLLME, EA-DLLME) (Shishov et al. 2020), and so on. In addition, the DLLME techniques can be classified based on the methods that are applied to separating the phases after extraction. Centrifugation, the addition of a salt (salt-induced), purging a gas (i.e., argon or nitrogen), and solidification of floating organic drop (DLLME-SFO) can be named as the routine methods (Šandrejová et al. 2016).

Recently, the LPME techniques based on DESs have gained much attention in the extraction of drugs and poisons from biological matrices. Most applications of DES-based LPME in biological analysis are summarized in Table 14.2.

In 2018, Yousefi et al. reported a two-phase HS-SDME method based on a hydrophobic magnetic Bucky gel for the extraction of volatile aromatic hydrocarbons from water and biological samples (Yousefi et al. 2018). The hydrophobic DES was formed by combining ChCl and chlorophenol, and the magnetic Bucky gel was prepared by mixing the DES and the magnetic multiwalled carbon nanotubes. A drop of magnetic Bucky gel was put in the space above the container of the sample solution. Compared to the conventional organic solvents, the DES demonstrated more suitability to create stable drops for HS-SDME owing to its lower volatility, higher viscosity, and adjustable miscibility.

Seidi and his co-workers presented the application of hydrophilic DES based on ChCl and urea as an acceptor phase in the three-phase HF-LPME procedure for the lead extraction from whole blood samples (Alavi et al. 2017). In this procedure the analyte (Pb^{2+}) was extracted from an aqueous sample

TABLE 14.2

Applications of DES-based liquid-phase microextraction (LPME) in biological analysis.

DES		DES Properties	Analytes	Biological Sample	Sample Treatment	Analytical Technique	Ref.
HBA	HBD						
Choline chloride	Chlorophenol	Hydrophobic	Volatile aromatic hydrocarbons	Human urine	Headspace single drop microextraction	GC-FID	(Yousefi et al. 2018)
Choline chloride	Urea	Hydrophilic	Lead	Human blood	Hollow-fibre liquid-phase microextraction	AAS	(Alavi et al. 2017)
Choline chloride, Methyltriphenylphosphonium bromide, Methyltriphenylphosphonium iodide	Ethylene glycol	–	Steroidal hormones	Human plasma, urine	Three-phase hollow-fibre liquid-phase microextraction	HPLC-UV	(Khataei et al. 2018)
Choline chloride	Phenylethanol	Relatively hydrophobic	Antiarrhythmic agents	Human plasma, urine	Hollow-fibre microextraction	HPLC-UV	(Rajabi et al. 2019)
Choline chloride	5,6,7,8-Tetrahydro-5,5,8,8-tetramethylnaphthalen-2-ol	Hydrophilic	Methadone	Human plasma, urine	Air-assisted emulsification liquid-liquid microextraction	GC-FID	(Lamei et al. 2017)
Choline chloride	Phenol, Ethylene glycol, urea	Hydrophilic	Anti-depressant drugs	Human plasma	Air-agitated emulsification microextraction	HPLC-UV	(Moghadam et al. 2018)
Borneol	Decanoic acid, Oleic acid, Thymol	Hydrophobic	Warfarin	Plasma, urine	Air-assisted liquid-liquid microextraction	HPLC-UV	(Majidi and Hadjmohammadi, 2020)
Methyltrioctylammoniumbromide	Decanoic acid	Hydrophobic	Malondialdehyde, formaldehyde	Human urine	Vortex-assisted liquid-liquid microextraction	HPLC-UV	(Safavi et al. 2018)
Trioctylmethylammonium chloride	Hydroquinone, 4-phenylphenol	Hydrophobic	Formaldehyde	Duck blood pig blood	Vortex-assisted liquid-liquid microextraction	HPLC	(Zhang et al. 2019b)
Trioctylmethylammonium chloride	Oleic acid	Hydrophobic	Nitrite	Saliva, urine	Vortex-assisted dispersive liquid-liquid microextraction	HPLC-UV	(Zhang et al. 2019a)

(Continued)

TABLE 14.2 (Continued)
Applications of DES-based liquid-phase microextraction (LPME) in biological analysis.

DES		DES Properties	Analytes	Biological Sample	Sample Treatment	Analytical Technique	Ref.
HBA	HBD						
Tetrabutylamonium chloride	Decanoic acid	–	Cu (II)	Bovine liver	Ultrasound-assisted liquid-phase microextraction	AAS	(Kanberoglu et al. 2018)
Choline chloride	Phenol	Hydrophobic	Mercury	Fish tissue	Ultrasound-assisted deep eutectic solvent-based on liquid-phase microextraction	AAS	(Thongsaw et al. 2019)
Choline chloride	Phenol, 1-Phenylethanol, 4-Chlorophenol, Ethylene glycol	–	Macrolides	Swine urine	Ultrasonic-assisted dispersive liquid-liquid microextraction	LC-MS/MS	(Liao et al. 2020)
Imidazolium chloride ionic liquids	1-Undecanol	Low hydrophilic	Mercury	Human blood	Vortex-assisted dispersive liquid-liquid microextraction	AAS	(Akramipour et al. 2018)
Choline chloride	Decanoic acid	Hydrophobic	As (III), As (V), Se (IV), Se (VI), Hg (II)	Human blood	Liquid-phase microextraction	AAS	(Akramipour et al. 2019a)
Choline chloride	Decanoic acid	Hydrophobic	Se (IV)/Se (VI)	Human blood	Solidified deep eutectic solvent microextraction	AAS	(Akramipour et al. 2019b)
Benzyltriphenylphosphonium bromide	Phenol	Hydrophobic	Cr (VI)	Human urine	Solidified deep eutectic solvent microextraction	AAS	(Seidi et al. 2019)
Menthol	Phenylacetic acid	–	Pesticides	Human plasma, urine	Dispersive liquid-liquid microextraction	GC-MS	(Jouyban et al. 2019a)

Menthol	Phenylacetic acid	–	Pesticides	Saliva, breath condensate samples	Organic droplets liquid-phase microextraction	GC-MS	(Jouyban et al. 2019b)
Choline chloride	Phenol	–	Neonicotinoid insecticide	Human urine	Cloud-point microextraction	HPLC-DAD	(Kachangoon et al. 2020)
Menthol	Ketoprofen, Diclofenac	–	Ketoprofen, Diclofenac	Human urine	Solidified deep eutectic solvent microextraction	HPLC-UV	(Shishov et al. 2018)
Menthol	Formic acid	–	Ketoprofen, Diclofenac	Beef liver	Effervescence-assisted dispersive liquid-liquid microextraction	HPLC-MS/MS	(Shishov et al. 2020)

solution into 1-octanol containing *N,N,N*-cetyltrimethylammoniumbromid (CTAB) as the extraction phase and then back-extracted into the DES containing $KClO_4$ as the acceptor phase. The method was followed by atomic absorption spectroscopy (AAS) for the determination of lead. In this respect, Yamini and his co-workers also reported another application of DESs in the three-phase HF-LPME method (Khataei et al. 2018). In this work, the suitability of three types of DESs as the extraction solvent in the HF-LPME method for the extraction of cyproterone acetate and dydrogesterone from biological samples was evaluated. The results demonstrated that the nature of the DES has a significant effect on the efficiency of the method. Another application of DES in this extraction method was reported by Rajabi et al. (Rajabi et al. 2019). In this report, a new relatively hydrophobic DES based on ChCl and 1-phenylethanol was introduced as an extraction solvent in the HF-LPME procedure for extraction of antiarrhythmic drugs (propranolol, carvedilol, verapamil, and amlodipine) in biological and environmental samples (urine, plasma, and wastewater). The results proved the good compatibility of DES to hollow fibre pores and its excellent capability for ionizable chemical extraction without any requirement for carrier agents.

In 2017, Lamei et al. developed an AA-DLLME procedure using a new hydrophilic DES for the pre-concentration and extraction of methadone from water, urine, and plasma samples (Lamei et al. 2017). The DES was formed by using choline chloride (ChCl) as an HBA component and 5,6,7,8-tetrahydro-5,5,8,8-tetramethylnaphthalen-2-ol as an HBD component. In this work, tetrahydrofuran (THF) was applied as a demulsifier solvent, which caused aggregating of the DES and formation of a turbid solution. Moreover, the dispersion of DES droplets into the sample solution was created by several times sucking and injecting the mixture of sample, extracting, and demulsifier agents. Finally, the extraction solvent was collected on the top of the solution by centrifugation, and then it was analyzed using the gas chromatography (GC) system. In 2018, Ghoochani Moghadam et al. employed a similar AA-DLLME procedure for the extraction of three anti-depressant drugs (escitalopram, desipramine, and imipramine) from human plasma and pharmaceutical wastewater (Ghoochani Moghadam et al. 2018). Moreover, in 2020, the AA-DLLME procedure was also used for the extraction of warfarin from plasma and urine samples (Majidi and Hadjmohammadi 2020).

Safavi et al. reported a simple VA-LLME procedure based on hydrophobic DES for the extraction of low molecular weight aldehydes (malondialdehyde and formaldehyde) from urine samples (Safavi et al. 2018). In this procedure, the water-immiscible DES based on methyltrioctylammonium bromide and decanoic acid was used as the acceptor phase. The method was followed by HPLC-UV for measuring the analytes. The VA-LLME based on the hydrophobic DES procedure was also proposed for the extraction of formaldehyde from blood samples (Zhang et al. 2019b) and nitrite from urine and saliva samples (Zhang et al. 2019a).

Kanberoglu et al. reported the application of DES in the digestion of the liver samples and also the UA-DLLME extraction of copper from the digested samples prior to its analysis by AAS (Kanberoglu et al. 2018). In this case, DES based on ChCl/lactic acid and tetrabutylammonium chloride/decanoic acid were applied in the digestion step and extraction step, respectively. Before extraction, copper ions were complexed with sodium dimethyl dithiocarbamate. In another work, the DES-based UA-DLLME method combined with AAS was used for pre-concentration and determination of mercury (Hg^{2+} and CH_3Hg^+) in water and fish samples (Thongsaw et al. 2019). Mercury in the form of CH_3Hg^+ was extracted directly to hydrophobic DES based on ChCl/phenol due to its hydrophobicity, but Hg^{2+} were complexed with dithizone before extraction. Liao et al. also developed a DES-based UA-DLLME procedure for macrolide antibiotics extraction from swine urine samples (Liao et al. 2020).

Three works have been published by Fattahi groups focusing on the application of hydrophobic DESs on the VA-DLLME based on the solidification of a floating organic drop (VA-DLLME-SFO) method for extraction of toxic metal cations (As(III), As(V), Se(IV), Se(VI), and Hg(II)) from blood samples (Akramipour et al. 2019a, 2019b, 2018). In these reports, DESs based on 1-octyl-3-methylimidazolium chloride/1-undecanol (Akramipour et al. 2018) and ChCl/Phenol (Akramipour et al. 2019a, 2019b) with a freezing point around ambient temperature were applied as the extraction solvents, and diethyl-dithiophosphoric acid was used as a chelating agent. In this procedure, the mixture of the DES and the chelating agent was rapidly injected into the sample solution with a syringe, and the mixture was shaken by a vortex agitator. After that, by centrifugation, the tiny droplets of DES separated and floated at the surface of the solution, and finally, it was solidified after cooling the solution. Seidi et al. also reported a similar hydrophobic DES-based DLLME procedure for the extraction of Cr(VI) from urine samples (Seidi et al. 2019). Cr(VI) contents of the urine samples were complexed with 1,5-diphenylcarbazone prior to the extraction step. Water-immiscible DES consisting of benzyltriphenylphosphonium bromide and phenol was chosen as an extraction solvent.

Nowadays, the importance of insecticide monitoring in biological samples is well known. In this regard, Jouyban et al. introduced a new, interesting procedure based on DLLME for the extraction of pesticides from biological samples such as human urine (Jouyban et al. 2019a), plasma (Jouyban et al. 2019a), saliva (Jouyban et al. 2019b), and exhaled breath condensate (Jouyban et al. 2019b). In these works, a novel extraction vessel was designed for extraction of pesticides based on the DLLME-SFO procedure (Figure 14.2). The designed vessel was a double-glazed, U-shaped glass equipped with a glass filter as shown in Figure 14.2. DES based on menthol and phenylacetic acid was selected as an extraction solvent with a density lower than water. In this procedure, the sample solution was transferred into the device on the glass filter, and then the extraction solvent was passed through the vessel filter by means of the nitrogen gas flow. In this step, the extractant was dispersed in the sample solution, and a cloudy solution was obtained with the aid of gas bubbles. Finally, the extraction solvent was collected on the surface of the aqueous sample solution without centrifugation, and it was solidified with circulating cool water. The pesticides were analyzed by using a gas chromatography-mass spectrometry (GC-MS) system. In another report, Kachangoon et al. developed a cloud-point extraction combined with in-situ metathesis reaction of DESs for extraction of neonicotinoid insecticides prior to HPLC analysis (Kachangoon et al. 2020). DES based on ChCl/phenol and Triton X-114 were used as a disperser and extraction solvent, respectively.

A new strategy for the LLME of non-steroidal anti-inflammatory drugs (NSAIDs) from biological samples was proposed by Shishov et al. (Shishov et al. 2018). This microextraction method was based on the in-situ formation of a eutectic mixture between menthol and NSAIDs, which results in the separation of NSAIDs from the aqueous phase. The proposed method was applied for HPLC-UV monitoring of ketoprofen and diclofenac in the biological samples. Two years later, the same group developed an EA-DLLME based on DES decomposition for extraction of NSAIDs from beef liver samples (Shishov et al. 2020). In the first step of this procedure, analytes were separated from a liver sample in a sodium carbonate solution. In the second step, decomposition of DES (menthol/formic acid) into the extractant (menthol) and proton donor agent (formic acid) occurred in the sample solution. Therefore, an effervescent reaction between sodium carbonate and formic acid produced carbon dioxide and consequently helped in a dispersion of extractant.

14.3.2 Utilization of Deep Eutectic Solvents in Solid-Based Microextraction Methods

Although most applications of DES in the microextraction of drugs and poisons from biological matrices are related to LPME techniques, there are some reports of the application of DESs in SPME (Table 14.3). The SPME technique is a simple and effective sample pre-treatment method that was introduced in 1990 (Arthur and Pawliszyn 1990). This methodology combines sampling, extraction, and

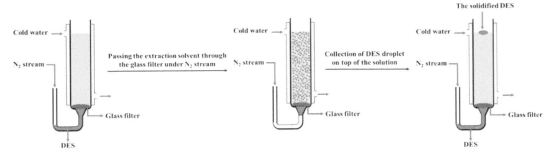

FIGURE 14.2 Scheme of the proposed DLLME-SFO procedure applied for the extraction of pesticides from biological samples such as human urine (Jouyban et al. 2019a), plasma (Jouyban et al. 2019a), saliva (Jouyban et al. 2019b), and exhaled breath condensate (Jouyban et al. 2019b).

TABLE 14.3
Applications of DES-Based Solid-phase Microextraction (SPME) in Biological Analysis.

DES		DES Properties	Analytes	Biological Sample	Sample Treatment	Analytical Technique	Ref.
HBA	HBD						
Choline chloride	Urea, ethylene glycol, Glycerol	Hydrophilic	Dopamine, epinephrine, norepinephrine	Human serum, urine	Dispersive micro-solid-phase extraction	HPLC-UV	(Khezeli and Daneshfar 2015)
Choline chloride	Itaconic acid	–	Anti-inflammatory drugs	Human plasma	Solid-phase microextraction	HPLC-UV	(Wang et al. 2018)
Choline chloride	Urea	Hydrophilic	Losartan	Human urine, plasma	In-tube solid-phase microextraction	HPLC-UV	(Asiabi et al. 2018)
Choline chloride	Thiourea	–	Ag (I)	Human hair	Hollow-fibre solid-phase microextraction	AAS	(Karimi et al. 2018)
Choline chloride	Ethylene glycol	Hydrophilic	Levofloxacin	Human plasma	Microextraction by packed sorbent	HPLC-UV	(Meng and Wang 2019)
Choline chloride	Ethylene glycol	Hydrophilic	Meloxicam	Human plasma, urine	Magnetic dispersive micro-solid-phase extraction	HPLC-UV	(Rastbood et al. 2020)

concentration operations in one step only, which considerably reduces solvent consumption and operating duration (Wang et al. 2018). In fact, the SPME process involves two basic steps, partitioning of analytes between the sorbent and the sample matrix and then the desorption of the concentrated analytes to the eluent solvent (Ho 2011). Therefore, selecting a suitable sorbent and eluent solvent is crucial in the design of SPME methods (Khezeli and Daneshfar 2015).

In 2015, Khezeli et al. developed a SPME procedure for the extraction of epinephrine, dopamine, and norepinephrine from biological samples prior to HPLC-UV analysis (Khezeli and Daneshfar 2015). In this process, the magnetic metal-organic framework core-shell named Fe_3O_4@MIL-100 (Fe) and the ChCl-based DESs were proposed as sorbent and eluent solvent, respectively.

In 2018, Wang et al. prepared a polymer monolithic column based on the DES for in-tube SPME of NSAIDs from water and plasma samples (Wang et al. 2018). A poly-(DES-ethylene glycol-dimethacrylate) monolith was synthesized inside the polydopamine-functionalized poly(ether ether ketone) tube by applying a DES (ChCl/itaconic acid) as a functional monomer. This column was coupled with a HPLC-UV and used for online analysis of three NSAIDs (ketoprofen, flurbiprofen, and diclofenac sodium). In the same year, Yamini and his co-workers described a new coating based on a DES for usage in electrochemically controlled in-tube SPME (EC-IT-SPME) (Asiabi et al. 2018). This coating was prepared by electrochemical deposition of the polypyrrole/DES nanocomposite on the inner walls of a stainless-steel tube. It was applied for the EC-IT-SPME of losartan from urine and plasma samples.

In 2018, Karimi et al. immobilized DES of ChCl and thiourea on the surface of graphene oxide nanosheets and used it as a new sorbent in the HF-SPME of silver ions from water and hair samples (Karimi et al. 2018).

In 2019, Meng et al. used a DES as a porogen for preparation of molecularly imprinted polymers with a pseudo template (Meng and Wang 2019). It was applied as an adsorbent material in the microextraction of levofloxacin from human plasma by packed sorbent.

In 2020, Rastbood et al. applied a DES as a carrier and disperser of ferrofluids in magnetic dispersive SPME of meloxicam from human plasma and urine samples (Rastbood et al. 2020). A ferrofluid was prepared by using silica-coated magnetic nanoparticles and a DES (ChCl/ethylene glycol) as an adsorbent and carrier, respectively. The DES carrier reduced the extraction time via increasing the surface contact between the adsorbent and the target analytes.

14.3.3 Utilization of Deep Eutectic Solvents in Combined Microextraction Methods

A combination of the sample pre-treatment methods is a suitable way to overcome the limitations of individual techniques and offer their synergistic benefits (Sajid and Płotka-Wasylka 2018). Additionally, combinations of extraction methods have their own unique merits. Combined extraction methods can be divided into two groups, conventional extractions combined with microextraction methods and binary microextraction techniques (Sajid and Płotka-Wasylka 2018). Some applications of the combined extraction methods based on DESs in biological analysis are summarized in Table 14.4.

Hemmati et al. introduced two consecutive dispersive SPME/AA-LLME procedures for the influential clean-up and enrichment of lorazepam and clonazepam from human plasma and urine samples (Hemmati et al. 2017). Polythiophene-sodium dodecyl benzene sulphonate/iron oxide nanocomposite was used as a sorbent in the SPME step. The magnetic nature of the proposed sorbent made the clean-up step quick and convenient. The purification step was followed by a highly rapid and efficient emulsification microextraction process based on the ChCl-based DESs for further enrichment. Finally, the instrumental analysis was feasible via HPLC-UV.

The dispersive solid-phase extraction (SPE) methods can be coupled with the other microextraction techniques, such as DLLME, in order to improve extraction performance parameters, such as enrichment factor (Mohebbi et al. 2018). In this regard, Mohebbi et al. proposed an SPE procedure coupled with DES-based AA-DLLME for extraction of some tricyclic antidepressant drugs from human plasma and urine samples prior to determination by GC-MS (Mohebbi et al. 2018). In this procedure, C_{18} was used as a sorbent, and ChCl/4-chlorophenol DES was applied as an elution and extraction solvent.

TABLE 14.4
Applications of DES-Based Combined Microextraction Methods in Biological Analysis.

DES		DES Properties	Analytes	Biological Sample	Sample Treatment	Analytical Technique	Ref.
HBA	HBD						
Choline chloride	Phenol, urea, ethylene glycol	Hydrophilic	Clonazepam, lorazepam	Human plasma, urine	Dispersive solid/liquid phase microextractions	HPLC-UV	(Hemmati et al. 2017)
Choline chloride	4-Chlorophenol	Hydrophobic	Tricyclic antidepressant drugs	Human plasma, urine	Dispersive solid-phase extraction and deep eutectic solvent-based air-assisted liquid-liquid microextraction	GC-MS	(Mohebbi et al. 2018)
Choline chloride	Benzyl ethylenediamine, 2,4-dichloro aniline p-chlorophenol	Hydrophobic	Anti-seizure drugs	Human urine	Dispersive liquid-liquid microextraction	GC-FID	(Soltanmohammadi et al. 2020)
Menthol	Acetic acid	Hydrophobic	Mefenamic acid	Human urine	Vortex-assisted liquid-phase microextraction	HPLC-UV	(Dil et al. 2019)
Menthol	Octanoic acid	Hydrophobic	Doxycycline	Human plasma, urine	Vortex-assisted dispersive liquid-liquid microextraction	HPLC-UV	(Alipanahpour Dil et al. 2020)
Choline chloride	Stearic acid	Hydrophobic	Polycyclic aromatic hydrocarbons	Human urine, saliva	Air-assisted liquid-liquid microextraction	GC-MS	(Jouyban et al. 2020)

Jouyban and his co-workers employed a salt-induced LLE method combined with a DES-based DLLME technique followed by GC for extraction and measurement of three anti-seizure drugs (carbamazepine, diazepam, and chlordiazepoxide) in urine samples (Soltanmohammadi et al. 2020). In the first step of this procedure, the analytes were extracted into iso-propanol as a water-miscible extraction solvent. In the second step, the iso-propanol, which contains the extracted analytes, was applied as a disperser solvent, and the water-miscible DESs were used as the extraction solvent in the DLLME process.

Recently, Ghaedi and his co-workers reported the successful application of ferrofluids based on hydrophobic DESs as an extraction phase in DLLME of mefenamic acid (Dil et al. 2019) and doxycycline (Alipanahpour Dil et al. 2020) from biological samples. Ferrofluid is typically the stable suspension of magnetic nanoparticles in the carrier liquid (Alipanahpour Dil et al. 2019). In these works, the ferrofluids were prepared by mixing the oleic acid-coated Fe_3O_4 magnetic nanoparticles in the hydrophobic DESs. A similar procedure was used by Jouyban et al. for the extraction of 16 polycyclic aromatic hydrocarbons from urine and saliva samples of tobacco smokers (Jouyban et al. 2020).

14.4 Conclusion and Future Perspectives

The complexity of the biological matrices as well as the low concentration of analytes in the biological samples has made it necessary to perform the extraction step before the biological analysis. Along with the advancement of green chemistry, much attention has been paid to the use of green solvents instead of toxic organic solvents in all fields of analytical chemistry, especially extraction. Recently, DESs as a novel group of alternative and eco-sustainable solvents have been applied in various extraction methods. By substituting traditional toxic organic solvents with DESs, the original merits of extraction techniques, such as low costs, ease in operation, and environmental safety, were enhanced. In this chapter, the main applications of DESs in the microextraction of drugs and poisons from biological matrices were overviewed.

There is no doubt about the further growth of DES applications in biological analysis in the close future. Future studies on the DESs maybe focus on the following topics; (1) introducing novel DESs with various properties and their utilization in the designing of new efficient extraction methods; (2) further development of DES applications as extracting solvents, eluent solvents, or chemical media for synthesis of different sorbents; and (3) designing snovel extraction methodologies based on DESs for analyte monitoring in biological matrices.

REFERENCES

Abbott, Andrew P., David Boothby, Glen Capper, David L. Davies, and Raymond K. Rasheed. "Deep Eutectic Solvents Formed between Choline Chloride and Carboxylic Acids: Versatile Alternatives to Ionic Liquids." *Journal of the American Chemical Society* 126, (2004): 9142–7.

Abbott, Andrew P., Glen Capper, David L. Davies, Raymond K. Rasheed, and Vasuki Tambyrajah. "Novel Solvent Properties of Choline Chloride/Urea Mixtures." *Chemical Communications*, (2003): 70–71.

Ahmadi, Raheleh, Bahram Hemmateenejad, Afsaneh Safavi, Zahra Shojaeifard, Maryam Mohabbati, and Omidreza Firuzi. "Assessment of Cytotoxicity of Choline Chloride-based Natural Deep Eutectic Solvents Against Human Hek-293 Cells: A Qsar Analysis." *Chemosphere* 209, (2018a): 831–8.

Ahmadi, Raheleh, Bahram Hemmateenejad, Afsaneh Safavi, Zahra Shojaeifard, Azin Shahsavar, Afshan Mohajeri, Maryam Heydari Dokoohaki, and Amin Reza Zolghadr. "Deep Eutectic–Water Binary Solvent Associations Investigated by Vibrational Spectroscopy and Chemometrics." *Physical Chemistry Chemical Physics* 20, (2018b): 18463–73.

Ahmadi, Raheleh, Ghasem Kazemi, Amir M. Ramezani, and Afsaneh Safavi. "Shaker-Assisted Liquid-Liquid Microextraction of Methylene Blue Using Deep Eutectic Solvent Followed by Back-Extraction and Spectrophotometric Determination." *Microchemical Journal* 145, 2019: 501–7.

Akhond, Morteza, Ghodratollah Absalan, Tayebe Pourshamsi, and Amir M. Ramezani. "Gas-Assisted Dispersive Liquid-Phase Microextraction Using Ionic Liquid as Extracting Solvent Forspectrophotometric Speciation of Copper." *Talanta* 154, (2016): 461–6.

Akramipour, Reza, Mohammad Reza Golpayegani, Mahmoud Ghasemi, Negar Noori, and Nazir, Fattahi. "OptimIzation of a Methodology for Speciation of Arsenic, Selenium and Mercury in Blood Samples Based on the Deep Eutectic Solvent." *MethodsX* 6, (2019a): 2141–7.

Akramipour, Reza, Mohammad Reza Golpayegani, Mahmoud Ghasemi, Negar Noori, and Nazir Fattahi. "Development of an Efficient Sample Preparation Method for the Speciation Ofse(iv)/se(vi) and Total Inorganic Selenium in Blood of Children with Acute Leukemia." *New Journal of Chemistry* 43, (2019b): 6951–8.

Akramipour, Reza, Mohammad Reza Golpayegani, Simin Gheini, and Nazir Fattahi. "Speciation of Organic/Inorganic Mercury and Total Mercury in Blood Samples Using Vortex Assisted Dispersive Liquid-Liquid Microextraction Based on the Freezing of Deep Eutectic Solvent Followed by GFAAS." *Talanta* 186, (2018): 17–23.

Alavi, Leila, Shahram Seidi, Ali Jabbari, and Tahmineh Baheri. "Deep Eutectic Liquid Organic Salt as a New Solvent for Carrier-mediated Hollow Fiber Liquid Phase Microextraction of Lead From Whole Blood Followed by Electrothermal Atomic Absorption Spectrometry." *New Journal of Chemistry* 41, (2017): 7038–44.

Alipanahpour Dil, Ebrahim, Mehrorang Ghaedi, and Arash Asfaram. "Application of Hydrophobic Deep Eutectic Solvent as the Carrier for Ferrofluid: A Novel Strategy For Pre-concentration and Determination of Mefenamic Acid in Human Urine Samples by High Performance Liquid Chromatography Under Experimental Design Optimization." *Talanta* 202, (2019): 526–30.

Alipanahpour Dil, Ebrahim, Mehrorang Ghaedi, Arash Asfaram, Lobat Tayebi, and Fatemeh Mehrabi. "A Ferrofluidic Hydrophobic Deep Eutectic Solvent for the Extraction of Doxycycline From Urine, Blood Plasma and Milk Samples Prior to its Determination by High-performance Liquid Chromatography-Ultraviolet." *Journal of Chromatography A* 1613, (2020): 460695.

Arthur, Catherine L., and Janusz Pawliszyn. "Solid Phase Microextraction with Thermal Desorption Using Fused Silica Optical Fibers." *Analytical Chemistry* 62, (1990): 2145–8.

Asiabi, Hamid, Yadollah Yamini, Maryam Shamsayei, and Jamshid Azarnia Mehraban. "A Nanocomposite Prepared From a Polypyrrole Deep Eutectic Solvent And Coated Onto the Inner Surface of a Steel Capillary for Electrochemically Controlled Microextraction of Acidic Drugs Such as Losartan." *Microchimica Acta* 185, (2018): 169.

Choi, Young Hae, Jaap van Spronsen, Yuntao Dai, Marianne Verberne, Frank Hollmann, Isabel W. C. E. Arends, Geert-Jan Witkamp, and Robert Verpoorte. "Are Natural Deep Eutectic Solvents the Missing Link in Understanding Cellular Metabolism and Physiology?" *Plant Physiology* 156, (2011): 1701–5.

Cunha, Sara C., and José O. Fernandes. "Extraction Techniques with Deep Eutectic Solvents." *TrAC Trends in Analytical Chemistry* 105, (2018): 225–39.

Farajzadeh, Mir Ali, and Mohammad Reza Afshar Mogaddam. "Air-Assisted Liquid-Liquid Microextraction Method as a Novel Microextraction Technique; Application in Extraction and Preconcentration of Phthalate Esters in Aqueous Sample Followed by Gas Chromatography-Flame Ionization Detection." *Analytica Chimica Acta* 728, (2012): 31–38.

Florindo, Catarina, F.S. Oliveira, L.P.N. Rebelo, Ana M. Fernandes, and I.M. Marrucho. "Insights into the Synthesis and Properties of Deep Eutectic Solvents Based on Cholinium Chloride and Carboxylic Acids." *Sustainable Chemistry & Engineering* 2, (2014): 2416–25.

Gill, Iqbal, and Evgeny Vulfson. "Enzymic Catalysis in Heterogeneous Eutectic Mixtures of Substrates." *Trends in Biotechnology* 12, (1994): 118–22.

Guthrie, Frederick. "On Eutexia." *Proceedings of the Physical Society* 6, (1884): 124–46.

Gutiérrez, María C., María L. Ferrer, C. Reyes Mateo, and Francisco del Monte. "Freeze-Drying of Aqueous Solutions of Deep Eutectic Solvents: A Suitable Approach to Deep Eutectic Suspensions of Self-assembled Structures." *Langmuir* 25, (2009): 5509–15.

Hemmati, Maryam, Maryam Rajabi, and Alireza Asghari. "A Twin Purification/Enrichment Procedure Based on Two Versatile Solid/Liquid Extracting Agents for Efficient Uptake of Ultra-Trace Levels of Lorazepam and Clonazepam From Complex Bio-matrices." *Journal of Chromatography A* 1524, (2017): 1–12.

Ho, Tien D., Anthony J. Canestraro, and Jared L. Anderson. "Ionic Liquids in Solid-Phase Microextraction: A Review." *Analytica Chimica Acta* 695, (2011): 18–43.

Huang, Jie, Xiuyun Guo, Tianyi Xu, Lanyan Fan, Xinpeng Zhou, and Shihua Wu. "Ionic Deep Eutectic Solvents for the Extraction and Separation of Natural Products." *Journal of Chromatography A* 1598, (2019): 1–19.

Jouyban, Abolghasem, Mir Ali Farajzadeh, and Mohammad Reza Afshar Mogaddam. "Dispersive Liquid–liquid Microextraction Based on Solidification of Deep Eutectic Solvent Droplets for Analysis of Pesticides in Farmer Urine and Plasma by Gas Chromatography–Mass Spectrometry." *Journal of Chromatography B* 1124, (2019a): 114–21.

Jouyban, Abolghasem, Mir Ali Farajzadeh, Maryam Khoubnasabjafari, Vahid Jouyban-Gharamaleki, and Mohammad Reza Afshar Mogaddam. "Development of Deep Eutectic Solvent Based Solidification of Organic Droplets-Liquid Phase Microextraction; Application to Determination of Some Pesticides in Farmers Saliva and Exhaled Breath Condensate Samples." *Analytical Methods* 11, (2019b): 1530–40.

Jouyban, Abolghasem, Mir Ali Farajzadeh, Mahboob Nemati, Ali Akbar Alizadeh Nabil, and Mohammad Reza Afshar Mogaddam. "Preparation of Ferrofluid From Toner Powder and Deep Eutectic Solvent Used in Air-assisted Liquid-Liquid Microextraction: Application in Analysis of Sixteen Polycyclic Aromatic Hydrocarbons in Urine and Saliva Samples of Tobacco Smokers." *Microchemical Journal* 154, (2020): 104631.

Kachangoon, Rawikan, Jitlada Vichapong, Yanawath Santaladchaiyakit, and Supalax Srijaranai. "Cloud-Point Extraction Coupled to In-situ Metathesis Reaction of Deep Eutectic Solvents for Preconcentration and Liquid Chromatographic Analysis of Neonicotinoid Insecticide Residues in Water, Soil and Urine Samples." *Microchemical Journal* 152, (2020): 104377.

Kalhor, Payam, and Khashayar Ghandi. "Deep Eutectic Solvents for Pretreatment, Extraction, and Catalysis of Biomass and Food Waste." *Molecules* 24, (2019): 4012.

Kanberoglu, Gulsah Saydan, Erkan Yilmaz, and Mustafa Soylak. "Usage of Deep Eutectic Solvents for the Digestion and Ultrasound-assisted Liquid Phase Microextraction of Copper in Liver Samples." *Journal of the Iranian Chemical Society* 15, (2018): 2307–14.

Karimi, Mehdi, Shayessteh Dadfarnia, and Ali Mohammad Haji Shabani. "Hollow Fibre-supported Graphene Oxide Nanosheets Modified with a Deep Eutectic Solvent to be Used for the Solid-Phase Microextraction of Silver Ions." *International Journal of Environmental Analytical Chemistry* 98, (2018): 124–37.

Khataei, Mohammad Mahdi, Yadollah Yamini, Ali Nazaripour, and Meghdad Karimi. "Novel Generation of Deep Eutectic Solvent as an Acceptor Phase in Three-Phase Hollow Fiber Liquid Phase Microextraction for Extraction and Preconcentration of Steroidal Hormones From Biological Fluids." *Talanta* 178, (2018): 473–80.

Khezeli, Tahera, and Ali Daneshfar. "Dispersive Micro-Solid-Phase Extraction of Dopamine, Epinephrine and Norepinephrine From Biological Samples Based on Green Deep Eutectic Solvents and Fe_3O_4@ MIL-100 (Fe) Core–Shell Nanoparticles Grafted with pyrocatechol." *RSC Advances* 5, (2015): 65264–73.

Lamei, Navid, Maryam Ezoddin, and Khosrou Abdi. "Air Assisted Emulsification Liquid-Liquid Microextraction Based on Deep Eutectic Solvent for Preconcentration of Methadone in Water and Biological Samples." *Talanta* 165, (2017): 176–81.

Li, Guizhen, and Kyung Ho Row. "Utilization of Deep Eutectic Solvents in Dispersive Liquid-Liquid Micro-Extraction." *TrAC Trends in Analytical Chemistry* 120, (2019): 115651.

Li, Guizhen, Tao Zhu, and Yingjie Lei. "Choline Chloride-based Deep Eutectic Solvents as Additives for Optimizing Chromatographic Behavior of Caffeic Acid." *Korean Journal of Chemical Engineering* 32, (2015): 2103–8.

Li, Linnan, Yamin Liu, Zhengtao Wang, Li Yang, and Huwei Liu. "Development and Applications of Deep Eutectic Solvent Derived Functional Materials in Chromatographic Separation." *Journal of Separation Science* 44, (2020): 1098–121.

Li, Na, Junjie Du, Di Wu, Jichao Liu, Ning Li, Zhiwei Sun, Guoliang Li, and Yongning Wu. "Recent Advances in Facile Synthesis and Applications of Covalent Organic Framework Materials as Superior Adsorbents in Sample Pretreatment." *TrAC Trends in Analytical Chemistry* 108, (2018): 154–66.

Li, Xiaoxia, and Kyung Ho Row. "Development of Deep Eutectic Solvents Applied in Extraction and Separation." *Journal of Separation Science* 39, (2016): 3505–20.

Liao, Qie Gen, Zhang Da Wen, Luo Lin Guang, and Zha Xi Ci Dan. "Ultrasonic-Assisted Dispersive Liquid–Liquid Microextraction Based on a Simple and Green Deep Eutectic Solvent for

Preconcentration of Macrolides From Swine Urine Samples." *Separation Science Plus* 3, (2020): 22–27.

Liu, Hanghui, and Purnendu K. Dasgupta. "Analytical Chemistry in a Drop. Solvent Extraction in a Microdrop." *Analytical Chemistry* 68, (1996): 1817–21.

Majidi, Seyedeh Maedeh, and Mohammad Reza Hadjmohammadi. "Hydrophobic Borneol-based Natural Deep Eutectic Solvents as a Green Extraction Media for Air-assisted Liquid-Liquid Micro-extraction of Warfarin in Biological Samples." *Journal of Chromatography A* 1621, (2020): 461030.

Makoś, Patrycja, Edyta Słupek, and Jacek Gębicki. "Hydrophobic Deep Eutectic Solvents in Microextraction Techniques–A Review." *Microchemical Journal* 152, (2020): 104384.

Malaei, Reyhane, Amir M. Ramezani, and Ghodratollah Absalan. "Analysis of Malondialdehyde in Human Plasma Samples Through Derivatization with 2,4-dinitrophenylhydrazine by Ultrasound-assisted Dispersive Liquid–Liquid Microextraction-GC-FID Approach." *Journal of Chromatography B* 1089, (2018): 60–69.

Martins, Mónia A. R., Simão P. Pinho, and João A. P. Coutinho. "Insights into the Nature of Eutectic and Deep Eutectic Mixtures." *Journal of Solution Chemistry* 48, (2019): 962–82.

Meng, Jia, and Xu Wang. "Microextraction by Packed Molecularly Imprinted Polymer Combined Ultra-High-Performance Liquid Chromatography for the Determination of Levofloxacin in Human Plasma." *Journal of Chemistry* 2019, (2019): 1–9.

Moghadam, Ahmad Ghoochani, Maryam Rajabi, and Alireza Asghari. "Efficient and Relatively Safe Emulsification Microextraction Using a Deep Eutectic Solvent for Influential Enrichment of Trace Main Anti-depressant Drugs From Complicated Samples." *Journal of Chromatography B* 1072, (2018): 50–59.

Mohebbi, Ali, Saeid Yaripour, Mir Ali Farajzadeh, Mohammad Reza Afshar Mogaddam. "Combination of Dispersive Solid Phase Extraction and Deep Eutectic Solvent–based Air–assisted Liquid–Liquid Microextraction Followed by Gas Chromatography–Mass Spectrometry as an Efficient Analytical Method for the Quantification of Some Tricyclic Antidep." *Journal of ChromatographyA* 1571, (2018): 84–93.

Płotka-Wasylka, Justyna, Małgorzata Rutkowska, Katarzyna Owczarek, Marek Tobiszewski, and Jacek Namieśnik. "Extraction with Environmentally Friendly Solvents." *TrAC Trends in Analytical Chemistry* 91, (2017): 12–25.

Rajabi, Maryam, Nooshin Ghassab, Maryam Hemmati, and Alireza Asghari. "Highly Effective and Safe Intermediate Based on Deep Eutectic Medium For Carrier Less-three Phase Hollow Fiber Microextraction of Antiarrhythmic Agents in Complex Matrices." *Journal of Chromatography B* 1104, (2019): 196–204.

Ramezani, Amir M., and Ghodratollah Absalan. "Employment of a Natural Deep Eutectic Solvent as a Sustainable Mobile Phase Additive for Improving the Isolation of Four Crucial Cardiovascular Drugs by Micellar Liquid Chromatography." *Journal of Pharmaceutical and Biomedical Analysis* 186, (2020): 113259.

Ramezani, Amir M., Ghodratollah Absalan, and Raheleh Ahmadi. "Green-modified Micellar Liquid Chromatography for Isocratic Isolation of Some Cardiovascular Drugs with Different Polarities Through Experimental Design Approach." *Analytica Chimica Acta* 1010, (2018): 76–85.

Ramezani, Amir M., Raheleh Ahmadi, and Ghodratollah Absalan. "Designing a Sustainable Mobile Phase Composition for Melamine Monitoring in Milk Samples Based on Micellar Liquid Chromatography and Natural Deep Eutectic Solvent." *Journal of Chromatography A* 1610, (2020): 460563.

Rastbood, Samira, Mohammad Reza Hadjmohammadi, and Seyedeh Maedeh Majidi. "Development of a Magnetic Dispersive Micro-Solid-Phase Extraction Method Based on a Deep Eutectic Solvent as a Carrier for the Rapid Determination of Meloxicam in Biological Samples." *Analytical Methods* 12, (2020): 2331–7.

Ribeiro, Bernardo D., Catarina Florindo, Lucas C. Iff, Maria A. Z. Coelho, and Isabel M. Marrucho. "Menthol-based Eutectic Mixtures: Hydrophobic Low Viscosity Solvents." *ACS Sustainable Chemistry & Engineering* 3, (2015): 2469–77.

Safavi, Afsaneh, Raheleh Ahmadi, and Amir M. Ramezani. "Vortex-assisted Liquid-Liquid Microextraction Based on Hydrophobic Deep Eutectic Solvent for Determination of Malondialdehyde and Formaldehyde by HPLC-UV Approach." *Microchemical Journal* 143, (2018): 166–74.

Sajid, Muhammad, and Justyna Płotka-Wasylka. "Combined Extraction and Microextraction Techniques: Recent Trends and Future Perspectives." *TrAC Trends in Analytical Chemistry* 103, (2018): 74–86.

Šandrejová, Jana, Natalia Campillo, Pilar Viñas, and Vasil Andruch. "Classification and Terminology In Dispersive Liquid–Liquid Microextraction." *Microchemical Journal* 127, (2016): 184–86.

Seidi, Shahram, Leila Alavi, Ali Jabbari. "Dispersed Solidified Fine Droplets Based on Sonication of a Low Melting Point Deep Eutectic Solvent: A Novel Concept for Fast and Efficient Determination of Cr(VI) in Urine Samples." *Biological Trace Element Research* 188, (2019): 353–62.

Shishov, Andrey, Artur Gerasimov, Daria Nechaeva, Natalia Volodina, Elena Bessonova, and Andrey Bulatov. "An Effervescence-assisted Dispersive Liquid–Liquid Microextraction Based on Deep Eutectic Solvent Decomposition: Determination of Ketoprofen and Diclofenac in Liver." *Microchemical Journal* 156, (2020): 104837.

Shishov, Andrey Y., Mvgeny V. Chislov, Daria V. Nechaeva, Leonid N. Moskvin, and Andrey V. Bulatov. "A New Approach for Microextraction of Non-steroidal Anti-inflammatory Drugs From Human Urine Samples Based on In-situ Deep Eutectic Mixture Formation." *Journal of Molecular Liquids* 272, (2018): 738–45.

Smith, Emma L., Andrew P. Abbott, and Karl S. Ryder. "Deep Eutectic Solvents (DESs) and Their Applications." *Chemical Reviews* 114, (2014): 11060–82.

Soltanmohammadi, Fatemeh, Mohammad Reza Afshar Moghadam, Maryam Khoubnasabjafari, and Abolghasem Jouyban. "Development of Salt Induced Liquid–Liquid Extraction Combined with Amine Based Deep Eutectic Solvent-dispersive Liquid–Liquid Microextraction; An Efficient Analytical Method for Determination of Three Anti-seizures in Urine Samples." *Pharmaceutical Sciences* 26, (2020): 323–31.

Tan, Ting, Mingliang Zhang, Yiqun Wan, and Hongden Qiu. "Utilization of Deep Eutectic Solvents as Novel Mobile Phase Additives for Improving the Separation of Bioactive Quaternary Alkaloids." *Talanta* 149, (2016):85–90.

Tang, Baokun, Heng Zhang, and Kyung Ho Row. "Application of Deep Eutectic Solvents in the Extraction and Separation of Target Compounds From Various Samples." *Journal of SeparationScience* 38, (2015): 1053–64.

Thongsaw, Arnon, Yuthapong Udnan, Gareth M. Ross, and Wipharat Chuachuad Chaiyasith. "Speciation of Mercury in Water and Biological Samples by Eco-friendly Ultrasound-assisted Deep Eutectic Solvent Based on Liquid Phase Microextraction with Electrothermal Atomic Absorption Spectrometry." *Talanta* 197, (2019): 310–8.

van Osch, Dannie J.G.P., Lawien F. Zubeir, Adriaan van den Bruinhorst, Marisa A.A. Rocha, and Maaike C. Kroon. "Hydrophobic Deep Eutectic Solvents as Water-Immiscible Extractants." *Green Chemistry* 17, (2015); 4518–21.

Wang, Rong, Wen Li, and Zilin Chen. "Solid Phase Microextraction with Poly(Deep Eutectic Solvent) Monolithic Column Online Coupled to HPLC for Determination of Non-steroidal Anti-inflammatory Drugs." *Analytica Chimica Acta* 1018, (2018): 111–8.

Yousefi, Seyedeh Mahboobeh, Farzaneh Shemirani, and Sohrab Ali Ghorbanian. "Enhanced Headspace Single Drop Microextraction Method Using Deep Eutectic Solvent Based Magnetic Bucky Gels: Application to the Determination of Volatile Aromatic Hydrocarbons in Water and Urine Samples." *Journal of Separation Science* 41, (2018): 966–74.

Zhang, Kaige, Shuangying Li, Chuang Liu, Qi Wang, Yunhe Wang, and Jing Fan. "A Hydrophobic Deep Eutectic Solvent-based Vortex-assisted Dispersive Liquid-Liquid Microextraction Combined with HPLC for the Determination of Nitrite in Water and Biological Samples." *Journal of Separation Science* 42, (2019a): 574–81.

Zhang, Kaige, Chuang Liu, Shuangying Li, and Jing Fan. "A Hydrophobic Deep Eutectic Solvent Based Vortex-assisted Liquid-Liquid Microextraction for the Determination of Formaldehyde From Biological and Indoor Air Samples by High Performance Liquid Chromatography." *Journal of Chromatography A* 1589, (2019b): 39–46.

15

Hyphenation of Derivatization with Microextraction Techniques in Analytical Toxicology

Abhishek Chauhan and Utkarsh Shukla
Polymer R&D, North Site, Atul Ltd. Atul, Valsad (Gujarat), India

CONTENTS

15.1	Introduction	239
15.2	Overview of Microextraction Techniques Coupled with Derivatization	240
	15.2.1 Solid-Phase Microextraction (SPME)	240
	15.2.2 Dispersive Liquid-Liquid Microextraction (DLLME)	241
	15.2.3 Single Drop Microextraction (SDME)	241
	15.2.4 Liquid-Phase Microextraction (LPME)	242
	15.2.5 Hollow-Fibre Liquid-Phase Microextraction Technique (HF-LPME)	242
15.3	Derivatization Techniques and Their Various Modes	242
	15.3.1 Derivatization Techniques for Gas Chromatography	242
	15.3.2 Derivatization Techniques for High-Performance Liquid Chromatography	243
	15.3.2.1 Fluorescent Derivatization	243
	15.3.2.2 UV Derivatization	243
15.4	Application of Combination of Derivatization with Microextraction Techniques in Analytical Toxicology	243
	15.4.1 Water Samples	243
	15.4.2 Urine Samples	244
	15.4.3 Other Biological Samples	245
	15.4.4 Application on Other Sample Matrixes	245
15.5	Conclusion	246
References		246

15.1 Introduction

In the present scenario of global development, people are exposed to numerous toxicants through various modes in their day-to-day lives. These chemicals move in the environment as well as in the food chain via different sources (water, soil, air, etc.) and accumulate in human beings in acute or chronic ways. This results in different type of toxicities. Identification of analytes or toxicants in complex matrices like blood, tissues, food, etc., is the most challenging task in the field of analytical toxicology.

Thus, it is necessary to identify these toxicants efficiently to find out the root cause and take corrective action to overcome the problem. With the advancement in analytical instrumentation for determining analytes, there has also been developments or advancements in sample preparation techniques that decrease the cost and time of analysis by reducing or eliminating the use of toxic organic solvents.

Various microextraction techniques have been implemented for the extraction and pre-concentration of analytes for their analysis using different chromatographic instruments. Examples of some popular types of microextraction techniques are solid-phase microextraction (SPME), dispersive liquid-liquid microextraction (DLLME), single drop microextraction (SDME), and liquid-phase microextraction (LPME).

In order to analyze a wide range of organic compounds by microextraction and chromatographic techniques, several improvements were done, for which chemical derivatization of polar organic compounds is an important aspect while dealing with their chromatographic analysis.

Generally, derivatization is performed to make organic compounds volatile enough to get analyzed in gas chromatography (GC) or to make UV/fluorescent active to get analyzed in high-performance liquid chromatography (HPLC). In the field of analytical toxicology, derivatization is preferred to take place at room temperature within a short span of time with good solubility of derivatives in organic solvents (Rutkowska et al. 2014; Lin et al. 2013; Risticevic et al. 2010).

The derivatization process can be classified into three modes:

1. Pre-column
2. Post-column
3. On-column

Additionally, on the basis of chromatographic instruments, the derivatization process can be further classified as follows:
For GC

1. Alkylation
2. Acylation
3. Silylation

For HPLC

1. UV derivatization
2. Fluorescent derivatization

Therefore, the coupling of derivatization with microextraction techniques results in fast, sensitive, high enrichment, and effective pre-concentration of analytes.

15.2 Overview of Microextraction Techniques Coupled with Derivatization

As advancements in the field of analytical toxicology, various microextraction techniques, combined with different chemical derivatization approaches, have been developed, which are discussed here (Kabir et al. 2017).

15.2.1 Solid-Phase Microextraction (SPME)

SPME is a well-established microextraction technique. It is a solvent-less extraction technique that involves a fibre (with polymeric coating) as an adsorption media (extracting phase). This technique was first established by Pawliszyn et al. (Risticevic et al. 2010;Kabir et al. 2017).

SPME is a two-step extraction technique:

1. Adsorption
2. Desorption

Here, first the analyte gets adsorbed into the fibre and then the fibre gets directly desorbed into the analytical instrument (GC, GC-mass spectrometry, etc.).

The two different modes of SPME are as follows:

1. Headspace SPME (HS-SPME) – Fibre is exposed to the vapours phase above the sample.
2. Direct immersion SPME (DI-SPME) – Fibre gets directly immersed in the sample.

The optimization parameters required for SPME are as follows:

1. Coating of the fibre
2. Extraction mode
3. Sample volume
4. Sample agitation
5. pH
6. Ionic strength
7. Extraction time and temperature
8. Desorption time and temperature
9. Derivatization in the case of polar analytes

15.2.2 Dispersive Liquid-Liquid Microextraction (DLLME)

DLLME is based on the ternary solvent system, i.e., extraction solvent, dispersion solvent, and sample. The process of DLLME involves the injection of the extraction and dispersion solvent system to an aqueous sample, resulting in the formation of a cloudy solution (Sajid 2018). A pre-requirement of traditional DLLME is as follows:

1. Extraction solvent must be immiscible with water and miscible with dispersive solvent.
2. Extraction solvent must show high affinity with target analytes.

Traditional DLLME follows the use of high-density extraction solvent possessing (e.g., dichloromethane, chloroform, carbon tetrachloride) in comparison to water, but advancement takes place, and the use of low-density extraction solvent (hexane, toluene, xylene, etc.,) takes place as an alternative mode of DLLME, as discussed in previous chapters.

Advantages of DLLME involve the following:

1. Instantaneous partitioning of analytes from the sample
2. Easy recovery of extraction solvent
3. Smaller amount of solvent required (in microliters)
4. High enrichment factor
5. Sensitive, rapid, and cost-effective

15.2.3 Single Drop Microextraction (SDME)

The SDME concept was first introduced by Cantwell This technique used an acceptor phase, where a small micro-droplet (upto 2 µL) of extraction solvent was suspended, which is compatible with the GC system.

Classification of SDME is as follows:

1. Two-phase SDME: This includes direct immersion SDME, continuous-flow SDME, drop-to-drop SDME, and directly suspended SDME.
2. Three-phase SDME: This include HS-SDME and liquid-liquid-liquid SDME.

Advantages of SDME include the following:

1. Fast and sensitive analysis
2. High distribution coefficient
3. Low consumption of sample and solvent

15.2.4 Liquid-Phase Microextraction (LPME)

LPME is a solvent-minimized sample preparation technique, alternative to liquid-liquid extraction (LLE). Here, solvent in microliters is used to extract the analytes rather than in millilitres to litres, as in LLE.

LPME is known for its rapidness, simplicity, low cost, low solvent consumption, need for a small amount of sample, reduced waste, and high enrichment of analytes.

15.2.5 Hollow-Fibre Liquid-Phase Microextraction Technique (HF-LPME)

Rasmussen et al. introduced the concept of HF-LPME. Here, a polypropylene-based hollow fibre is utilized as a membrane. This should ensure that the solvent is immiscible with water to establish the retentivity of solvent within the pores of the membrane. A thin layer is formed by organic solvent within the wall of the hollow fibre. The type of acceptor phase decides the mode of the system, i.e., two-phase or three-phase system.

For the two phase system, the generally used solvents are toluene and n-octanol as the organic phase.
For the three-phase system, n-octanol and dihexyl ether are used as supported liquid membrane (SLM).

15.3 Derivatization Techniques and Their Various Modes

As discussed earlier, derivatization is the process of chemically modifying the analyte to make it amenable for the chromatographic analysis.

Different modes of derivatization are performed based on chromatographic or analytical instruments.

15.3.1 Derivatization Techniques for Gas Chromatography

As discussed earlier, the most commonly used derivatization procedures for analysis of polar analytes by GC include alkylation, acetylation, and silylation:

Alkylation is the widely used derivatization process to protect -OH (carboxylic), -SH, and -NH groups to enhance the volatility of the compound and improve its chromatographic properties (Lin et al. 2013; Moldoveanu and David 2018; Halket and Zaikin 2004). Alkyl halides (alkyl bromide, alkyl iodide, etc.) are the most widely used reagents for alkylation.

Some of the other used reagents for alkylation are tetramethylammonium hydroxide, diazomethane, etc.

Acetylation refers to the introduction of an acetyl group into a chemical compound. In this process, a number of analytical steps get minimized due to direct derivation of analytes in the water sample within a minute. Various acetylating reagents are used for derivatization, such as acetic anhydride and chloroformates (methyl chloroformates, ethyl chloroformates, etc.). The derivatization process takes place at room temperature. Acetylation has been used as a derivatization process for the analysis of environmental, pharmaceutical, forensic, and clinical samples.

Silylation is the most widely used technique of the derivatization approach for GC analysis of polar analytes. Most of the functional groups get covered and easily derivatized using the silylation technique. Here, the hydrogen of the polar group of analytes gets replaced with the trimethylsilyl group. The nucleophilic substitution reaction (S_N2) mechanism takes place in the presence of a strong leaving group.

Various silylating reagents are available as derivatizing reagents, such as Bis(trimethylsilyl)trifluoroacetamide (BSTFA) and N-methyl-N-(ter-butyldimethylsilyl) trifluoroacetamide (MTBSTFA).

15.3.2 Derivatization Techniques for High-Performance Liquid Chromatography

In HPLC, derivatization is performed in order to introduce groups in a molecule to increase sensitivity for a UV or fluorescent detector (Moldoveanu and David 2018). Here, pre- and post-column derivatization are performed for detection of a UV or fluorescent detector.

15.3.2.1 Fluorescent Derivatization

This technique is used to derivatize non- or weekly fluorescing compound to enhance the sensitivity of the analysis. It also increases the linear dynamic response range and selectivity.

Some of the fluorescent derivatizing reagents are 9-chloromethyl anthracene, 9-(2-hydroxy ethyl)-carbazole, 5-(4-pyridyl)-2-tiophenmethanol, dansyl chloride, etc. (Xie et al. 2012).

15.3.2.2 UV Derivatization

Similar to fluorescent derivatization, various chromophores are used to convert UV inactive molecule into UV active. Some of the commonly used derivatizing reagents are o-phthalaldehyde, Ninhydrin, 2,4-Dinitro phenyl hydrazine etc., (Adegoke 2012).

15.4 Application of Combination of Derivatization with Microextraction Techniques in Analytical Toxicology

Combination of microextraction technique with derivatization proves itself a very promising approach in analytical toxicology due to its high enrichment factor, making analyte suitable for chromatographic analysis (GC, HPLC, etc.), low solvent consumption, eco-friendly, rapid analysis, etc. Various applications of coupling these two approaches in different matrices have been explored in the field of analytical toxicology. Some of them are described in this section.

15.4.1 Water Samples

Varanusupakul et al. developed a method for analyzing haloacetic acids in water using in-situ derivatization coupled with a hollow-fibre membrane-based microextraction technique (Varanusupakul et al. 2007). In this work, analytes were derivatized using acidic methanol and were simultaneously extracted by using hollow-fibre membrane in HS mode. The method achieved a good limit of detection, below 18 µg/L for all haloacetic acid.

Zhang et al. utilized an LPME technique followed by injector-port derivatization for the determination of acidic pharmaceutically active compounds (clofibric acid, ibuprofen, naproxen, and ketoprofen) using GC-MS in water samples (Zhang and Lee 2009). Trimethylanilinium hydroxide (TMAH) was used as a derivatizing reagent. The limit of detection that was achieved ranged between 0.01 and 0.05 µg/L.

Rodriiguez et al. explored the concept of SPME coupled with on-fibre derivatization followed by GC-MS analysis for the analysis of anti-inflammatory drugs in sewage water samples (Rodriiguez et al. 2004). N-methyl-N-(tert-butyldimethylsilyl)-trifluoroacetamide (MTBSTFA) was used as a reagent for the derivatization of drugs directly inside the heated injection port of GC-MS. The limit of quantification of this reported method ranged from 12 to 40 ng/L.

Basheer et al. determined the endocrine-disrupting alkylphenols, chlorophenols, and bisphenol A by using hollow-fibre protected LPME coupled with injector-port derivatization followed by GC-MS analysis (Basheer and Lee 2004). Derivatization was performed by using bis(trimethylsilyl)trifluoroacetamide (BSTFA). This method showed a good enrichment factor, up to 162 fold, and the limit of detection ranged from 0.005 to 0.015 µg/L in the selective ion monitoring mode.

Esaghi et al. developed a method for determining non-steroidal anti-inflammatory drugs by using in-situ derivatization coupled with HF-LPME in municipal waste water samples (Eshaghi 2009). Here, the drugs taken into consideration for analysis were ibuprofen, naproxen, and ketoprofen. In this work, sealed polypropylene hollow-fibre was used for extraction purposes and tetra butyl ammonium sulphate as a derivatizing reagent. The detection limit of method ranged from 1 to 2 ng/L.

In one work, Loua et al. developed a method for determining octylphenol and nonylphenol in water samples by using the concept of simultaneous derivatization coupled with DLLME (Luoa et al. 2010). In this method, a mixture of dispersion solvent and catalyst (methanol and pyridine mixture) was added to the known amount of sample. In continuation, the derivatizing reagent and the extraction solvent (i.e., methyl chloroformate and chloroform) was injected rapidly to perform simultaneous derivatization and extraction. The limit of detection for both the analytes was found to be 0.03 and 0.02 μg/L, respectively.

15.4.2 Urine Samples

A method was developed by Nabil et al. to determine antidepressants (fluvoxamine, nortriptyline, and maprotiline) in urine samples by coupling simultaneous derivatization with temperature-assisted DLLME followed by GC-FID analysis (Nabil et al. 2015). In this method, dimethylformamide was used as a disperser solvent, 1,1,2,2-tetrachloroethane as an extraction solvent, and acetic anhydride as a derivatizing reagent. The enrichment factor was found to be 820–1070 fold, whereas the detection limit was found to be in the range of 2 to 4 ng/mL.

Jain et al. developed a rapid, simple, and cost-effective method for determining quinine by using ultrasound-assisted DLLME followed by injector-port silylation in urine samples (Jain et al. 2013). Ethanol was used as a disperser solvent, dichloromethane as an extraction solvent, and N,O-bis (trimethylsilyl)trifluoroacetamide (BSTFA) as a derivatizing reagent. The method achieved good linearity ($R^2 = 0.999$), and the detection limit was 5 ng/mL.

In another work, Gupta et al. utilized the same concept for determining polycyclic aromatic hydrocarbons (PAHs) metabolites {1-napthol (NAP), 9-phenanthrol (PHN), 1-hydroxypyrene (1-OHP), 2-hydroxyfluorene (2-HF), 3-hydroxyfluorene (3-HF), 9-hydroxyfluorene (9-HF), and 6-hydroxychrysene (OH-CHRY)} in urine samples using DLLME coupled with injector-port silylation followed by GC-MS/MS analysis (Gupta et al. 2015). Trichloroethylene was used as an extraction solvent, ethanol as a disperser solvent, and BSTFA as a derivatizing reagent. The limit of detection of the defined method ranged from 1 to 9 ng/mL and showed good linearity with R^2 ranges from 0.987 to 0.999.

Lin et al. extracted pyrethroid metabolites such as 3-phenoxybenxoic acid, 4-fluoro-3-phenoxybenzoicacid, and cis and trans-3-(2,2-dichlorovinyl)-2,2-dimethylcyclopropane-1-carboxylic acid by using LPME coupled with in-syringe derivatization followed by GC-ECD analysis of urine samples (Lin et al. 2015). The method showed high enrich factor ranges from 69.8 to 154.6. The detection limit ranged from 1.6 to 17 ng/mL.

Racamonde et al. demonstrated the application of SPME coupled with in-sample derivatization for the extraction and determination of amphetamines and ecstasy-related stimulants (methamphetamine, 3,4-Methylendioxyamphetamine, 3,4-Methylendioxymethamphetamine, and 3,4-Methylendioxyethylamphetamine) in water and urine samples. Isobutyl chloroformate was used as a derivatizing reagent, and PDMS-DVB (polydimethylsioloxane-divinylbemzene) fibre was used for extraction purposes (Racamonde et al. 2013). The detection limit of method was found to be less than1 μg/L in both the matrixes.

Further, Klimowska et al. developed a technique where off-line microextraction was performed using packed sorbent followed by a solid support derivatization process to determine pyrethroid metabolites (cis-2,2-dimethyl-3-(2-chloro-3,3,3-trifluoro-1-propenyl)-cyclopropanecarboxylic acid, cis/trans-3-(2,2-dichlorovinyl)-2,2-dimethylcyclopropane-1-carboxylic acids, cis-(2,2-dibromovinyl)-2,2-dimethylcyclopropane-1-carboxylic acid, and 3-phenoxybenzoic acid) in urine samples using GC-MS (Klimowska and Wielgomes 2018). Microextraction was performed using a manually operated semiautomatic syringe and derivatization using 1,1,1,3,3,3-hexafluoroisopropanol and diisopropylcarbodiimide. The developed method showed quantification limit ranges from 0.06 to 0.08 ng/mL/.

A method was developed for the analysis of various metabolites of fluorene by Geimner et al. (2002). In this method, SPME was used for the extraction of analytes, whereas HS silylation was performed for on-fibre derivatization purposes using BSTFA and MTBSTFA. The method was found suitable for profile analysis of polycyclic aromatic hydrocarbon metabolites in a urine sample of an occupationally exposed person. Separation was found satisfactory for all metabolites without matrix interference.

15.4.3 Other Biological Samples

In the field of analytical toxicology, biological samples like blood, milk, tissues, etc., play an important role where xenobiotics are analyzed to find out the severity of exposure. Several literatures are available where application of simultaneous derivatization coupled with microextraction showed a significant outcome of the study.

A solvent-free extraction methodology coupled with on-fibre silylation was explored by Aresta et al. to determine phytoestrogens (equol, enterodiol, daidzen, genistein, and glycitein) in different milk samples of cows, goats, and soy-rice (Aresta et al. 2019). Direct immersion SPME was used for extraction, whereas BSTFA was used for on-fibre derivatization followed by GC-MS analysis. The detection limit ranged from 0.02 to 0.28 ng/mL, whereas the quantification level ranged from 0.08 to 0.95 ng/mL.

Bustamante et al. utilized the concept of packed sorbent, LLME, and derivatization for the analysis and determination of diet-derived phenolic acids (e.g., phenylacetic acid, butanedioic acid) in gerbil plasma samples (Bustamante et al. 2017). In this method, optimization and comparison of extraction techniques and derivatization processes were performed. Derivatization was performed using BSTFA. Multivariate analysis was also performed to derive the optimum method condition for the extraction and derivatization of the analytes.

Erarpat et al. developed a method for combining ultrasound-assisted derivatization and switchable solvent LPME for identifying l-methionine in a human plasma sample (Erarpat et al. 2019). Here, ethyl chloroformate was used as a derivatizing reagent followed by LPME. N,N-Dimethyl benzylamine was used for the synthesis of a switchable solvent. The method possesses a good detection and qualification limit of 3.3 ng/g and 11.0 ng/g, respectively.

15.4.4 Application on Other Sample Matrixes

In another report, Sun et al. developed a method by using HS-LPME coupled with needle-based derivatization for the extraction and determination of volatile organic acids (e.g., formic acid, propionic acid, caproic acid) in tobacco. Derivatization was performed using BSTFA (Sun et al. 2008). Here, the mixture of BSTFA and decane was used as a solvent for HS-LPME. The method found good applicability in tobacco samples, with good linearity and repeatability.

Xu et al. proposed an analytical method for the analysis of 3-monochloropropane-1,2-diol (3-MCPD) and 2-monchloropropane-1,3-diol (2-MCPD) fatty acid ester and glycidyl esters in edible oil by using in-situ derivatization and HS-SPME (Xu et al. 2020). The central composite design was implemented to optimize the extraction time and derivatization temperature. The method possesses detection limits of 3.9 µg/L and 5.3 µg/L for 3-MCPD and 2-MCPD, respectively.

A sensitive and rapid method was developed by Lee et al. to determine 4-methylimidazole in a red ginseng product possessing caramel colours by using coupling of DLLME with in-situ derivatization followed by GC-MS analysis (Lee et al. 2018). In this method, chloroform and acetonitrile were used as extraction and disperser solvents, whereas isobutyl chloroformate was used as a derivatizing reagent. The detection limit of the method for 4-methylimidazole was found to be 0.96 µg/L.

A technique based on DLLME coupled with 2-Naphthalenthiol was developed by Faraji et al. for the analysis and determination of acrylamide in bread and biscuit samples using HPLC (Faraji et al. 2018). In this technique, ultrasound-assisted DLLME followed the extraction of analyte, whereas derivatization was done by using 2-Naphthalenthiol. The method was found to be simple, sensitive, and rapid, with a detection limit of 3µg/L and quantification limit of 9 µg/L.

Farajzadeh et al. developed a method for the extraction and derivatization of parabens by using a technique that was lighter than water, air-assisted LLME, by using a homemade device for extraction purposes (Farajzadeh et al. 2018). Here, an organic solvent was used, which was lighter than water for the paraben (methyl, ethyl, and propyl parabens) extractions from different samples (cosmetics, hygiene, and food samples). Acetic anhydride was used as a derivatizing reagent, and p-xylene was used as an extraction solvent. An inverse funnel with capillary tube was used as a homemade device. This method showed an enrichment and enhancement factor in the range of 370 to 430 and 489 to 660, respectively. The detection limit was found in the range of 0.90 to 2.7 ng/mL.

A microextraction technique coupled with derivatization was developed by Jouyban et al. to determine salbutamol in exhaled breath condensate (EBC) samples by using GC-MS (Jouyban et al. 2020). Extraction was performed by using microwave-enhanced air-assisted LLME. In this method, 1-fluoro-2,4-dinitrobenzene was used as a derivatizing reagent, whereas a mixture of N,N-diethylethanolammonium chloride, dichloroacetic acid, and octanoic acid deep eutectic solvent was used as an extraction solvent. The method showed a detection limit of 0.074 and 0.37 μg/L in water and EBC, respectively.

Norouzi et al. proposed a method for determining morphine and oxymorphone by coupling derivatization with microwave-enhanced three-component deep eutectic solvent-based air-assisted LLME followed by GC-MS analysis. The method was based on a ternary deep eutectic solvent, which was used as an extraction solvent (Norouzi et al. 2020). The method was found to be cost-effective, rapid, efficient, and green. The extraction solvent was a mixture of choline chloride-menthol-phenylacetic acid deep eutectic solvent, whereas butyl chloroformate was used as a derivatization agent and picoline was used as a catalyst for derivatization. The detection limit for morphine and oxymorphone was found to be 2.1 and 1.5 ng/mL, respectively.

One application of SDME coupled with in-syringe derivatization was explored by Saraji et al. in 2006 for determining organic acids (Cinnamic acid, o-coumaric acid, caffeic acid, and p-hydroxybenzoic acid) in fruits and juices by using GC-MS (Saraji and Mousavinia 2006). Here, hexyl acetate was used as an extraction solvent and BSTFA as a derivatizing reagent. Only 2.5 μl of solvent was used to extract the analytes from the matrix. The detection limit for different acids was found to be in the range of 0.6 to 164 ng/mL.

Recently, Jain et al. reported a simple, rapid, and cost-effective analytical procedure based on coupling of DLLME with injection-port silylation for quantitative analysis of morphine in illicit opium samples. After performing DLLME with chloroform and acetone as extraction and disperser solvents, the sedimented phase along with the derivatizing reagent, i.e., N,O-Bis(trimethylsilyl)acetamide (BSA), was injected into the heated injection port of GC-MS. Unlike in-vial silylation, this simple approach resulted in instantaneous derivatization of morphine without the need for any external derivatization apparatus, lengthy reaction time, and moisture-free conditions (Jain et al.).

15.5 Conclusion

Many xenobiotic compounds possess polar functional groups, such as carboxyl, hydroxyl, amine, carbonyl, etc., which undergo hydrogen bonding that contributes significantly to intermolecular interactions. Additionally, the metabolism of these xenobiotics in the body also adds polar functional groups in order to convert them into more hydrophilic compounds. The primary reason for derivatization of such xenobiotics is to increase their volatility, reduce their polarity, and improve their chromatographic behaviour, mainly for GC analysis. The combination of microextraction strategies with derivatization methods not only results in an overall analytical method that is rapid, eco-friendly, cost-effective, and sensitive but also provides a versatile analytical platform for analysis of xenobiotics and their metabolites, which has a high sample throughput and is less laborious in comparison to traditional sample preparation approaches.

REFERENCES

Adegoke, Olajire A. "Chemical Derivatization Methodologies for UV-Visible Spectrophotometric Determination of Pharmaceuticals." *International Journal of Pharmaceutical Sciences Review and Research* 14, (2012): 6–24.

Aresta, Antonella, Pietro Cotugno, and Carlo Zambonin. "Solid-Phase Microextraction and On-fiber Derivatization for Assessment of Mammalian and Vegetable Milks with Emphasis on the Content of Major Phytoestrogens." *Scientific Reports* 9-6398, (2019): 1–8.

Basheer, Chanbasha H., and Hian Kee Lee. "Analysis of Endocrine Disrupting Alkylphenols, Chlorophenols and Bisphenol-a Using Hollow Fiber-protected Liquid-Phase Microextraction Coupled with Injection Port-Derivatization Gas Chromatography–Mass Spectrometry." *Journal of Chromatography A* 1057, (2004): 163–9.

Bustamante, Luis, Diana Cardenas, Dietrich von Baer, Edgar Pastene, Daniel Duran- Sandoval, Carola Vergara, and Claudia Mardones. "Evaluation of Microextraction by Packed Sorbent, Liquid-Liquid Microextraction and Derivatization Pretreatment of Diet-Derived Phenolic Acids in Plasma by Gas Chromatography with Triple Quadrupole Mass Spectrometry." *Journal of Separation Science* 40, (2017): 3487–96.

Erarpat, Sezin, Süleyman Bodur, Elif Öztürk Er, and Sezgin Bakirdere. "Combination of Ultrasound-Assisted Ethyl Chloroformate Derivatization and Switchable Solvent Liquid-Phase Microextraction for the Sensitive Determination of L-methionine In Human Plasma by GC-MS." *Journal of Separation Science* 43, (2019): 1100–06.

Eshaghi, Zarrin. "Determination of Widely Used Non-Steroidal Anti-Inflammatory Drugs in Water Samples by In Situ Derivatization, Continuous Hollow Fiber Liquid-Phase Microextraction and Gas Chromatography-Flame Ionization Detector." *AnalyticaChimicaActa* 641, (2009): 83–88.

Faraji, Mohammad, Mohammadrezza Hamdamali, Fezzeh Aryanasab, and Meisam Shabanian. "2-Naphthalenthiol Derivatization Followed by Dispersive Liquid-Liquid Microextraction as an Efficient and Sensitive Method for Determination of Acrylamide in Bread and Biscuit Samples Using High-Performance Liquid Chromatography." *Journal of Chromatography A* 1558, (2018): 14–20.

Farajzadeh, Mir Ali, Mehri Bakhshizadeh Aghdam, Mohammad Reza Afshar Mogaddam, and Ali Akbar Alizadeh Nabil. "Simultaneous Derivatization and Lighter-Than-Water Air-Assisted Liquid-Liquid Microextraction Using a Homemade Device for the Extraction and Preconcentration of Some Parabens in Different Samples." *Journal of Separation Science* 41, (2018): 3105–12.

Geimner, Günter, Peter Gartner, Christian Krassnig, and H. Tausch. "Identification of Various Urinary Metabolites of Fluorene Using Derivatization Solid-Phase Microextraction." *Journal of Chromatography B* 766, (2002): 209–18.

Gupta, Manoj Kumar, Rajeev Jain, Pratibha Singh, Ratnasekhar Ch, and Mohana Krishna Reddy Mudiam. "Determination of Urinary PAH Metabolites Using DLLME Hyphenated to Injector Port Silylation and GC–MS-MS." *Journal of Analytical Toxicology* 39, (2015): 365–73.

Halket, John M., and Vladimir V. Zaikin. "Derivatization in Mass Spectrometry– 3. Alkylation (Arylation)." *European Journal of Mass Spectrometry* 10, (2004): 1–19.

Jain, Rajeev, Meenu Singh, Aparna Kumari, and Rohitshva Mani Tripathi. "A Rapid and Cost-Effective Method Based on Dispersive Liquid-Liquid Microextraction Coupled to Injection Port Silylation-gas Chromatography-Mass Spectrometry for Determination of Morphine In Illicit Opium." *Analytical Science Advances* (2020): 1–10

Jain, Rajeev, Mohana Krishna Reddy Mudiam, Ratnasekhar Ch, Abhishek Chauhan, Haider A. Khan, and R.C. Murthy. "Ultrasound Assisted Dispersive Liquid-Liquid Microextraction Followed by Injector Port Silylation: A Novel Method for Rapid Determination of Quinine in Urine by GC-MS." *Bioanalysis* 5, (2013): 2277–86.

Jouyban, Abolghasem, Mir Ali Farajzadeh, Maryam Khoubnasabjafari, Vahid Jouyban-Gharamaleki, and Mohammad Reza Afshar Mogaddam. "Derivatization and Deep Eutectic Solvent-Based Air-Assisted Liquid-Liquid Microextraction of Salbutamol in Exhaled Breath Condensate Samples Followed by Gas Chromatography-Mass Spectrometry." *Journal of Pharmaceutical and Biomedical analysis* 191, (2020): 113572.

Kabir, Abuzar, Marcello Locatelli, and Halil I. Ulusoy. "Recent Trends in Microextraction Techniques Employed in Analytical and Bioanalytical Sample Preparation." *Separations* 3, (2017): 1–15.

Klimowska, Anna, and Bartosz Wielgomes. "Off-line Microextraction by Packed Sorbent Combined with on Solid Support Derivatization and GC-MS: Application for the Analysis of Five Pyrethroid Metabolites in Urine Samples." *Talanta* 176, (2018): 165–71.

Lee, Jangho, Yunyeol Lee, Tae Gyu Nam, and Hae Won Jang. "Dispersive Liquid-Liquid Microextraction with In Situ Derivatization Coupled with Gas Chromatography and Mass Spectrometry for the Determination of 4-methylimidazole in Red Ginseng Products Containing Caramel Colors." *Journal of Separation Science* 41, (2018): 3415–423.

Lin, Chiu-Hwa, Cheing-Tong Yan, Ponnusamy Vinoth Kumar, Hong-Ping Li, and Jen-Fon Jen. "Determination of Pyrethroid Metabolites in Human Urine Using Liquid Phase Microextraction

Coupled In-syringe Derivatization Followed by Gas Chromatography/Electron Capture Detection." *Analytical and Bioanalytical Chemistry* 401, (2015): 927–37.

Lin, Huaqing, Jiale Wang, Lingjie Zeng, Gang Li, Yunfei Sha, Da, Wu, and Baizhan, Liu. "Development of Solvent Micro-extraction Combined with Derivatization." *Journal of Chromatography A* 1296, (2013): 235–42.

Luoa, Shusheng, Ling Fanga, Xiaowei Wanga, Hongtao Liub, Gangfeng Ouyanga, Chongyu, Lana and Tiangang Luan. "Determination of Octylphenol and Nonylphenol in Aqueous Sample Using Simultaneous Derivatization and Dispersive Liquid–liquid Microextraction Followed by Gas Chromatography–Mass Spectrometry." *Journal of Chromatography A* 1217, (2010): 6762–8.

Moldoveanu, Serban C., and Victor David. Derivatization Methods in GC and GC/MS 2018. https://www.intechopen.com/books/gas-chromatography-derivatization-sample-preparation-application/derivatization-methods-in-gc-and-gc-ms.

Nabil, Ali Akbar Alizadeh, Nina Nouri, and Mir Ali Farajzadeh. "Determination of Three Antidepressants in Urine Using Simultaneous Derivatization and Temperature-assisted Dispersive Liquid-Liquid Microextraction Followed by Gas Chromatography-Flame Ionization Detection." *Biomedical Chromatography* 29, (2015): 1094–102.

Norouzi, Fatemeh M.A., Maryam Khoubnasabjafari, Vahid Jouyban-Gharamaleki, Jafar Soleymani, Abolghasem Jouyban, Mir Ali Farajzadeh, and Mohammad Reza Afshar Mogaddam. "Determination of Morphine and Oxymorphone in Exhaled Breath Condensate Samples: Application of Microwave Enhanced Three-Component Deep Eutectic Solvent-Based Air-Assisted Liquid-Liquid Microextraction and Derivatization Prior to Gas Chromatography-Mass Spectrometry." *Journal of Chromatography B* 1152, (2020): 122256.

Racamonde, Inés, Rosario Rodil, José Benito Quintana, and Rafael Cela. "In-Sample Derivatization-Solid-Phase Microextraction of Amphetamines and Ecstasy Related Stimulants From Water and Urine." *AnalyticaChimicaActa* 770, (2013): 75–84.

Risticevic, Sanja, Heather Lord, Tadeusz Gorecki, Catherine L. Arthur, and Janusz Pawliszyn. "Protocol for Solid Phase Microextraction Method Development." *Nature Protocol* 5, (2010): 122–39.

Rodrıiguez, Isaac, J. Carpinteiro, J.B. Quintana, Antonia M. Carro, R.A. Lorenzo and Rafael Cela. "Solid-Phase Microextraction with On-Fiber Derivatization for the Analysis of Anti-Inflammatory Drugs in Water Samples." *Journal of Chromatography A* 1024, (2004): 1–8.

Rutkowska, Małgorzata, Kinga Dubalska, Piotr Konieczka and Jacek Namiesnik. "Microextraction Technique Used in The Procedures for Determining Organomercury and Organotin Compounds in Environment Samples." *Molecules* 19, (2014): 7581–609.

Sajid, Muhammad. "Dispersive Liquid-Liquid Microextraction Coupled with Derivatization: A Review of Different Modes, Applications, and Green Aspects." *TrAC Trends in Analytical Chemistry* 106, (2018): 169–82.

Saraji, Mohammad, and Fakhraldin Mousavinia. "Single-Drop Microextraction Followed by In-syringe Derivatization and Gas Chromatography-Mass Spectrometric Detection for Determination of Organic Acids in Fruits and Fruit Juices." *Journal of Separation Science* 29, (2006): 1223–9.

Sun, Shi-Hao, Jian-Ping Xie, Fu-Wei Xie, and Yong-Li Zong. "Determination of Volatile Organic Acids in Oriental Tobacco by Needle-Based Derivatization Headspace Liquid-Phase Microextraction Coupled to Gas Chromatography/Mass Spectrometry." *Journal of Chromatography A* 1179, (2008): 89–95.

Varanusupakul, Pakorn, Narongchai Vora Adisak, and Bancha Pulpoka. "In Situ Derivatization and Hollow Fiber Membrane Microextraction for Gas Chromatographic Determination of Haloacetic Acids in Water." *AnalyticaChimicaActa* 598, (2007): 82–86.

Xie, Zhenming, Lin Yu, Huijuan Yu, and Qinying Deng. "Application of a Fluorescent Derivatization Reagent 9-Chloromethyl Anthracene on Determination of Carboxylic Acids by HPLC." *Journal of Chromatographic Science* 50, (2012): 464–8.

Xu, Minwei, Zhao Jin, Zhongyu Yang, Jiajia Rao, and Bingcan Chen. "Optimization and Validation of In-situ Derivatization and Headspace Solid-Phase Microextraction for Gas Chromatography-Mass Spectrometry Analysis of 3-MCPD Esters, 2-MCPD Esters and Glycidyl Esters in Edible Oils Via Central Composite Design." *Food Chemistry* 307, (2020): 125542.

Zhang, Jie, and Hian Kee Lee. "Application of Dynamic Liquid-Phase Microextraction and Injection Port Derivatization Combined with Gas Chromatography–Mass Spectrometry to the Determination of Acidic Pharmaceutically Active Compounds in Water Samples." *Journal of Chromatography A* 1216, (2009): 7527–32.

Index

β-agonist 126
β-blockers 97, 126
π-stacking 97

3, 4-methylenedioxymethamphetamine (MDMA) 159, 165, 167
3-phenoxybenzoic acid (3-PBA) 166, 244
6-monoacetylmorphine 16

acylation 240, 242
air assisted-dispersive liquid-liquid microextraction (AA-DLLME) 224, 228, 231
alkaloids 153
alkylation 240, 242
amitriptylene 137, 153, 154
amphetamine 136, 137, 150, 153, 163, 165, 167, 180, 244
amphetamine 14, 136, 137, 150, 153, 163, 165, 167, 180, 244
analytical chemistry 159, 160, 172, 200
analytical toxicology 2, 11, 23, 120, 125, 154, 160, 186, 195, 239
antemortem 2
antidepressant drugs 98, 123, 153, 164, 165, 189, 228, 244
aromatic inhibitors 188
arsenic 162
atomic absorption spectrometry (AAS) 228
atomic force microscopy 125
auxillary dispersion solvent 163
azithromycin 137
azole antimicrobial drug residue 186

barbiturates 17, 153
benzodiazepines 14, 150, 154, 162, 163, 166, 186
benzylecgonine 15
bioanalysis 61, 65
biological fluids 62
biological samples 61, 133, 162, 172, 221
biomarkers 76, 122
bisphenol-A 140, 199, 243
blood 2, 123, 160
budesonide 126
bupivacaine 137

cabon nanomaterial 97
cannabinoids 15, 126, 150, 153, 154
capillary electrophoresis (CE) 63, 132, 137, 172, 198
capillary zone electrophoresis 154
captopril 136
carbamate 136
carbon nanotube 16, 224

carbon-ceramic material 65
carbowax 13
central composite design 212, 245
chemical adhesion technique 65
chlordiazepoxide 166
chlorinated diphenyl ether 86
chlorophenol 181
citalopram 123, 179, 180
clinical toxicology 2, 76
cocaethyl 15
cocaine 15, 77, 150, 163
continuous-flow microextraction 132
cortisol 126
cotinine 153
covalent imprinting 195
cyanide 17
cyclic voltametry 125
cypermethrin 166

deep eutectic solvents (DES) 9, 135, 137, 221, 246
derivatization 205, 240
designer solvent 206
desipramine 153, 228
desorption corona beam ionization mass spectrometry 123
dichloroacetic acid 17
diffusion coefficient 173
dip coating 119
direct immersion SDME (DI-SDME) 135, 140, 149
dispersive liquid-liquid microextraction (DLLME) 12, 25, 37, 133, 147, 149, 153, 160, 161, 162, 207, 209, 210, 225, 229, 233, 240, 241, 245
dispersive micro solid-phase extraction (Dispersive μ-SPE) 24
divenylbenzene (DVB) 13, 210
draw-eject cycle 73, 76, 77
drop protection 133
drop-to-drop electromembrane extraction 176
drugs of abuse 150, 180

ecstasy 15
effervescence-assisted dispersive liquid-liquid microextraction (EA-DLLME) 224, 229
electromembrane extraction (EME) 4, 7, 153, 165, 172
electrospinning coating 120
electrospray ionization-ion mobility mass spectrometry 153
electrospun nanofiber 43
endocrine disrupting chemicals 86
enrichment factor 6, 74, 123, 150, 212, 241
environmental toxicology 86
ethyl alcohol 17

249

ethylene glycol dimethacrylate 197
eutectic point 222
exhaled breadth condensate 126, 246
extraction voltage 174

fabric phase sorptive extraction (FPSE) 4, 8, 183, 184
fentanyl 16, 150
fluoroquinone 86
fluoxetine 98, 123
food toxicology 86
forensic analysis 23
forensic samples 46
forensic toxicology 2, 86, 139
free liquid membrane 175, 178

gas assisted-dispersive liquid-liquid microextraction (GA-DLLME) 224
gas chromatography (GC) 61, 62, 72, 131, 133, 172, 198, 210, 233, 240
gas chromatography-electron capture detection (GC-ECD) 166
gas chromatography-mass spectrometry (GC-MS) 12, 63, 125, 137, 229, 231
gel electromembrane extraction 175, 178
grapheme oxide 65, 97
green analytical chemistry (GAC) 61, 147, 184, 190, 195, 205, 211, 221

hair 3, 160
hallucinogens 150, 154, 165
haloacetic acid 243
headspace SDME (HS-SDME) 136, 137, 140, 147, 149
heavy metal 181
heptafluorobutyric anhydride 15
heptaflurobutyric chloride 15
heroin 154
high performance liquid chromatography (HPLC) 13, 61, 186, 188, 231, 240
hollow-fibre liquid-phase microextraction 147, 150, 172, 200, 224, 228, 242
hormones 121, 126
hydrogen bond acceptor 9, 222
hydrogen bond donor 9, 222
hypnotic drugs 153

inflammatory bowel treatment drugs 188
injection port silylation 166, 244, 246
ionic liquids (IL) 4, 9, 64, 97, 135, 162, 206, 221
ionic liquid-stir cake sorptive extraction 210
ionic strength 13

kaempferol 197
ketamine 150, 154, 167

levofloxacin 231
liquid chromatography 63, 72, 122, 125, 133, 153, 172, 198
liquid-liquid extraction (LLE) 4, 11, 12, 38, 73, 161, 147, 160, 172, 183, 207, 209, 212

liquid-liquid microextraction (LLME) 160
liquid-phase microextraction (LPME) 4, 8, 25, 131, 132, 147, 150, 183, 200, 224, 240, 242
lithium 180
loperamide 179
low-density solvent-dispersive liquid-liquid microextraction 162
lysergic acid diethylamide (LSD) 150, 154, 165

magnetic ionic liquid 137
magnetic ionic liquid 207, 210
magnetic molecularly imprinted polymer 197
magnetic solid-phase dispersion 199
magnetic three-phase single drop microextraction 137
mass transfer 71
matrix interferences 74, 123
mebendazole 179
metabolomics 76, 126
metal organic framework 35, 35, 43, 98, 122
methacrylic acid-197
methadone 16, 77, 126, 150, 166, 179, 180
methyl parathion 14
methylenedioxyamphetamine (MDA) 165
micro-electromembrane extraction 178
microextraction with packed sorbent (MEPS) 4, 7, 71, 197
microextraction 4, 133, 205
microfluidic 150
micro-LLE 62
micro-solid phase extraction (μ-SPE) 23, 195
miniaturization 193, 195, 205
molecularly imprinted polymer (MIP) 4, 8, 25, 33, 65, 66, 73, 98, 164, 194
monomer 195
monomer-template complex 195
monomer-template interaction 195
morphine 16, 246
multiwall carbon nanotube 33

nails 3
nalmefenac 179
naltrexone 179
nano-electrospray ionization-mass spectrometry 136
narcotics 126, 150
natural deep eutectic solvent 222
nerve agent 181
nicotine 153
nitrofurantoin 106
non-covalent imprinting 195
non-steroidal anti-inflammatory drugs 229, 231, 243

ochratoxin A 86
octadecyl silica (C18) 65, 76, 77, 86, 125
on-chip electromembrane extraction 175
on-fibre derivatization 243, 245
opiates 16, 153
opium alkaloids 16, 153, 154, 166, 246
organochlorine pesticide 13
oxeladin 137

Index

oxymorphine 246

parabens 98, 140, 246
parallel electromembrane extraction 178
paroxetine 179
penicillin 190
pentrafluorobenzylbromide (PFBBr) 15
pesticides 13, 14, 86, 97, 136, 162, 166
pethidine 179
phencyclidine 159, 165
phenothiazine 17
phthalate 86
physical adhesion technique 65
phytoestrogen 245
pipette tip µ-SPE 38
placket-burman design 212
poisoning 11
polyacrylate (PA) 13, 64, 199
polyacrylonitrile 106
polycyclic aromatic hydrocarbons (PAH) 13, 76, 86, 197, 198, 210, 212, 244
polydimethylsiloxane (PDMS) 13, 62, 63, 118, 199, 210
polyethylene glycol (PEG) 13, 199
polytetrafluoroethane (PTFE) 66, 132, 135
porous graphitic carbon 73
porous membrane protected µ-SPE 25, 37
postmortem 2
propranolol 137
propyl chloroformate 15
protein precipitation 76, 86, 189
pulsed electromembrane extraction 177
pyrethroid 166, 244

quantum dot 136
quercetine 125
quinine 244

repaglindine 125, 126
rotating disk sorptive extraction 199

saliva / oral fluid 3, 14, 15, 86, 160
salt assisted liquid-liquid extraction 164
sample preparation 2, 11, 61, 147, 160, 172, 183, 239
sample preparation 193
scanning electron microscope (SEM) 125
sedative drugs 153
semi-covalent imprinting 195
sertraline 123, 179, 180
signal-to-noise ratio 133
silylation 66, 240, 242
single drop microextraction (SDME) 8, 12, 131, 132, 133, 147, 149, 160, 183, 209, 212, 224, 240, 241, 246
single-wall carbon nanotube 33
sol-gel 8, 65

solid phase extraction (SPE) 4, 11, 23, 38, 62, 73, 147, 160, 163, 172, 183, 194, 195, 197, 209, 231
solidification floating organic droplet microextraction 132
solidification of floating organic droplet-dispersive liquid-liquid microextraction (DLLME-SFO) 153, 165, 225, 229
solid-phase microextraction (SPME) 4, 6, 12, 23, 38, 62, 63, 73, 117, 160, 161, 194, 198, 210, 212, 231, 240
soxhlet extraction 11, 160
spectrophotometry 137
spin coating 120
spin column µ-SPE 43
steroids 126
stir bar 6
stir-bar sorptive extraction (SBSE) 4, 7, 62, 183, 194, 198, 210
supported liquid membrane (SLM) 8, 147, 172, 176, 177
surface-displayed amino group 77
surfactant assisted dispersive liquid-liquid microextraction 163

template 195, 199
testosterone 126
tetracycline 97
thebaine 179
therapeutic drug monitoring 75, 77
thermal desorption 6
thermal desorption 63
thin-film microextraction (TFME) 4, 7, 198
thin-film SPME 117, 119
toxicology 65, 66, 75, 147, 150
tramadol 16
triazine herbicide 98, 186
trichloroacetic acid 17
trichloroethanol 17
trichloroethylene 17
triclocarban 140
trimipramine 153, 154, 165

ultrasound-assisted dispersive liquid-liquid microextraction 164, 224
ultrasound-assisted emulsification microextraction (USEME) 25, 37, 178
ultrasound radiation 24
urine 2, 121, 160, 165

van der Waals interaction 97, 207
viscera 166
vitrous humour 2, 160
volatile organic chemical (VOC) 13
vortex-assisted dispersive liquid-liquid microextraction (VA-DLLME) 224, 228

xenobiotics 2, 11, 86